U.Seeliger C.Odebrecht J.P.Castello *Subtropical Convergence Environments*

Springer
Berlin
Heidelberg
New York
Barcelona
Budapest
Hong Kong
London
Milan
Paris
Santa Clara
Singapore
Tokyo

U. Seeliger C. Odebrecht J. P. Castello (Eds.)

Subtropical Convergence Environments

The Coast and Sea in the Southwestern Atlantic

With 66 Figures

Springer

Dr. Ulrich Seeliger
Dr. Clarisse Odebrecht
Lic. Jorge Pablo Castello

Department of Oceanography
Universidade do Rio Grande
96201 Rio Grande, RS
Brazil

ISBN-13: 978-3-642-64418-4 Springer-Verlag Berlin Heidelberg New York

Library of Congress Cataloging-in-Publication Data

Subtropical convergence environments : the coast and sea in the southwestern Atlantic / Ulrich Seeliger, Clarisse Odebrecht, Jorge Pablo Castello (eds.).
p. cm.
Includes bibliographical references and index.
ISBN-13: 978-3-642-64418-4 e-ISBN-13: 978-3-642-60467-6
DOI: 10.1007/978-3-642-60467-6

1. Coastal ecology--South Atlantic Ocean. 2. Marine ecology--South Atlantic Ocean. 3. Marine biology--South Atlantic Ocean. 4. Biological diversity--South Atlantic Ocean. I. Seeliger, U. (Ulrich), 1944- . II. Odebrecht, Clarisse, 1952- III. Castello, Jorge Pablo.
QH93.6.S83 1997
574.5'2636'091636--dc20 96-32225

Cover design: Design & Production, Heidelberg
Typesetting: K. Darms, Bevern
SPIN 10519611 31/3137 5 4 3 2 1 0 - Printed on acid-free papier

To our parents for helping
to make our professional dreams
come true

Preface

Our understanding of coastal and marine ecosystems, like coasts and seas themselves, is continuously evolving. Earlier work on these environments and their biota in the warm-temperate southwestern Atlantic has amassed useful data, which now need to be analysed as the evaluation, estimation, and management of these resources depend on comprehending the system as a whole. The considerable amount of information now dispersed in scientific papers, reports, and dissertations requires new ways of organization. To encompass the pertinent knowledge, to deal with conflicting interests, to identify research needs, and to meet future demands, an attempt has been made to synthesize the rapidly expanding database, although this effort can only be viewed as one stage in the process of understanding these systems. It is hoped that the book will be useful to many different users, including other scientists, resource managers, students, and interested laypersons.

First the book describes the environmental setting and the biota of the Patos Lagoon estuary, the coastal region, and the continental shelf. Following sections relate these components in space and time to illustrate functional aspects within and the integrated character between systems. The last sections of the book emphasize the impact of fisheries and other human activities on the coast and sea and suggest socio-economically feasible, multifaceted management strategies. An inventory of all marine taxa in the warm-temperate southwestern Atlantic is added as an appendix.

The supportive funding for the book was provided by the Brazilian Government (Program for the Support of Scientific and Technological Development in Environmental Science-PADCT/CIAMB/CNPq/CAPES) and the World Bank. We are grateful to all of our colleagues who authored individual contributions, for extracting the pertinent information from the enormous body of original literature upon which this book is based, and for generously sharing their unpublished data. Special thanks go to Drs. Jose Tundisi and Wolfgang Pfeiffer for their encouragement to edit this book.

Rio Grande, Brazil
October 1996

Ulrich Seeliger

Contents

List of Contributors

Abreu, P.C., Dept. of Oceanography, Universidade do Rio Grande, 96201-900 Rio Grande, RS, Brazil

Asmus, M.L., Dept. of Oceanography, Universidade do Rio Grande

Baumgarten, M.G., Dept. of Chemistry, Universidade do Rio Grande

Bemvenuti, C.E., Dept. of Oceanography, Universidade do Rio Grande

Bianchini, A., Dept. of Physiological Sciences, Universidade do Rio Grande

Calliari, L.J., Dept. of Geological Sciences, Universidade do Rio Grande

Capítoli, R.R., Dept. of Oceanography, Universidade do Rio Grande

Castello, J.P., Dept. of Oceanography, Universidade do Rio Grande

Costa, C.S.B., Dept. of Oceanography, Universidade do Rio Grande

Cousin, J.C.B., Dept. of Morpho-Biological Sciences, Universidade do Rio Grande

Duarte, A.K., Dept. of Oceanography, Universidade do Rio Grande

Fillmann, G., Dept. of Oceanography, Universidade do Rio Grande

Garcia, C.A.E., Dept. of Physics, Universidade do Rio Grande

Garcia, V.M.T., Dept. of Oceanography, Universidade do Rio Grande

Gianuca, N.M., Dept. of Oceanography, Universidade do Rio Grande

Gloeden, I.M., Dept. of Oceanography, Universidade do Rio Grande

Haimovici, M., Dept. of Oceanography, Universidade do Rio Grande

Levy, J.A., Dept. of Chemistry, Universidade do Rio Grande

Montú, M., Dept. of Oceanography, Universidade do Rio Grande

Muelbert, J.H., Dept. of Oceanography, Universidade do Rio Grande

Niencheski, L.F., Dept. of Chemistry, Universidade do Rio Grande

Odebrecht, C., Dept. of Oceanography, Universidade do Rio Grande

Phonlor, G., Dept. of Oceanography, Universidade do Rio Grande

Pinedo, M.C., Dept. of Oceanography, Universidade do Rio Grande

Santos, E.A., Dept. of Physiological Sciences, Universidade do Rio Grande

Seeliger, U., Dept. of Oceanography, Universidade do Rio Grande

Sinque, C., Dept. of Oceanography, Universidade do Rio Grande

Tagliani, P.R.A., Dept. of Oceanography, Universidade do Rio Grande

Vieira, J.P., Dept. of Oceanography, Universidade do Rio Grande

Vooren, C.M., Dept. of Oceanography, Universidade do Rio Grande

Klein, A.H.F., Faculty of Marine Sciences, Universidade do Vale do Itajaí, 88302-202 Itajaí, SC, Brazil

Resgalla, Jr.C., Faculty of Marine Sciences, Universidade do Vale do Itajaí

1 Introduction and Overview

U. Seeliger and C. Odebrecht

The South Atlantic between Africa and South America is under the domain of the high pressure centre of the Atlantic anticyclone which controls the climate and determines the large-scale oceanographic circulation. The northward flowing eastern boundary Benguela Current, the South Equatorial Current, and the western boundary currents, represented by the southward and northward flowing Brazil and Malvinas Currents, respectively, are the main currents of the anticyclonic subtropical gyre between 15 and 40° S, with the centre at surface between 25 and 30° S (Reid 1989). The western boundary currents converge between approximately 32 and 40° S (Castello and Möller 1977) and are forced offshore, giving rise to the Subtropical Convergence in the southwestern Atlantic (Fig. 1.1). Western boundary currents also receive seasonally large continental freshwater runoff (Castello and Möller 1977; Hubold 1980a; Ciotti et al. 1995) from the La Plata River drainage basin (3 170 000 km^2; Urien 1972) and, through the Patos Lagoon estuary, from the Patos-Mirim Lagoon watershed (201 626 km^2; Herz 1977).

The northern reaches of the Subtropical Convergence directly influence about 700 km of southwestern Atlantic coast, comprising approximately 100000 km^2 of southern Brazilian inshore, nearshore, continental shelf, and slope regions between Sta. Marta Grande Cape, Brazil (28° 40′ S) and Uruguay (34° 40′ S) (Fig. 1.1). The variability of the western boundary current circulation, the position of the Subtropical Convergence, and the volume of continental runoff are the principal forcing factors in this area. Together, they provide the foundation for physical and chemical characteristics which determine the biological structure and ecological processes. In consequence, shelf and slope environments are not dissociated from tropical and subantarctic waters nor do they function in isolation from the La Plata River, the huge Patos Lagoon and its estuary, or the fringing shorelines; thus, the coast and sea are integrated and interdependent. Interactions between the oligotrophic Brazil Current, the nutrient-rich waters of the Malvinas Current, and continental runoff make the region an important nursery and feeding area as well as reproduction grounds for fishery stocks of subtropical and antarctic origin which utilize the Brazil and Malvinas Currents for long distance transport. These interactions are also responsible for the high biological production of the coastal zone between the coastal plain and the continental shelf break, which is clearly demonstrated by satellite images (Longhurst et al. 1995).

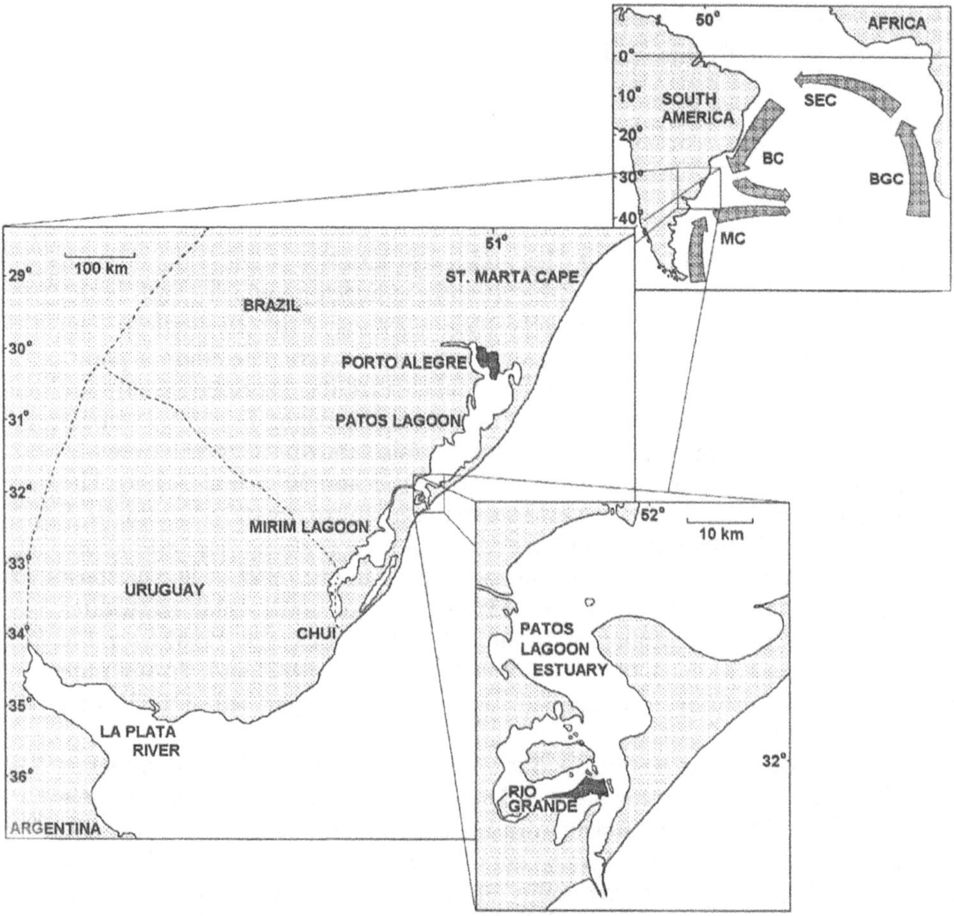

Fig. 1.1. Location of warm-temperate in- and offshore regions under the influence of the Subtropical Convergence in the southwestern Atlantic. Brazil (*BC*), Malvinas (*MC*), Benguela (*BGC*), and South Equatorial (*SEC*) Current

On a global scale, coastal zones are affected by climate and sea-level changes, whereas exploitation by man leads to a gradual decline of resources at a regional level. Initially, the plentiful resources of the southwestern Atlantic coast and sea were exploited by indigenous human populations and early European settlers without harm to the environment. Since it was first colonized the coastal zone has served as a natural channel of export, promoting the establishment of human settlements and industrial development, and the considerable resource potential has invited exploitation by national and foreign fishing fleets. During the last few decades, accelerated demographic growth and increasing economic demands have caused a decline of the seemingly limitless natural resources, and human activities might be on the verge of irreversibly changing the nature of the coast and sea in this region. Unfortunately, regional impacts tend to accumulate and eventually might become significant on a global scale. Studies of large marine ecosystems point to a global depletion of regional seas (Sherman et al. 1993).

The basis required for predicting the role of natural and man-mediated processes contributing to changes in the coastal zone is a detailed knowledge of biodiversity. Biodiversity is expressed by the sum of genes, individuals, species, communities, and ecosystems within a geographically defined region and is the biological response of an ecosystem to past environmental conditions. Tampering with biodiversity might thus severely curtail the capacity of ecosystems to reply with viable solutions to natural changes and man's impact. Owing to the great size and relative inaccessibility of the marine ecosystems, answers to the causes of loss and mechanisms of preservation of biodiversity are severely hampered by a lack of comprehensive surveys of marine taxa and overviews of the processes that control their diversity. New interdisciplinary approaches are needed, existing regional inventories of biodiversity should be capitalized on, and on-site monitoring of marine biodiversity is essential. The first testimony of marine biodiversity (see Appendix) of the Subtropical Convergence marine ecosystem in the warm-temperate southwestern Atlantic not only provides the necessary inventory to help trace changes in biodiversity with time but also furnishes essential background information for urgent management actions to support sustainability.

Despite an increasing commitment to solve current coastal management problems and to reduce the threat of diminishing fishery stocks, a comprehensive appraisal of the biological resources, an assessment of their sustainable exploration, and an evaluation of oceanographic processes in the warm-temperate southwestern Atlantic are lacking in the international literature. Contrary to the historic situation in other parts of the world, which suffered consequences of uncontrolled resource exploitation decades ago, the information now available may still prevent further deterioration of these marine ecosystems and their resources.

2 Regional Climate

A.H.F. Klein

The climate of coastal regions between latitudes 29 and 34° S is under the control of the high pressure centre of the Atlantic anticyclone (Monteiro 1968; Nobre et al. 1986). The latitudinal migration of the centre and the passage of polar frontal systems (Fig. 2.1) in 6-10-day intervals (Hartmann et al. 1986; Stech 1990) have a seasonally modifying influence on the climate (Kousky and Ferreira 1981; Fortune and Kousky 1983; Paz 1985; Vieira and Rangel 1988; Gan 1992). The proximity of the Subtropical Convergence and the stabilizing influence of the extensive inshore Patos-Mirim Lagoon system have imprinted warm-temperate characteristics on the coast and sea (Semenov and Berman 1977), while the distribution patterns of flora (Coutinho and Seeliger 1986; Cordazzo and Seeliger 1987) and fauna (Semenov 1978) classify the area as a warm-temperate biogeographical transition zone.

The marked influence of the Atlantic anticyclone leads to dominant NE winds (5 m s^{-1} mean velocity) throughout the year, followed by SW winds (8 m s^{-1} mean velocity) during a passage of a cold front (Stech and Lorenzetti 1992) which are more common in winter than in summer (Delaney 1965; Godolphin 1976; Zeltzer 1976; Tommazelli 1993). The region is subject to an average of 223 days of sunshine and 404 cal cm^{-2} mean daily solar radiation, with mean monthly radiation of 603 cal cm^{-2} day^{-1} in December and 213 cal cm^{-2} day^{-1} in June. The regional temperature regime is a function of season and number and intensity of cold front passages (Nobre et al. 1986). Mean annual temperatures vary between 19°C and 17°C in the north and south of the region, while monthly mean low and high temperatures vary between 13°C and 24°C in July and January, respectively (IBGE 1986). During the winter, some cold front passages may cause frost in the early morning hours of clear and dry days (Fortune and Kousky 1983; Nobre et al. 1986). Total mean annual precipitation (1200-1500 mm) may strongly vary from year to year and is principally related to the path and frequency of cold front passages (Paz 1985; Hartmann et al. 1986; Nobre et al. 1986; Gan 1992). Mean monthly rainfall is highest during the winter and spring (June to October), but a second peak may occur in summer (Castello and Möller 1978), when daily precipitation occasionally surpasses 100 mm (Gomes et al. 1987). The summer months are associated with a seasonal water deficit (Fig. 2.2), although precipitation and evaporation result in an average annual water surplus of 200-300 mm (IBGE 1986).

U. Seeliger et al.
Subtropical Convergence Environments

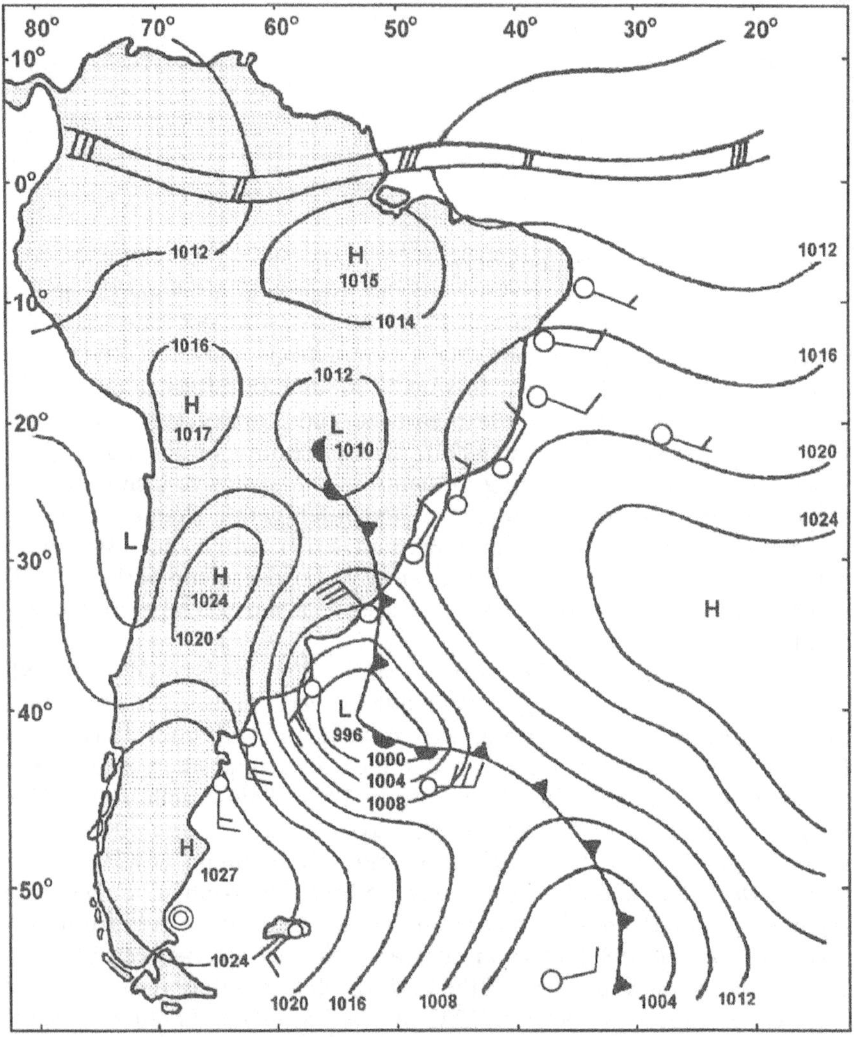

Fig. 2.1. Barometric pressure chart with approaching polar frontal system

In the southwestern Atlantic, distinct interannual variations in precipitation, with either a high amount of rainfall or dry periods, seem to be a consequence of the effect of the El Niño-Southern Oscillation cycle on the global climate (Nobre et al. 1986; Gan 1992), but the processes involved are still not well understood. This phenomena directly influences the amount of continental freshwater runoff and thus the biogeochemical processes in coastal and marine ecosystems of the Southwest Atlantic (Ciotti et al. 1995).

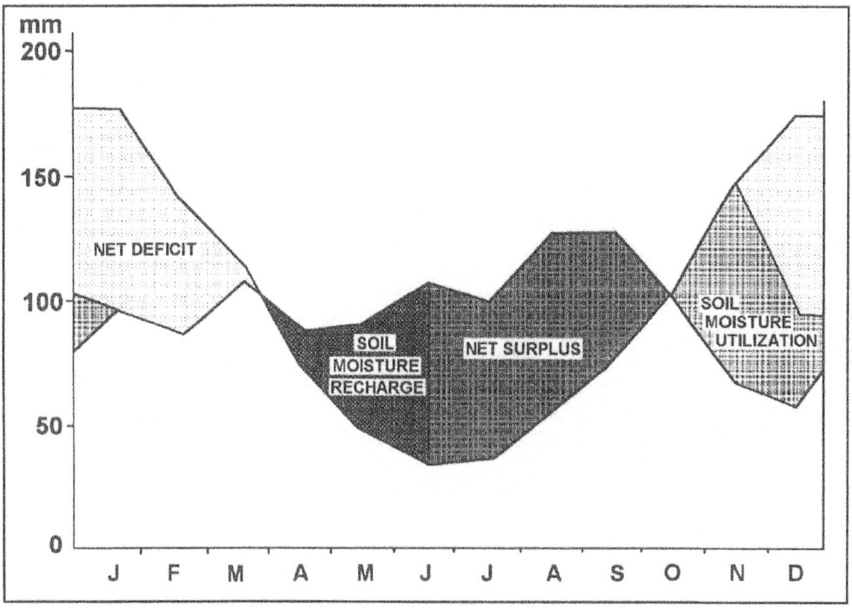

Fig. 2.2. Annual water balance of coastal regions at 32° S

3 Coastal Plain and Patos Lagoon

M.L. Asmus

The Patos Lagoon as a Natural Unit

Wetland and deep-water habitats (*sensu* Cowardin et al. 1979) of different types characterize the coastal plain of Brazil's extreme south, though the Patos-Mirim Lagoon complex is the dominant feature. With a surface area of 10 227 km^2, the Patos Lagoon classifies as the world's largest choked lagoon (Kjerfve 1986). It stretches in a NE-SW direction from 30^0 30'S to 32^0 12'S near the city of Rio Grande where the lagoon connects to the Atlantic Ocean. The lagoon receives waters from a 201 626 km^2 drainage basin, either directly from tributaries or through the São Gonçalo Channel from the Mirim Lagoon watershed (Fig. 3.1). Elevated precipitation and complex river flow patterns in the drainage basin result in highly dynamic and rapidly changing hydrographic processes.

The Patos Lagoon can be divided into five biological units (Fig. 3.2), represented by Guaíba Bay, Tapes Bay, Casamento Lagoon, the central lagoon body, and the estuary. Each unit is composed of deeper open and shallow protected waters. The Guaíba River is the major freshwater tributary. River flow dynamics and water quality fluctuate significantly. A delta is formed at the lower reaches of the river as a result of sediment deposition when the river flow energy dissipates upon entering upper Guaíba Bay. Both Tapes Bay and Casamento Lagoon are semi-open water bodies with distinctive hydrodynamic characteristics. The central open water body of the Patos Lagoon itself is composed of four elliptic cells which represent approximately 80% of the lagoon area. The shorelines are dominated by freshwater marshes and sandy beaches. There are 971 km^2 of estuarine area (approximately 10% of the lagoon) in the south which exchanges waters with the Atlantic Ocean through a 20 km long and 0.5-3 km wide inlet. Shallow, protected, marginal bays and a deeper, central, open water body constitute two ecologically distinct estuarine environments (Bonilha and Asmus 1994).

U. Seeliger et al.
Subtropical Convergence Environments

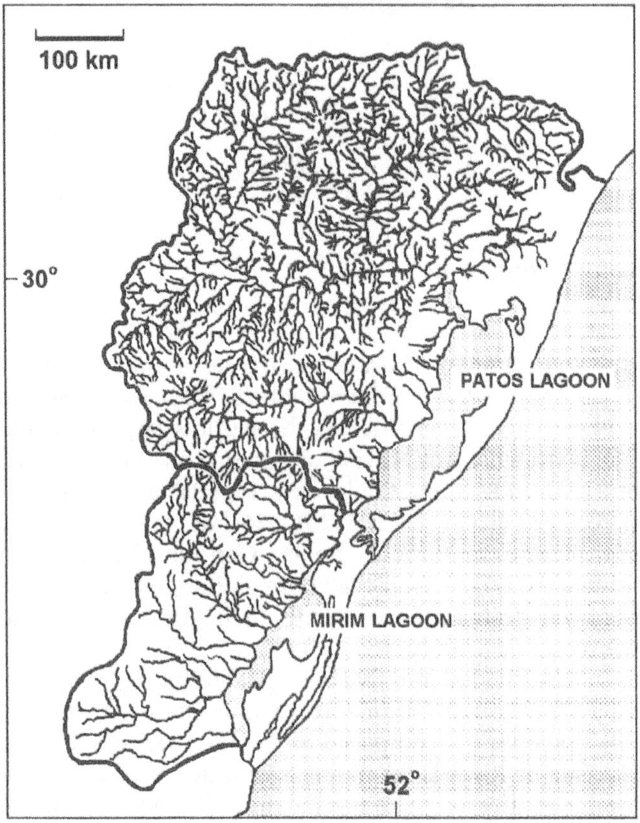

Fig. 3.1. The Patos and Mirim Lagoon watersheds

Settlement and Land Use History

Prior to the arrival of the first European colonists, the indigenous Chana and Tupi-Guarani Indians occupied the coastal plain. The Indians settled along the lagoon and ocean shores, where they exploited abundant fish, shellfish, and shrimp resources (Vieira 1983). The modern occupation history of the coastal plain began with the arrival of politically opposed Portuguese and Spanish Jesuits in 1605 and 1626, respectively. Following the Treaty of Madrid in 1750, which ratified the agreement of a status quo after the foundation of the town of Rio Grande in 1737, the coastal region became the centre of intense Portuguese colonization efforts. In order to reinforce its diplomatic achievements and to warrant sovereignty, Portugal occupied the territory in 1747 with settlers from the Azores, who entered through the port of Rio Grande.

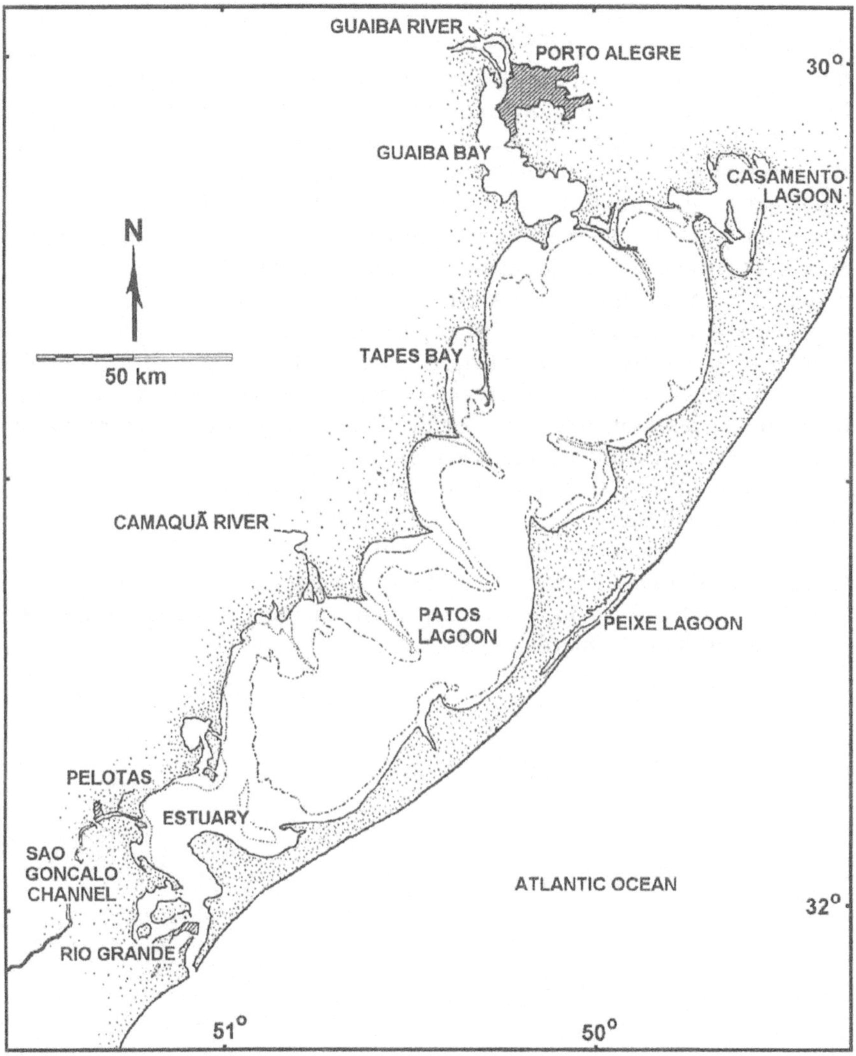

Fig. 3.2. The Patos Lagoon with major biological units

Colonization quickly radiated along the littoral towards Porto Alegre in the north and Chuí in the south (Fig. 1.1). The new settlers were endowed with land for cultivation and cattle breeding, which became the principal economic activity of the region (Viera and Rangel 1988). Agglomerations of settlements along the Patos Lagoon margins eventually gave rise to the present 16 counties, which together represent 90% of the population of the State of Rio Grande do Sul.

From the very beginning, these settlements not only exploited the abundant fishery resources of the Patos Lagoon as a food source, but also used the lagoon as a cultural, social, and commercial link. Advances in agricultural techniques, principally irrigated rice cultivation, as well as the implantation of modern means of transport, caused profound modifications in occupation and development of urban centres around the lagoon (Habiaga et al., unpubl.). Counties which in the past depended on local lacustrine ports for their economic growth were forced to exploit tourism as a new means of development. Today, virtually all port activities are assumed by the ports of Porto Alegre, Pelotas, and Rio Grande, which move approximately 22 500 000 tons of goods yearly. Over the last few decades, rapid and uncontrolled demographic and industrial growth around the lagoon has altered its natural processes and magnified environmental conflicts. Elevated runoff from the Guaíba River tributaries is responsible for adding ever increasing sediment loads, nutrients, hazardous heavy metals (Baisch et al. 1989), and agrotoxins, whilst navigation and port activities and the establishment of fertilizer and fish processing plants and petroleum refining have led to deterioration of the waters in the lower estuary (Almeida et al. 1993). Escalating conflicts between resource exploration and human impact call for urgent management of the Patos Lagoon and the coastal ecosystem (Asmus et al. 1984).

4 Environment and Biota of the Patos Lagoon Estuary

4.1 Geological Setting

L.J. Calliari

Regional Geological History

The Patos Lagoon is part of the coastal province which is geologically defined by the Pelotas Basin and the basement (Villwock 1978). The basement is characterized by the Sul-Rio Grandense shield, composed of crystalline Precambrian rock formations, and by Paleozoic and Mesozoic sedimentary and volcanic sequences of the Paraná Basin, both of which provided the sediments for the marginal Pelotas Basin. Owing to successive transgressions and regressions during the Cenozoic, clastic continental, transitional, and marine sediments of about 8000 m thickness accumulated in the Pelotas Basin. The emersed portion comprises the present flat coastal plain between 29 and 35° S which can geomorphologically be divided into an inner alluvial plain, the Lombas Barrier, the Guaíba-Gravataí Lagoon system, a multiple barrier complex, the Patos-Mirim Lagoon system, and the continental shelf (Fig. 4.1.1; Delaney 1965; Villwock 1984). Geomorphologically, the estuary of the Patos Lagoon is surrounded by tertiary and quaternary deposits of the low lands of the Patos-Mirim Lagoon system and the multiple barrier complex, which are characterized by eolian deposits, beach ridges, sand plains, stable and active dune fields, and lagoonal terraces (Fig. 4.1.2; Godolphin 1976; Zeltzer 1976; Paim et al. 1987).

Estuarine Evolution

The evolutionary history of the Patos-Mirim Lagoon system was decisively influenced by the development of a multiple sand barrier complex as a result of eustatic sea-level changes during the Quaternary (Villwock 1978). At least four successive transgression-regression cycles deposited presumably discontinuous barrier islands parallel to the coast. The isolation

U. Seeliger et al.
Subtropical Convergence Environments

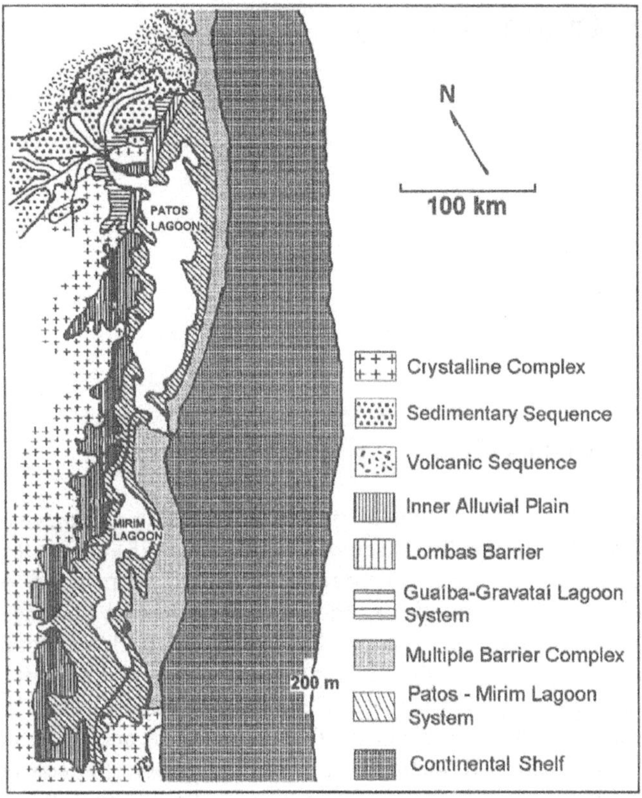

Fig. 4.1.1. Geology and geomorphology of the coastal province. (Modified after Villwock 1984)

of the Patos-Mirim Lagoon system was initiated by the formation of barrier III during the late Pleistocene (120 000 B.P.; Villwock et al. 1986). Coastal waves and wave-generated longshore currents progressively closed the barrier, except for a channel through which waters of the extensive drainage basin forced their way to the ocean. Thus, the entire system can be classified as a partially closed lagoon (Nichols and Allen 1981). The geomorphological characteristics support the model of lagoon evolution postulated by Zenkovitch (1967) and Rosen (1975). However, although geomorphological events during the late Pleistocene created the original estuarine conditions, erosion and deposition processes of Holocene morphodynamics were responsible for the evolution of the modern Patos Lagoon estuary. The last Holocene transgression (5500 B.P.) moulded barrier III deposits into a continuous scarp and deposited an external barrier (IV), composed of recent beach ridges and adjacent dunes (Villwock et al. 1986). The progradation of barrier IV, through deposition of seven regressive

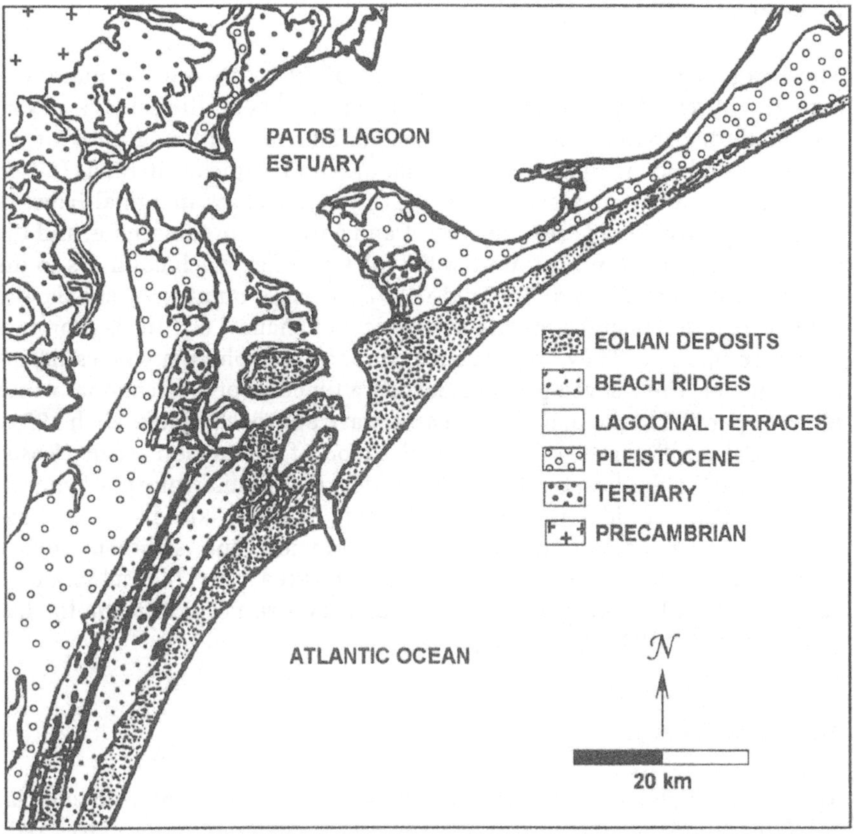

Fig.4.1.2. Geology and geomorphology of the Patos Lagoon estuary. (Modified after Paim et al. 1987)

beach ridges (Godolphin 1976) over a 13 km wide and 63 km long area immediately south of the estuary, has given rise to two different theories of local holocenic estuarine evolution. Godolphin (1976) postulated that interactions between lagoonal outflow and coastal hydrodynamics, like waves and longshore currents, formed the main process influencing the evolution of the Patos Lagoon estuary, whilst Long and Paim (1987) suggested that cyclic migrations of the main channel in relation to coastal orientation were responsible.

Estuarine Morphodynamics

The Patos Lagoon has a maximum width of 59.8 km, and the main axis extends 180 km. Average depth in the lagoon body and the estuary is 5 m, but depth distributions are highly variable. The bottom topography of the main lagoon body is characterized by natural and artificial channels (8-9 m depth), large adjacent areas (>5 m depth), and shallow marginal bays. In the estuary, large shoal areas between 1 and 5 m deep predominate, and the maximum depth (18 m) is found in the inlet channel which connects to the Atlantic Ocean (Calliari 1980; Toldo 1991). A large number of shoals and spits, which define embayments, are typical of shallow (<1 m) lagoon and estuarine margins. They are evidence of recent morphodynamic estuarine adjustments. Their development is largely a function of bidirectional winds blowing parallel to the main lagoon axis, wave approach (Zenkovitch 1959), microtidal oscillations, and the initial shape of the lagoon basin (Rosen 1975). Spits in the lower estuary are evidence of similar morphodynamic adjustments, but tidal inflow and flushing patterns are likely to have influenced their development. Geomorphological and sedimentological data indicate that lagoonal spits were deposited over the substrate of submerged banks, owing to a gradual water level rise (1 m) during the last 3 centuries (Toldo 1991).

Sediment Sources

The region's high level of precipitation causes considerable fluvial silt and clay transport from the large drainage basin, and thus the estuary receives sediments from a variety of sources. Owing to a gradual decrease (1 m/120 km) in elevation towards the estuarine inlet, the reduced flushing current velocities of the Patos-Mirim Lagoon complex lead to large sediment deposition from the water column during transport (Martins 1963; Hartmann et al. 1980). Apart from fluvial sediment input, the hydrodynamic erosion of estuarine margins, especially those comprised of Pleistocene and eolic Holocene formations, lagoonal terraces, and marshes, also contribute significant amounts of sediments. Additionally, sandy sediments of marginal dunes along the eastern shore are added during periods of strong NE winds to adjacent estuarine bays. Finally, high current velocities of inflowing seawater promote the deposition of sandy marine sediments in the lower estuary, and the benthic estuarine fauna and flora represent important sources of biodetritus.

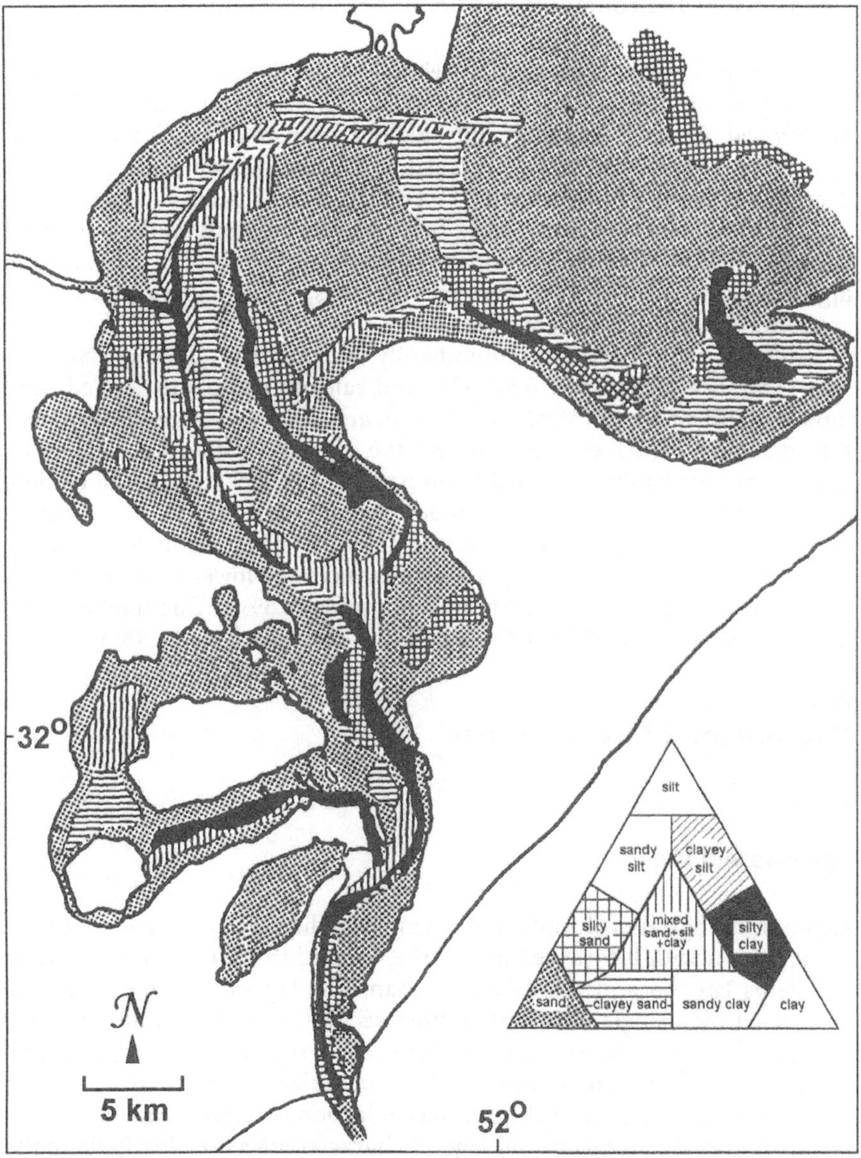

Fig. 4.1.3. Distribution of bottom types in the Patos Lagoon estuary

Depositional Environments and Sediment Characteristics

Six different bottom types can be related to depth and to hydrodynamic energy levels in the estuary (Calliari 1980; Fig. 4.1.3). Silty clay sediments are typical of deep channel and shallow protected bay environments (Martins 1963, 1966, 1971). Transitional environments change with decreasing depth from silty clay, clayey silt, mixed sand-silt-clay, clayey sand to silty sand bottoms. Sandy bottoms are predominantly associated with the shallow marginal portions of large shoals but occasionally occur in high current channels. Also, marginal marshes, sandy shores, and islands are common depositional environments.

Estuarine sediments are predominantly composed of quartz (90%), feldspar (5%), heavy minerals (0.5-1.5%), and calcium carbonate derived from molluscs and cirripeds (Martins 1963; Calliari 1980). Owing to the sedimentary diversity of the source area and the complex estuarine circulation, significant heavy mineral distribution patterns are lacking, though epidote and hornblend prevail in areas under the influence of the São Gonçalo Channel. Furthermore, anthropogenic activities have modified the sand, silt, and clay fraction of superficial sediments in the lower estuary through the introduction of organic carbon compounds (1.5% in fine fraction) and heavy metals (Cu, Zn, Cd, Pb, Cr, Mn, Fe, Ni, Co; Baisch et al. 1988).

4.2 Hydrographic Characteristics

C.A.E. Garcia

Freshwater Inflow

Geomorphological evolution, implemented by high coastal wave energy in combination with littoral sediment drift, formed the barrier that separates the Patos Lagoon and the Atlantic Ocean. The lagoon and the ocean are connected by a single inlet with a cross-section (0.5-3 km) which is small compared with the lagoon surface area. Both regional geomorphological events and the nature of the inlet characterize, more than any other parameter, the system as a choked coastal lagoon (Kjerfve 1986).

Hydrographic estuarine and lagoon characteristics are closely linked to and dominated by hydrological/riverine cycles. About 85% of the water from the immense Patos-Mirim Lagoon drainage basin originates from the Guaíba river, the Camaquã river, and from São Gonçalo Channel falling into the upper, central, and lower lagoon, respectively (Fig. 3.2). Freshwater discharge varies considerably between the summer/fall and winter/spring (6 to 5300 m^3 s^{-1} from the Camaquã basin and 41 to 25 000 m^3 s^{-1} from the Guaíba basin; Herz 1977). Especially during years under the influence of

El Niño, runoff greatly exceeds average values. Water flow through the São Gonçalo Channel is controlled by artificial locks to impede saltwater penetration from the estuary into the Mirim Lagoon. Only during periods of elevated water accumulation in the Mirim Lagoon are these locks opened. As a consequence, freshwater discharge into the upper estuary is variable (700-3000 $m^3 s^{-1}$), usually being more intense during the winter and spring.

Tides and Currents

The southern Brazilian coast is located in a region of minimal tidal influence (Defant 1961). The mean tidal amplitude is 0.47 m, and tides are predominantly mixed-diurnal with an *F*-parameter of 2.42. The tidal wave (O_1 = 25.8h), with an amplitude of 10.8 cm, represents the principal component (Herz 1977). Under typical summer conditions, the energy spectrum of the water level time series in the estuarine inlet reflects significant signals of diurnal (T=25.6 h), semi-diurnal (T=12.5 h), and quarter-diurnal (T=6.2 h) tidal components (Möller et al. 1996). As is typical for choked lagoons, the inlet channel of the Patos Lagoon acts as a set-down filter and therefore largely confines tidal influence to the channel and strongly attenuates amplitudes as the tidal wave advances into the estuary and lagoon. Owing to a large surface to volume ratio and to a gradual decrease in elevation of seawater, maximum current velocities in the main lagoon body are about 0.3 $m s^{-1}$, with a frequent inversion of direction (Möller, unpubl.). However, in the inlet flushing current velocities may reach 1.7-1.9 $m s^{-1}$ after prolonged periods of heavy rainfall (DNPVN 1941), whilst peak inflowing seawater current velocities approach 1.3 $m s^{-1}$.

Forcing Factors of Lagoon and Estuarine Circulation

Most choked coastal lagoon systems are wind forced, rather than tidal driven, and experience limited short-term marine variability (Kjerfve 1986). The Patos Lagoon region is influenced by a predominantly NE-SW wind regime. Winds from the NE quarter (mean velocity 3.6-5.1 $m s^{-1}$) occur during 22% of the year and winds from the SW quarter (mean velocity 5.7-8.2 $m s^{-1}$), during 12%. Seasonal wind patterns are represented by frequent and strong NE quarter winds between September and April and by SW quarter winds from May to October (Delaney 1965; DHN 1974). These winds blow along the NE-SW main axis of the lagoon body and were identified early on as the principal forcing factor in the Patos Lagoon system (Bicalho 1883; von Ihering 1885; Malaval 1922). They decisively control circulation, salinity distribution, and water levels, though fluvial discharge may also generate seasonal pressure gradients with elevated water levels after the onset of the rainy period.

Wind direction influences both local and large-scale circulation patterns. Under NE winds a pressure gradient rises along the main lagoon axis. A pressure gradient between the inlet channel and retreating adjacent coastal waters favours flushing of the lagoon water (Motta 1969). In contrast, SE and SW winds cause inversion of flow in the main lagoon body and raise the water level at its northern limits (Ferreira, unpubl.). Preliminary data indicate a time lag of 4-6 h for wind-induced seiching and set-up/set-down cycles (Möller, unpubl.). However, since systematic wind-driven circulation patterns are poorly developed and highly variable, the time-lag responses are likely to be unstable (Rosauro and Endres 1991). Also, modelling attempts (Bonilha 1974; Rosauro 1986) indicate that in the major part of the lagoon, water levels and vortex formations are strongly influenced by wind direction and velocity, respectively (Almeida et al. 1990). During the summer, SW winds tend to generate a stationary wave with a nodal line along the main axis of the lagoon and several cells with inverse current flow proximate to the margins (Möller et al. 1996). In the inlet channel, outflowing currents are forced along the western margin under the influence of NE winds, whilst SE and SW winds direct inflowing currents towards the shallower eastern bank (Gafrée 1927).

As a consequence of reduced tidal influence in the inlet and in the estuary, the salinity distribution lacks tidal variability but does correlate with wind forcing and variations in freshwater input on scales of hours to weeks. During periods of low fluvial discharge (i.e. summer/fall), onshore SE and SW winds force seawater through the inlet into the lower estuary and occasionally as far as 150 km into the lagoon. In contrast, NE quarter winds together with high fluvial discharge significantly decrease estuarine salinities (Calliari 1980; Costa et al. 1988a). Fluvial discharge in excess of 3000 $m^3 s^{-1}$ causes pronounced salinity stratification in the inlet, and higher values extend the estuarine mixing zone into coastal waters (Möller et al. 1991). Although our understanding of the system's hydrodynamics is still limited, it is evident that both the regional climate and the hydrological cycles are the principal forcing factors controlling lagoon and estuarine circulation patterns and salinity variations.

4.3 Environmental Chemistry

L.F. Niencheski and M. G. Baumgarten

Temperature and Salinity

The time-space variability of physico-chemical estuarine parameters and their component concentrations are a result of interactions between forcing factors, regional sediment characteristics, and anthropogenic activities. In a general manner, estuarine water temperature and salinity, with higher values during the summer, are related to the region's seasonal temperature cycles and wind and precipitation patterns, respectively (Vilas Boas 1990). Temperature and salinity profiles characterize different estuarine conditions. Homogeneous conditions, ranging from freshwater to full-strength seawater, are associated with high fluvial discharge and NE quarter winds or minimal freshwater discharge and SW quarter winds, respectively. In contrast, periods of strong freshwater runoff and concomitant SW wind-induced seawater penetration cause vertical stratification because freshwater advances in the deeper channels over the saltwedge. Also, lateral temperature and salinity stratification may occur in the inlet during reduced freshwater discharge (Kantin 1983). Therefore, rather than being under tidal influence, the extent of the saltwedge or unpredictable short-term changes from stratified to homogeneous conditions are principally controlled by the combined effects of highly variable runoff and wind patterns (Baptista 1984; Möller et al. 1991). Since the saltwedge strongly influences mixing processes of estuarine waters and the formation of salinity gradients, which directly act upon the liberation and/or removal of elements in the water column and the interfaces (Niencheski and Windom 1994), the estuary of the Patos Lagoon is chemically speaking highly unstable (Niencheski et al. 1986). A general absence of conservative mixing between lagoon and ocean waters increases this instability even further (Baptista 1984) because water masses are frequently a result of prior mixing (Vilas Boas 1990). Furthermore, owing to the physiographic heterogeneity of the estuary with deep channels, open shallow waters, and different types of marginal bays, each environment displays site-specific physico-chemical characteristics which strongly vary with the season and from year to year.

Suspended Matter and Dissolved Oxygen

The principal sources of suspended matter are the rivers in the northern (Jacuí, Sinos, Gravataí, Caí, Taquarí) and central (Camaquã) portion of the Patos Lagoon, the São Gonçalo Channel, and erosion processes in the southern part. Furthermore, interactions between geomorphological features of the narrow inlet channel and inflowing seawater during strong SW winds favour the resuspension of deposited sediments in the lower estuary (Niencheski and Windom 1994). As a result, suspended matter loads in the estuary tend to increase towards the ocean and depend strongly on precipitation patterns in the Patos-Mirim watershed. Despite high mean suspended matter loads in the channel (50 mg l^{-1}) and shallow water environments (30 mg l^{-1}), dissolved oxygen in the water tends to be close to saturation or even at supersaturation levels, especially in the inlet channel with pronounced hydrodynamic activity. The shallow depth (<1 m) of the larger part of the estuary favours water oxygen levels in equilibrium with the atmosphere.

Nutrients

Calculations of the annual estuarine nutrient budget demonstrate high dissolved nitrogen, phosphate, and silicate concentrations (Niencheski and Windom 1994) without any clear seasonal pattern. Since fluvial nutrient levels, added to the upper Patos Lagoon, suffer substantial reduction during transport before reaching the estuary, high dissolved and particulate nutrient concentrations are likely to be a result of industrial effluents and remobilization of bottom sediments in the lower estuary (Niencheski and Windom 1994).

The average phosphate concentration in estuarine channels and marginal bays north of the city of Rio Grande (Fig. 3.2) is approximately 1 μM, but may reach 3 μM during the summer (Kantin 1983). The higher phosphate concentrations in semi-enclosed bays south of Rio Grande (Fig. 4.3.1) are probably due to emission of industrial effluents, especially those from the local

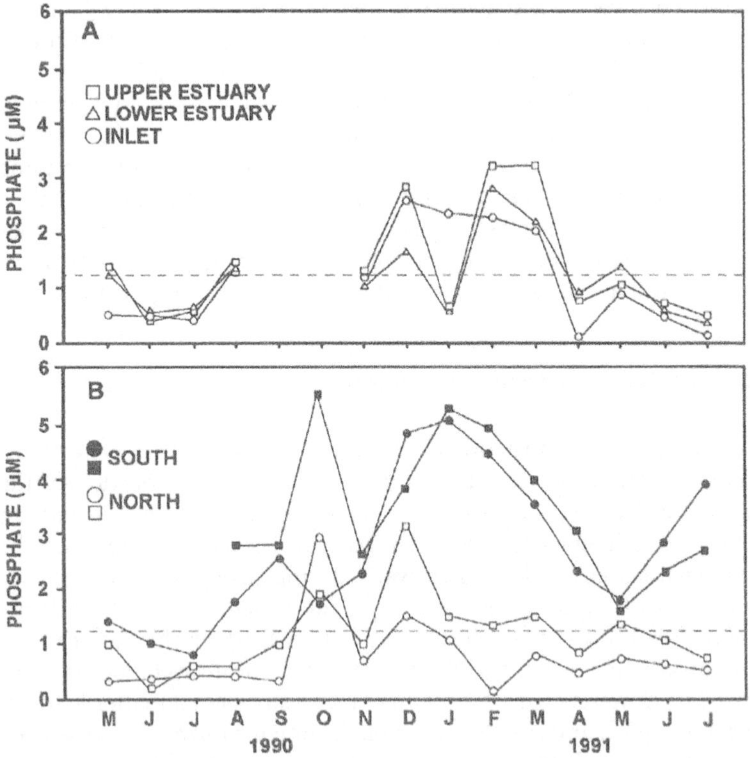

Fig. 4.3.1. Phosphate distribution in the inlet and the main channel of the upper and lower estuary (**A**) and in shallow estuarine waters north and south of Rio Grande city (**B**). *Dotted line* represents maximum levels at which phosphate appears to be buffered by sediments according to Liss (1976)

fertilizer plants (Baumgarten et al. 1995). Phosphate and suspended matter concentrations in the estuary are low during freshwater influence but tend to increase with salt water intrusion (Niencheski and Windom 1994), either because phosphate adsorption onto particulate matter decreases, owing to increased competition with Cl^- and SO_4^{2-} seawater ions for similar exchange sites (Liss 1976), or because desorption processes liberate phosphates from suspended matter. Several sources add nitrate and ammonia to the estuary. Elevated nitrate and ammonia concentrations in the upper estuary are partially a consequence of aquatic macrophyte and algal decomposition and mineralization processes in the lagoon (Kantin 1983). The concentrations in waters of the São Gonçalo Channel derive from agricultural activities in the Mirim Lagoon watershed and from raw domestic sewage discharge by the city of Pelotas. High surface water concentrations in the estuarine inlet (Vilas Boas 1990) and in waters around the port of Rio Grande (Baumgarten et al. 1995) seem to be a consequence of atmospheric input by local fertilizer industries and urban sewage. The Patos Lagoon water is the dominant source of dissolved silicate (approx. 175 µM) to the estuary. Although silicate concentrations oscillate along salinity gradients, they tend to decrease in a conservative manner towards the ocean, except when freshwater dominates the estuary.

Metals

Particulate metal concentrations in estuarine waters strongly depend on resuspension processes of bottom sediments. Sediment resuspension in the inlet channel by inflowing seawater leads to salinity, suspended matter, and particulate metal concentration gradients which decrease towards the upper estuary, whilst wind and wave action may locally contribute to high suspended matter and consequently to high particulate metal concentrations; thus, a clear seasonal pattern of metal distribution is lacking. Particulate metals are either a result of natural weathering or anthropogenic contribution. Their origin can be traced by using aluminium as a geochemical normalizer to establish metal/Al ratios and to determine enrichment factors. The enrichment factor of Ni, V, Ba, and Fe in estuarine waters with high suspension loads (>50 mg l^{-1}) is relatively constant, which suggests that these metals occur at natural levels. However, the enrichment factors of other metals (Cu, Zn, Pb, Li, Cr, Mn, Cd, As, Ag, P) tend to be higher in water with low suspended solids, indicating an anthropogenic contribution (Niencheski et al. 1994). Specifically elevated estuarine copper concentrations might be introduced by the Patos Lagoon which receives drainage waters from mining activities on the Camaquã River. Sporadic increases of suspended copper and lead levels may also reflect metal input by industrial activities in the Patos Lagoon watershed (Baumgarten 1987; Vilas Boas 1990; Niencheski and Windom 1994).

4.4 Tidal Marsh and Wetland Plants

C.S.B. Costa

Species Composition

As in Uruguay and Argentina (Costa and Davy 1992), the intertidal marshes of the islands and margins of the Patos Lagoon estuary are essentially flooded by brackish water and are occupied by annual and perennial tidal marsh and wetland plants. There are approximately 70 species in the lower estuarine marsh flora (Appendix). The tropical (*Paspalum vaginatum, Acrosticum aureum*) and cold temperate (*Atriplex patula, Limonium brasiliensis*) elements of the flora characterize southern Brazilian marshes as part of a warm-temperate biogeographical transition zone. Although species distribution occurs over a wide vertical range with a high degree of overlap, the distribution of dominant species is related to topography (Fig. 4.4.1). Sites with more than 50% cover of the marsh grasses *Spartina alterniflora, Scirpus americanus, Scirpus maritimus, Spartina densiflora,* and *Juncus effusus* occur more frequently in areas flooded during 64.0, 37.4, 25.5, 20.1, and 3.1% of the year, respectively. The local occurrence of monospecific patches additionally suggests interspecific competion as a determinant of plant distribution. Even though dominant species cover more than 50% of the marsh surface, secondary species (*Typha domingensis, Bacopa monnieri, Alternanthera phyloxeroides, Ischaemum minus, Acrosticum aureum, Polypogon monspeliensis*) may locally reach up to 90% cover.

Plant Community Structure

The vertical inundation gradient, the horizontal salinity gradient, and the presence of indicator species characterize six marsh plant communities (Costa and Copertino, unpubl.). Despite their proximity to the water table, upper marshes between 0.8 to 1.2 m above mean low water (MLW) only inundate during exceptionally high water levels in the estuary. Two communities, represented by the indicator species *Ischaemum minus, Polypogon monspeliensis, Cyperus polystachyus, Eleocharis* spp., *Alternanthera phyloxeroides* (Schafer 1985; Cordazzo and Seeliger 1988a), and by *Cyperus obtusatus, Panicum repens, Bacopa monierii, Paspalum vaginatum, Hydrocotyle bonariensis, Juncus acutus* (Reitz 1961; Cordazzo and Seeliger 1988a; Rocha and Costa 1988), characterize transitions to wetlands and dune slacks, respectively. In upper marsh areas (0.3-0.8 m above MLW), which are rarely inundated by either fresh or brackish water, a third community of dense stands of *Juncus effusus* becomes dominant (Capítoli et al. 1978).

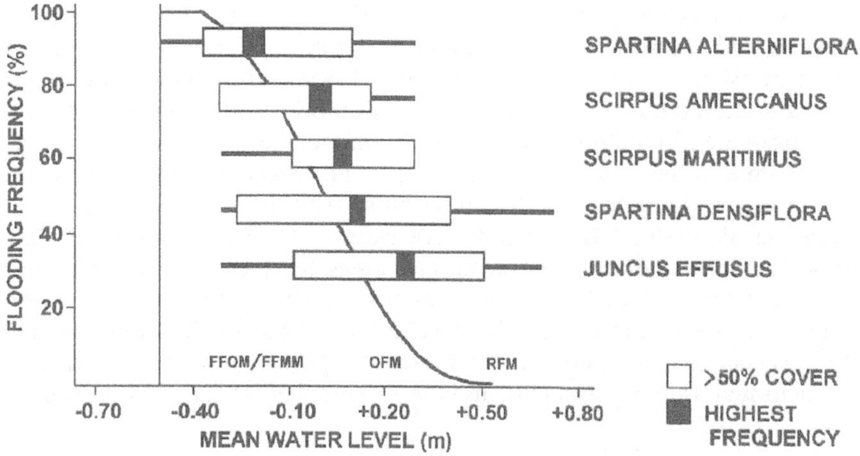

Fig. 4.4.1. Vertical range of distribution of dominant salt marsh species in the Patos Lagoon estuary. FFOM Frequently flooded oligohaline marsh; FFMM frequently flooded mesohaline marsh; OFM occasionally flooded marsh; RFM rarely flooded marsh

The majority of small islands and erosive banks in the estuary are comprised of occasionally flooded mid-marshes (0.1-0.3 m MLW) dominated by a dense *Spartina densiflora* community, which is also typical of nutrient-rich coastal marshes in Uruguay and northeastern Argentina (Parodi 1940; Mobberley 1956; Carnavale et al. 1987; Cordazzo and Seeliger 1988a). A combination of structural, physiological, and life-cycle characteristics permits *Spartina densiflora* to thrive in the mid-marshes despite high ion concentrations, low water and redox potential, impeded gas exchange during submersion, and perturbation by currents (Davy and Costa 1992). During drought periods the lignified leaves partially obtain support from a large number of siliceous bodies. A thick foliar epidermis, sunken stomates, vascular bundles protected by a lignified cell layer, and a C_4 metabolism effectively control the plant's water budget, thus avoiding the toxic and osmotic effects of excessive NaCl accumulation. Apart from active excretion by foliar salt glands, NaCl uptake is partially impeded by several suberized root cell layers. Finally, aerenchymatic leaf, stem, rhizome, and root tissues oxidate the rhizosphere, thereby preventing accumulation of CO_2 and toxic substances during anaerobic respiration in flooded sediments (Perazzolo and Pinheiro 1991; Bastos et al. 1993a,b). The perennial *Salicornia gaudichaudiana* co-occurs with *Spartina densiflora* but becomes particularly dominant in adjacent mud flats where prolonged inundation and large salinity fluctuations impede growth of the latter. Other accompanying species include *Vigna luteola* and *Senecio tweediei* which, owing to their vine-like growth, avoid light attenuation by the *Spartina densiflora* canopy (>1.5 m).

As elsewhere along the southwestern Atlantic coast (Chebataroff 1952; Oliveira and Nhunch 1986; Soriano-Sierra 1990), the frequently inundated lower marshes (between 0.10 and 0.50 m MLW) can be divided into oligo- (<6) and mesohaline (>6) salinity zones (Coutinho and Seeliger 1984) and are dominated by mono-specific *Scirpus americanus* and *Spartina alterniflora* communities, respectively (Fig. 4.4.2). Bluegreen and macroalgae are common components of both communities (Coutinho and Seeliger 1986; Yunes et al. 1990). Highly productive stands of *Scirpus maritimus* may opportunistically dominate transitional areas between mid-marshes and lower mesohaline marshes (Costa, in press). Owing to wave and wind deposited debris, these sites also favour the growth of the annual nitrophilous species *Chenopodium album*, *Atriplex patula*, and *Atriplex hastata*, the biannual *Apium graveolens*, and the perennial grass *Eriochloa punctata*.

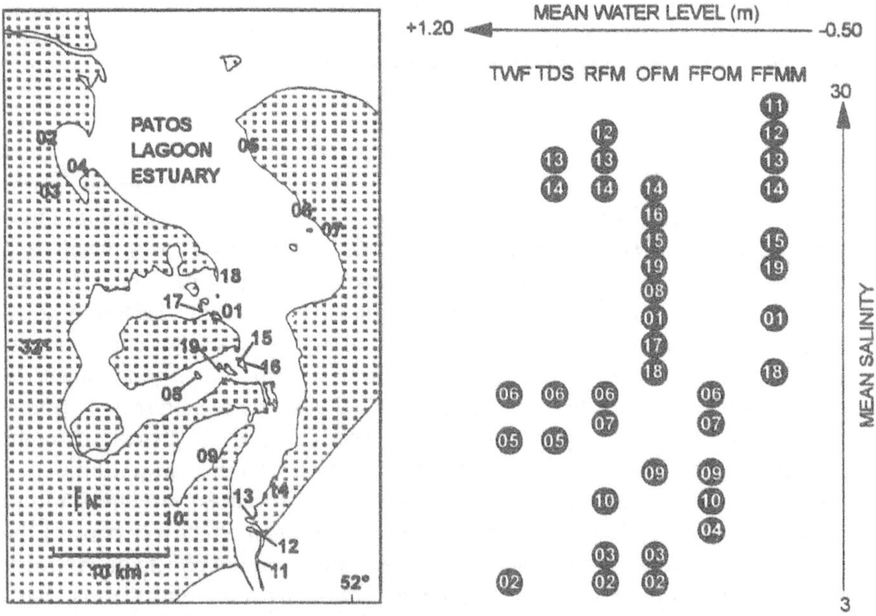

Fig. 4.4.2. Spatial distribution of different salt marsh communities in the Patos Lagoon estuary and their ordination according to mean water level and average salinity. *TWF* Transition to wet fields; *TDS* transition to dune slacks; *RFM* rarely flooded marsh; *OFM* occasionally flooded marsh; *FFOM* frequently flooded oligohaline marsh; and *FFMM* frequently flooded mesohaline marsh

4.5 Submersed Spermatophytes

U. Seeliger

Ecology of *Ruppia maritima* L.

The larger part of the shoals and shallow (<1.5 m) nearshore areas, with their reduced circulation and sandy sediments, is colonized by submersed higher vegetation. The principal colonizer is *Ruppia maritima* (Fig. 4.5.1), though *Zannichellia palustris, Potamogeton striatus, Myriophyllum brasiliense*, and *Ceratophyllum demersum* may become locally important during prolonged periods of low salinity (Cafruni et al. 1978; Cafruni 1983; Moreno 1994; Appendix).

The dispersal mechanisms of *Ruppia maritima* involve seeds and dislodged plants. Germination of seeds is regulated by environmental conditions acting upon the seeds after shedding. Since germination responses differ significantly between sites, populations appear to have adapted their life history to site-specific environmental conditions. In ephemeral marsh pools, for example, exposure to desiccation stress breaks the seed coat and stimulates the action of the seed's enzyme systems, thus preparing seeds for immediate germination during subsequent flooding. In permanently flooded estuarine shoals, low salinity and increasing temperatures may have a synergistic effect on seed germination. Since lower salinities occur in the winter and spring and higher temperatures in the spring and summer, the seeds will tend to germinate in spring when temperature (15^{0}C) and salinity conditions are optimal (Seeliger et al. 1984; Koch and Seeliger 1988), though low germination rates persists during the entire growth cycle.

Incident light, water temperature, and salinity support perennial growth of *Ruppia maritima* in the estuary, but annual growth cycles are more common. Growth may commence any time between the early spring and early summer and cease between the late summer and late fall. The highest growth, leaf density, and biomass are attained during the summer (Cafruni 1983; Moreno 1994). The formation of reproductive shoots with flowers and fruits may continue in the perennial population, but generally initiates in annual populations during peak growth in the summer (Fig. 4.5.2). Resource allocation into reproductive structures is highest in intertidal populations under exposure stress during summer. In lower subtidal populations flower and fruit formation is absent, suggesting the existence of depth-related sexual and vegetative reproductive strategies (Costa and Seeliger 1989).

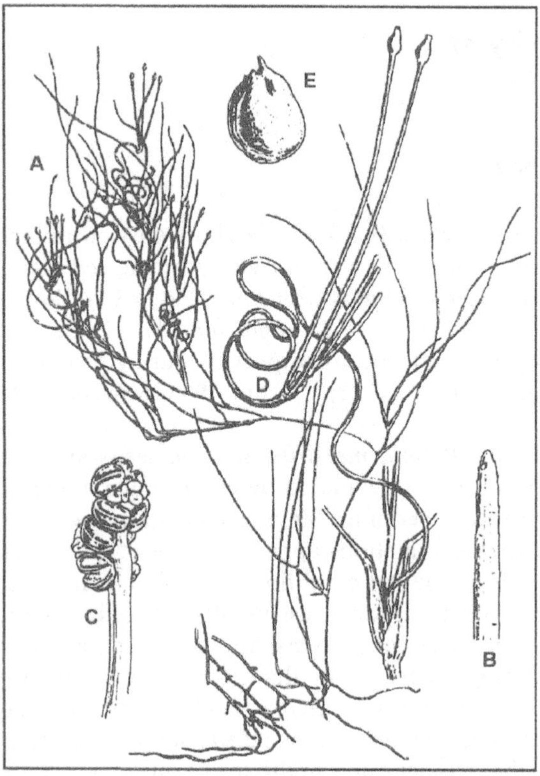

Fig. 4.5.1. Habit of *Ruppia maritima* plant (*A*), leaf tip (*B*), inflorescence (*C*), cluster of fruits on long peduncle (*D*), and seed (*E*)

Several factors, acting either alone or in concert, cause die-off of annual *Ruppia* populations in the late summer/fall. Experimental culture studies suggest that underwater light regimes, rather than water temperature and salinity, may limit growth and production of *Ruppia* populations in the estuary (Knak 1986). Additionally, desiccation may interfere with continued growth since the depth distribution of populations is restricted to a narrow range. Upper growth limits are controlled by air exposure (>20% $year^{-1}$) and lower limits, at about 70 cm below mean low water (MLW), by low underwater irradiance. After dry summers and during NE winds in the fall, estuarine water levels tend to remain for prolonged periods 50-60 cm below MLW (Costa et al. 1988a), which causes exposure of large areas and subsequent mortality of *Ruppia* beds (Costa and Seeliger 1989). Furthermore, shading by algal mats and epiphytes may induce the death of populations. Owing to the basal growth of *Ruppia* leaves, epiphytic colonization

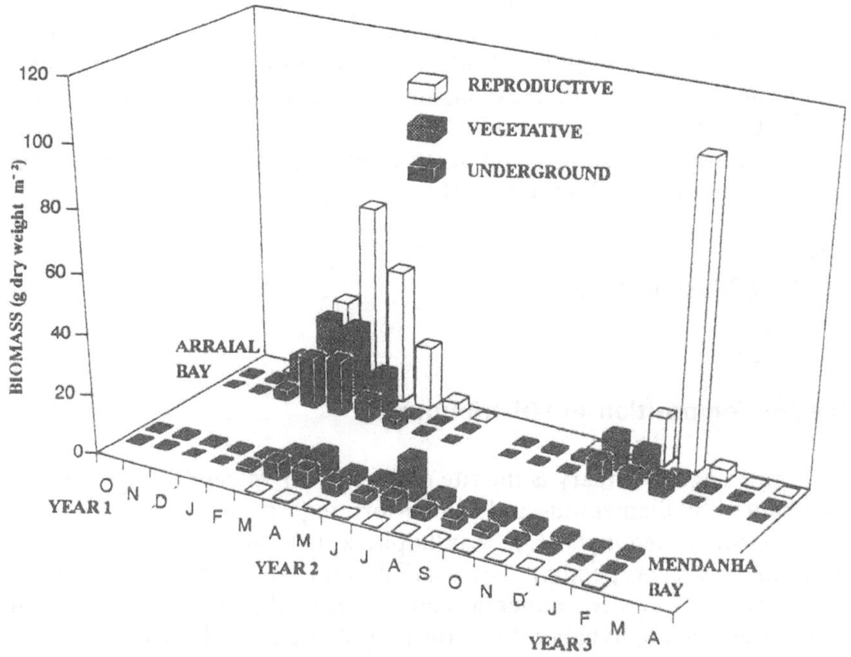

Fig. 4.5.2. Partitioning of mean monthly *Ruppia maritima* biomass in beds with annual (Arraial Bay) and perennial (Mendanha Bay) growth

is densest on the older leaf apex during the summer and fall (Ferreira and Seeliger 1985), resulting in sub-optimal light for continued growth. Also, drift algae tend to become entangled in *Ruppia* leaves and flowering stems. As a consequence, the drag force on the plants increases, making them susceptible to dislodging by waves and currents. These stressful conditions are likely to promote thinning of beds and may contribute to population die-off (Teixeira da Silva 1995).

Biomass and Production

Average total biomass production of *Ruppia* beds is about 25 g dry weight m^{-2}, though summer peaks can exceed 120 g dry weight m^{-2}, owing largely to the contribution of reproductive shoot biomass (Cafruni 1983; Moreno 1994). The average belowground biomass represents approximately 30% of the total biomass, but below/aboveground biomass ratios decrease during meadow development and especially after the formation of reproductive shoots (Fig. 4.5.2). The occurrence, permanence, and size of *Ruppia* beds as well as their biomass and production depend on whether populations follow annual or perennial growth cycles, probably as a function of underwater light conditions and hydrodynamic patterns. Large and unpredictable natural variations of these factors may cause the total extension of *Ruppia* beds in the estuary to differ between years by more than 40%, with total interannual standing plant stock estimates oscillating between 3200 and 5200 metric tons (Moreno 1994).

4.6 Benthic Macroalgae

U. Seeliger

Species Composition and Distribution

The Patos Lagoon estuary is the site of growth for 94 benthic algal species, represented by filamentous and colony-forming Cyanophyta (40), Chlorophyta (26), Phaeophyta (3), Xanthophyta (1), and Rhodophyta (24) (Coutinho 1982; Appendix). The spatially and temporally unpredictable, and thus stressful environmental conditions in the estuary are likely to account for Cyanophyta becoming the largest and most diverse component of the flora. Red and brown algae of marine origin progressively disappear towards the upper estuary, but species diversity remains approximately constant because the contribution of euryhaline green and bluegreen alga elements increases (Coutinho and Seeliger 1984). The horizontal distribution patterns define three algal groups (Fig. 4.6.1) corresponding to annual salinities of 1-12 in the upper estuary, 4-24 in the lower estuary, and 13-34 in the inlet. Although these salinity ranges may not express the salinity tolerances of species, they do indicate salinity preferences of different species groups. Superimposed on the salinity-dependent distribution are algal affinities to substrate stability. Algal colonization and growth in the lower estuarine shoals depend largely on the temporary presence of more or less stable substrates, though some species also thrive as floating mats in bays with reduced circulation. In contrast, the boulders of the jetties at both sides of the estuarine inlet provide a permanent substrate for growth and development. Substrate permanence clearly separates the local flora into two species groups (Coutinho and Seeliger 1984). The permanent rocks of the jetties and the influence of less turbid coastal waters also support short vertical distribution gradients of resident algal populations. At estuarine sites, algae lack vertical distribution patterns because high suspension loads cause strong attenuation of the incident light (Coutinho 1982; Moreno 1994).

Fig. 4.6.1. Similarity based on frequency of occurrence of 64 algae in different annual salinity regimes in the Patos Lagoon estuary.

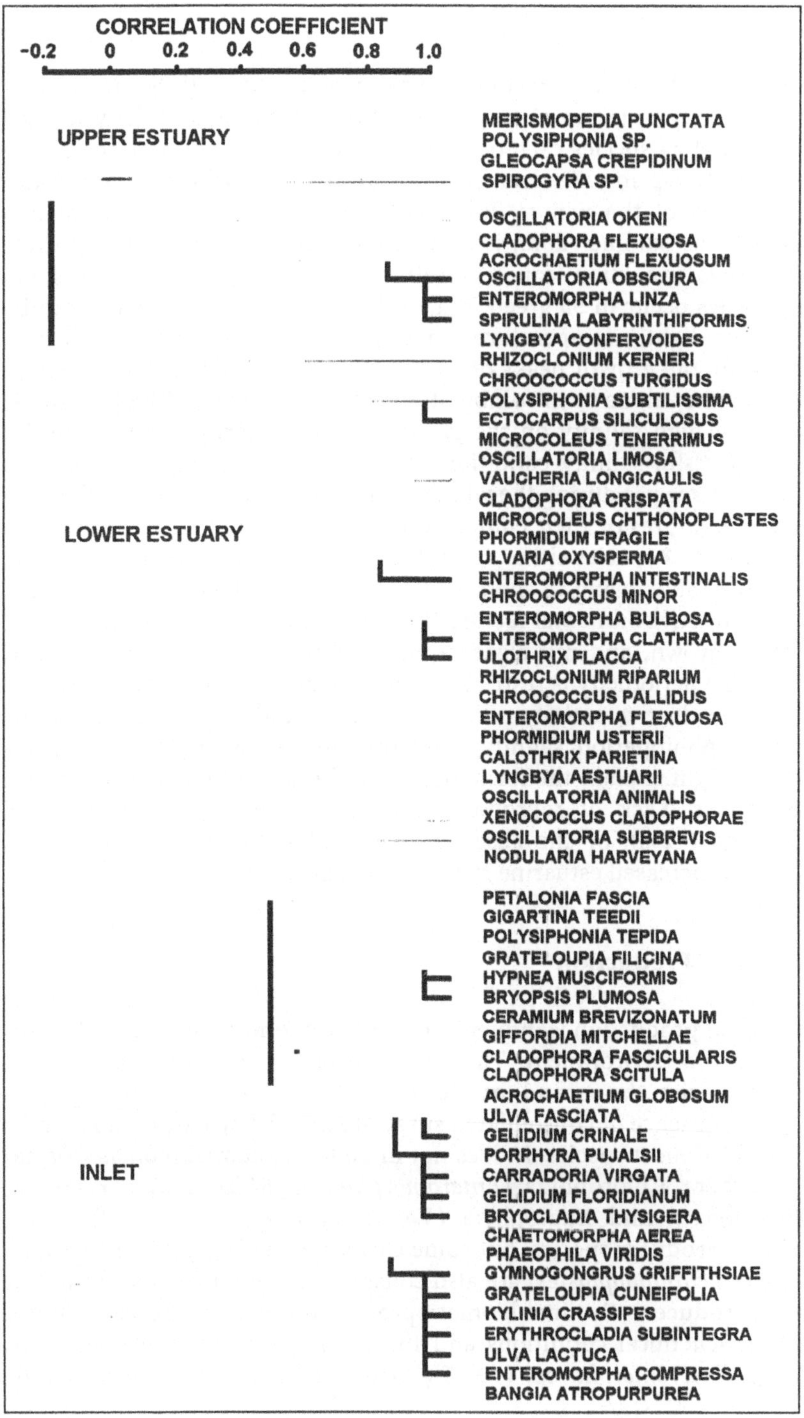
CORRELATION COEFFICIENT
-0.2
0
0.2
0.4
0.6
0.8
1.0
UPPER ESTUARY
LOWER ESTUARY
INLET
MERISMOPEDIA PUNCTATA
POLYSIPHONIA SP.
GLEOCAPSA CREPIDINUM
SPIROGYRA SP.
OSCILLATORIA OKENI
CLADOPHORA FLEXUOSA
ACROCHAETIUM FLEXUOSUM
OSCILLATORIA OBSCURA
ENTEROMORPHA LINZA
SPIRULINA LABYRINTHIFORMIS
LYNGBYA CONFERVOIDES
RHIZOCLONIUM KERNERI
CHROOCOCCUS TURGIDUS
POLYSIPHONIA SUBTILISSIMA
ECTOCARPUS SILICULOSUS
MICROCOLEUS TENERRIMUS
OSCILLATORIA LIMOSA
VAUCHERIA LONGICAULIS
CLADOPHORA CRISPATA
MICROCOLEUS CHTHONOPLASTES
PHORMIDIUM FRAGILE
ULVARIA OXYSPERMA
ENTEROMORPHA INTESTINALIS
CHROOCOCCUS MINOR
ENTEROMORPHA BULBOSA
ENTEROMORPHA CLATHRATA
ULOTHRIX FLACCA
RHIZOCLONIUM RIPARIUM
CHROOCOCCUS PALLIDUS
ENTEROMORPHA FLEXUOSA
PHORMIDIUM USTERII
CALOTHRIX PARIETINA
LYNGBYA AESTUARII
OSCILLATORIA ANIMALIS
XENOCOCCUS CLADOPHORAE
OSCILLATORIA SUBBREVIS
NODULARIA HARVEYANA
PETALONIA FASCIA
GIGARTINA TEEDII
POLYSIPHONIA TEPIDA
GRATELOUPIA FILICINA
HYPNEA MUSCIFORMIS
BRYOPSIS PLUMOSA
CERAMIUM BREVIZONATUM
GIFFORDIA MITCHELLAE
CLADOPHORA FASCICULARIS
CLADOPHORA SCITULA
ACROCHAETIUM GLOBOSUM
ULVA FASCIATA
GELIDIUM CRINALE
PORPHYRA PUJALSII
CARRADORIA VIRGATA
GELIDIUM FLORIDIANUM
BRYOCLADIA THYSIGERA
CHAETOMORPHA AEREA
PHAEOPHILA VIRIDIS
GYMNOGONGRUS GRIFFITHSIAE
GRATELOUPIA CUNEIFOLIA
KYLINIA CRASSIPES
ERYTHROCLADIA SUBINTEGRA
ULVA LACTUCA
ENTEROMORPHA COMPRESSA
BANGIA ATROPURPUREA

Fig. 4.6.1.

Seasonal Patterns

The seasonal distribution of the 94 algal species shows three patterns of algal periodicity. A group of 46 species is present throughout the year, and two groups of 24 species each occur either during the summer/fall or during the winter/spring. A seasonal alternation of eurythermal cool-water algae, as part of the winter flora, with eurythermal tropical elements, comprising the summer flora, is characteristic of floras in warm-temperate biogeographic transition zones (Coutinho and Seeliger 1986). Seasonal growth cycles occur in 55 green, red, and brown benthic algae and can be related to light, temperature, photoperiod, and salinity. During summer/fall, the growth peaks of 15 species significantly correlate to high salinities, either alone or in combination with other factors. The favourable influence of high salinity on the growth of the majority of these species suggests stressful salinity conditions during the remaining seasons. Peak growth of 11 species during the winter/spring largely depends on either low water temperatures or their combined effect with low light. The growth of the remaining 27 species either varies little or presents several peaks during he year, probably owing to site-specific conditions rather than to seasonal variations in environmental factors (Coutinho and Seeliger 1986). None of the green, brown, and red algae reproduces throughout the entire year; reproductive periodicity is restricted to one or two seasons. The lowest and highest reproductive activity occurs in January and May, respectively. As for growth, photoperiod, salinity, and temperature (either alone or in combination) might explain the formation of different reproductive structures in 45 species. Lack of reproduction in some Chlorophyta (23%) and Rhodophyta (12%), or the absence of one life-cycle phase, might be a response to increased estuarine stress (Coutinho 1982).

Biomass and Production

Benthic algal production is influenced by abiotic conditions and the heterogeneity of the environment. The first selects for the principal primary producer species, the second for the size and site of production. Although the general absence of consolidated substrate in the estuary probably limits the growth of most species, it does not impede substantial biomass formation of others. Chlorophyta (*Enteromorpha* sp., *Rhizoclonium riparium*, *Ulothrix flacca*) and Xantophyta (*Vaucheria longicaulis*) are the most prominent producer species, but some Cyanophyta (*Lyngbya confervoides*, *Microcoleus chthomoplastes*) are also conspicuous and important estuarine primary producers. Optimal biomass production of each species is linked to physico-chemical conditions and the permanence of stable substrate. Although *Enteromorpha* sp. (10-35 g dry weight m^{-2}) and *Rhizoclonium riparium* (47-817 g dry weight m^{-2}) form biomass throughout the year, desiccation during exposure tends to reduce biomass formation during some months. Biomass formation of *Lyngbya confervoides* (221 g dry weight m^{-2}), *Microcoleus chthomoplastes* (10 g dry weight m^{-2}), and *Vaucheria longicaulis* (21-161 g dry weight m^{-2}) is largely restricted to the summer, fall, and winter, respectively (Fig. 4.6.2). Large quantities of macroalgae which grow on sediments, shell debris, pebbles, and *Ruppia* plants are often dislodged by waves and currents and occasionally aggregate into extensive floating mats. As long as drift algae remain in the photic zone, continuous growth results in further biomass formation (Coutinho 1982).

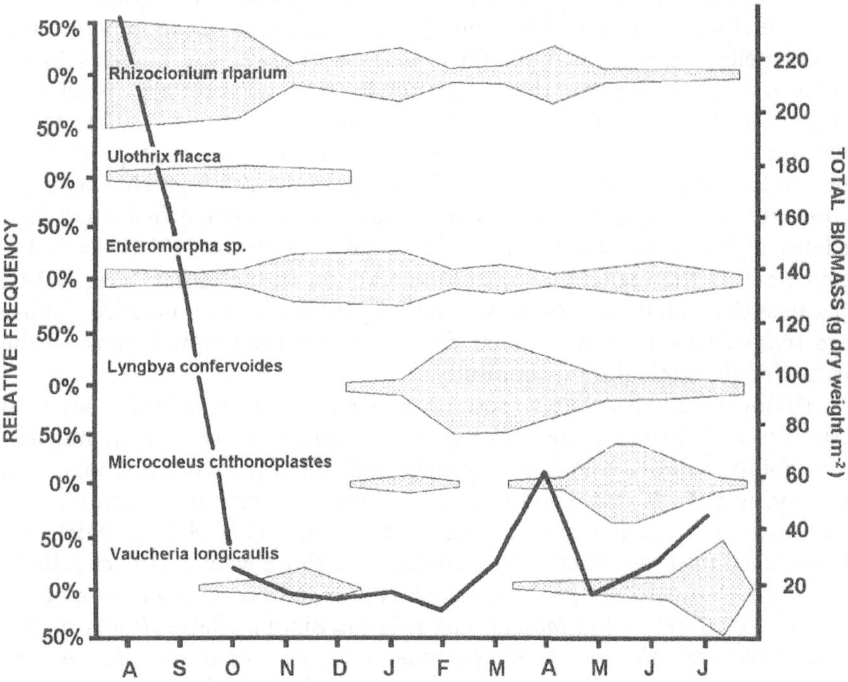

Fig. 4.6.2. Total average monthly biomass and relative monthly frequency of occurrence of principal algal producer species

4.7 Microalgae

C. Odebrecht and P. C. Abreu

Taxonomic Composition and Cell Abundance

Quantitative studies of the Patos Lagoon estuary phytoplankton have been done only recently (Abreu 1987; Odebrecht et al. 1987; Bergesch 1990; Torgan and Garcia 1990; Abreu 1992; Persich 1993; Abreu et al. 1994a; Yunes et al. 1994; Bergesch et al. 1995). Throughout the year, diatoms, monads, and a taxonomically heterogeneous group of small (<20 μm) flagellates (Prasinophyceae, Cryptophyceae, Haptophyceae, Chlorophyceae) are the principal organisms (Appendix). Cyanobacteria (Cyanophytes) and dinoflagellates are more abundant in periods of lower and higher salinity, respectively. Owing to their functional role as primary producers (Lindholm 1985), the number and biomass of the autotrophic ciliate *Mesodinium rubrum* indicate that it also makes up an important part of the phytoplankton community. Except during periods with high chlorophyll *a* concentrations owing to diatoms or cyanobacteria, unidentified small flagellates and monads are the most abundant organisms (10^6-10^9 cells l^{-1}), representing the major phytoplankton fraction in the estuary (Fig. 4.7.1). Because their identification poses special problems, more detailed studies are required to understand their dynamics and their contribution to the estuarine phytoplankton community.

Diatoms, dinoflagellates, and cyanobacteria follow distinct seasonal patterns as a result of meteorological conditions which govern nutrient availability, salinity, and light regimes. High phytoplankton abundance in the spring (10^6-10^9 cells l^{-1}) is related to blooms of euryhaline diatoms, like *Skeletonema costatum* and *Cerataulina daemon* (Fig. 4.7.1), which are introduced into the estuary from coastal nearshore waters. Higher salinity (>20) and temperature in the summer and/or fall favour maximum abundance (10^4-10^5 cells l^{-1}) of *Mesodinium rubrum*, dinoflagellates (*Prorocentrum minimum*, *Peridinium quinquecorne*), and the diatoms *Chaetoceros*, *Rhizosolenia*, *Coscinodiscus*, and *Odontella*. High numbers of the limnic cyanobacteria *Microcystis aeruginosa* (10^9 cells l^{-1}) and the oligohaline diatom *Skeletonema subsalsum* (10^6 cells l^{-1}) occur during flushing and low salinity conditions (<5). Also, interannual differences in species composition are largely due to haline conditions which vary with precipitation and runoff.

Large-scale meteorological factors control the phytoplankton composition in the entire estuary in addition to local hydrographic factors and winds, which strongly interact with benthic and planktonic microalgal communities in the extensive shoals (Bergesch et al. 1995). The brackish pennate diatoms *Cylindrotheca* (=*Nitzschia*) *closterium*, *Bacillaria paradoxa*,

Terpsinoe americana, *Melosira* spp., *Surirella* spp., and the centric *Pleurosira laevis* are frequently resuspended from the sediments into the water column. Phytoplankton cell concentrations per volume tend to be higher in shoals than in the deeper open estuarine waters and are about three orders of magnitude lower than those of benthic microalgae (Bergesch et al 1995).

Epiphytic microalgae are important colonizers on submersed plants of *Ruppia maritima* and on other available substrates. Colonization on *Ruppia* starts in the spring with the diatom *Cocconeis placentula*, which adheres to the leaves through mucus strands liberated by the raphe. Subsequently, *Synedra fasciculata* becomes fixed on live or empty frustules of *Cocconeis*, forming a three-dimensional mosaic which permits colonization of *Amphora*, *Nitzschia*, and naviculoid diatoms. During the final stage of colonization at the end of the summer, diatoms like *Pleurosira laevis*, *Melosira*, *Navicula*, *Rhopalodia*, and *Mastogloia* completely cover the leaf apex (Ferreira and Seeliger 1985).

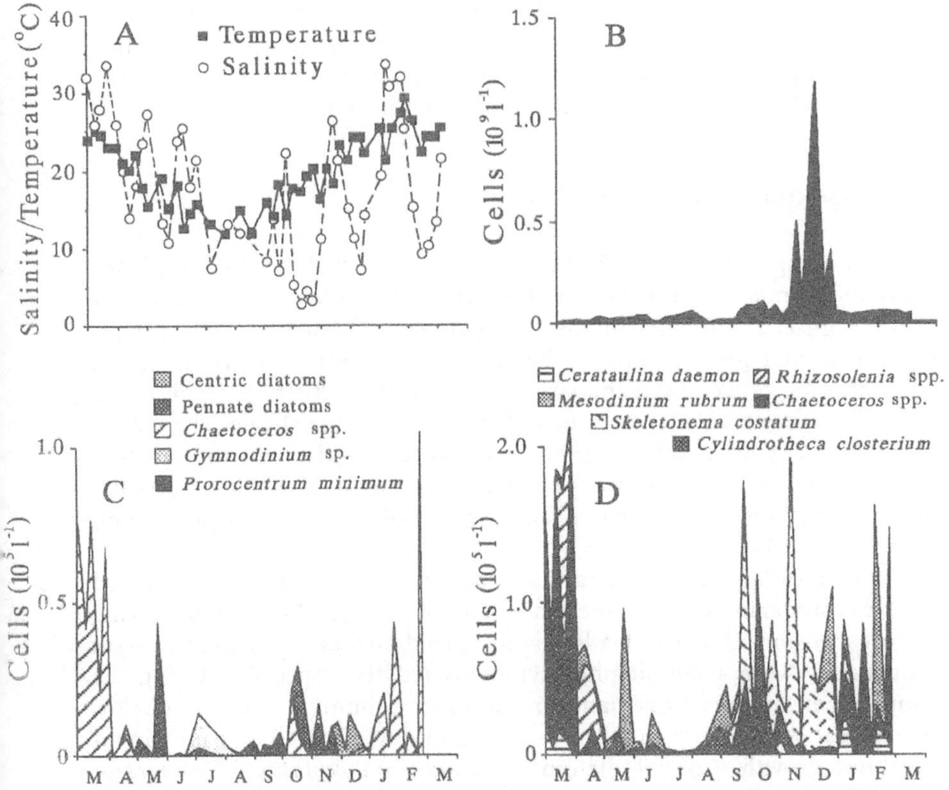

Fig. 4.7.1. **A** Water temperature and salinity and **B** cell concentration of autotrophic nanoflagellates (3-4 μm); **C** diatoms and dinoflagellates (<20 μm); and **D** *Mesodinium rubrum* and diatoms (>20 μm) in the Patos lagoon estuary (1989-1990)

Chlorophyll *a*

Pronounced daily and weekly variability and annual and interannual oscillations of chlorophyll *a* are common in the estuary. Distinct seasonal pulses with high concentrations (30-75 mg m^{-3}) occur in the spring and summer, but occasionally high concentrations are also observed in winter (Bergesch 1990; Proença et al. 1994; Abreu et al. 1994a). While annual and interannual variability is principally related to seasonal light regimes and meteorological conditions, like precipitation and wind direction and speed, small-scale variability is largely a function of direct wind impact. Similarly, wind-induced resuspension together with high productivity in shallow areas (<2 m) generates concentrations of phytoplankton biomass twice as high as those in the deeper channels (>3 m) (Bergesch 1990; Proença 1990), resulting in marked horizontal variability of chlorophyll *a*. In both deep water and shoal environments, pico- (<2 μm) and nanoplankton (2-20 μm) represent an annual mean of more than 70% of the total chlorophyll *a* concentration (Bergesch and Odebrecht 1987; Odebrecht et al. 1988; Proença 1990; Persich 1993; Abreu et al. 1994a). During spring/summer, the contribution of microphytoplankton (>20 μm) rises and the mean contribution may approximate 30% of the total phytoplankton chlorophyll *a*.

Phytoplankton Primary Production

Lower (2-5 mg C m^{-3} h^{-1}) and higher (160-350 mg C m^{-3} h^{-1}) values of phytoplankton primary production in the estuary are associated with winter (June-August) and spring/summer, respectively (Abreu et al. 1994a). During winter, net production may be negative, suggesting a prevalence of heterotrophic processes in the water column. Productivity per volume in channel environments tends to be a third of that in the shoals. The largest portion of the assimilated carbon is taken up by the <20 μm fraction (annual mean = 70.3% of total particulate production). Dissolved production represents 5-60% (annual mean = 22.6%) of total primary production (Abreu et al. 1994a).

Seasonal differences in primary production are linked to irradiance and temperature regimes, and to the availability of nitrogen compounds. Owing to resuspension of seston loads, highly turbid waters (coefficient of light reduction >10 m^{-1}) inhibit photosynthetic activity, especially during the winter when the mean irradiance in the water column is frequently below the critical value of 7.8 μmol m^{-2} s^{-1} (40 gcal day^{-1}, Riley 1967) for phytoplankton growth. Coastal diatoms (*Skeletonema costatum*, *Chaetoceros* spp.), dinoflagellates (*Prorocentrum minimum*), and *Mesodinium rubrum* seem to be adapted to high light conditions (>400 μmol m^{-2} s^{-1}), while nanoflagellate communities appear to photosynthesize more efficiently at low light (<50 μmol m^{-2} s^{-1}; Abreu 1992). Nitrogen compounds for primary production are introduced into the estuary either during freshwater outflow or during intrusion of nutrient-rich seawater (Abreu et al. 1995a).

4.8 Bacteria and Protozooplankton

P. C. Abreu and C. Odebrecht

Bacteria

During the last decade increased research efforts and modern analytical techniques have drastically changed our understanding of microbial ecology in aquatic environments. New facts have led to the "microbial loop" concept which postulates that the dissolved fraction of organic matter becomes available to higher trophic levels through bacterial uptake of dissolved organic carbon, which in turn is grazed upon by flagellates. Energy and material are channelled through ciliates and microzooplankton to higher consumers (Azam et al. 1983). Recent investigations into the ecology of microbial organisms of the Patos Lagoon estuary and adjacent coastal waters have shown that bacteria contribute significantly to the carbon budget of the estuarine foodweb (Abreu et al. 1992). Bacterial biomass exceeds phytoplankton biomass during several months of the year, but together they represent approximately 58% of the particulate organic carbon. Microbes are likely to play an important role in the lagoon, estuarine, and coastal foodwebs (Abreu 1992), and preliminary data suggest that the phytoplanktonic, bacterial, heterotrophic flagellate, and ciliate biomasses are closely related (Fig. 4.8.1).

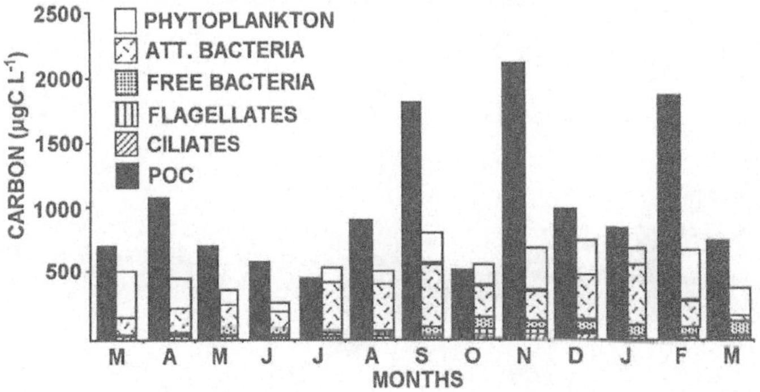

Fig. 4.8.1. Seasonal variation of particulate organic carbon (*POC*) and carbon-based biomass of phytoplankton, free and attached bacteria, heterotrophic flagellates, and ciliates in the Patos Lagoon estuary

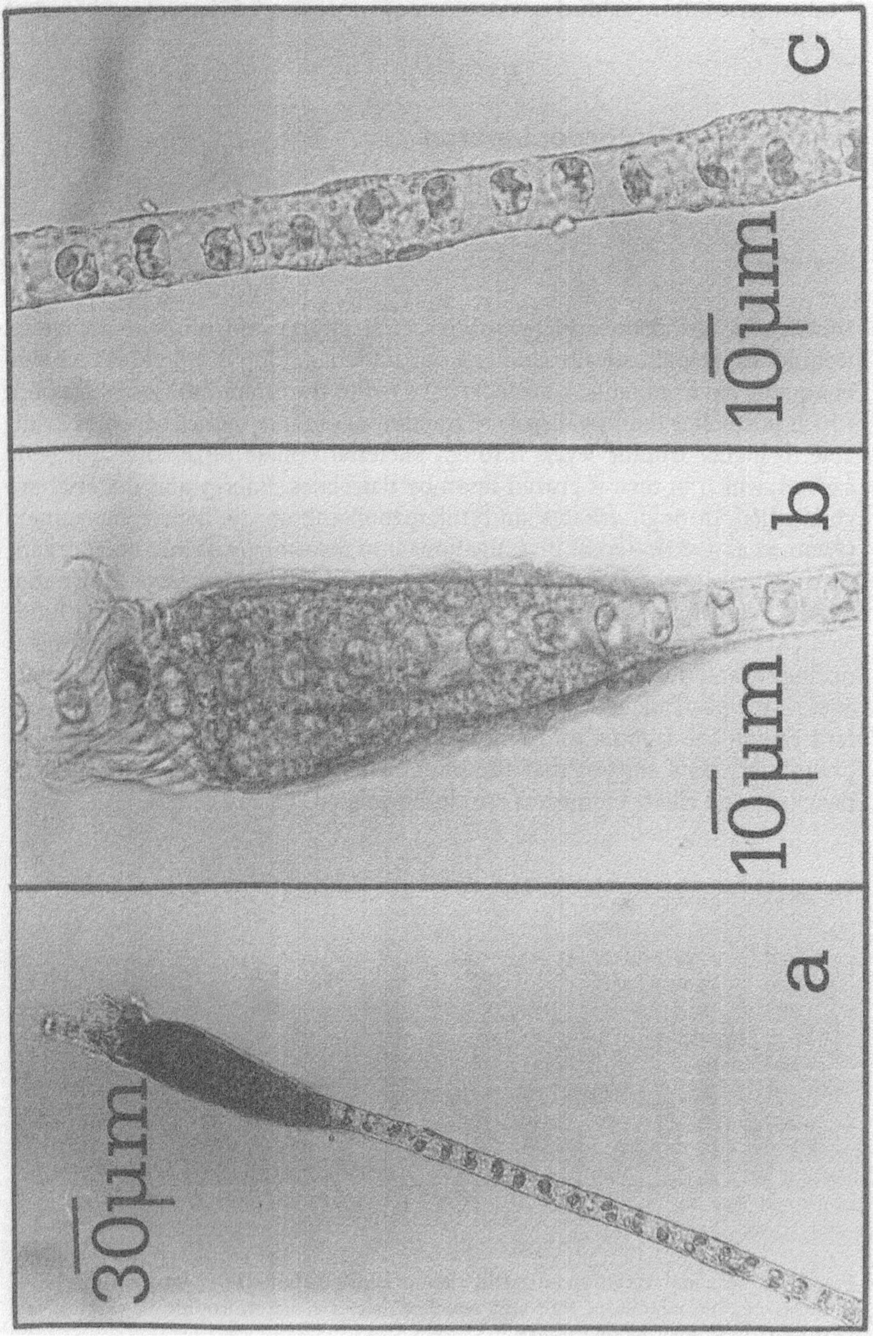

Fig. 4.8.2. Feeding of *Strombidium* sp. (75 x 25 µm) by engulfing a chain of *Skeletonema* (a,b) and details of damaged *Skeletonema* cells (c)

A first evaluation of bacterial production in the coastal plume of turbid estuarine waters and in immediately adjacent clear coastal waters demonstrated that although in turbid waters bacterial production (2.09-6.55 mg C m^{-3} h^{-1}) is much lower than phytoplankton production (79.02 mg C m^{-3} h^{-1}), bacteria may consume as much as 60% of the phytoplankton-produced organic matter (Abreu et al. 1995b). In order to evaluate the importance of the detrital chain in the estuarine foodweb, future studies will have to focus on the role of bacterial succession and activity during degradation processes of the large estuarine macrophyte and macroalgal biomass and on grazing and transfer rates of microbial organic matter to higher trophic levels.

Protozooplankton

Protozooplanktonic organisms in the estuary are represented by taxonomically diverse groups of heterotrophic flagellates (2-3 μm), dinoflagellates and ciliates, like loricate (tintinnids) and aloricate oligotrichids (Strombidiidae) (Appendix). The abundance of flagellates (195-3800x10^3 cells l^{-1}) and ciliates (1-62 cells l^{-1}) oscillates during the year (Abreu et al. 1992), but large numbers (10^{4-5} cells l^{-1}) of loricate ciliates (principally *Leprotintinnus* and *Tintinnopsis*) and aloricates of various sizes (*Strombidium* spp.) also occur. Heterotrophic flagellate (21 μg C l^{-1}) and ciliate (34 μg C l^{-1}) peak carbon biomass estimates are highest during the spring, while values between 3 and 4 μg C l^{-1} occur during the winter (Abreu 1992; Abreu et al. 1992). The availibility of bacterial biomass directly controls the biomass of heterotrophic flagellates. However, microcosm experiments indicate that flagellate biomass is insufficient to sustain continuous growth of ciliate predators because ciliate production (168.9 mg C m^{-3} day^{-1}) exceeds flagellate production (39.3 mg C m^{-3} day^{-1}). It is likely that ciliates, rather than exclusively grazing on flagellates, occasionally ingest bacteria, small phytoplankton cells, and organic detritus to satisfy their nutritional requirements (Abreu et al. 1992). Ciliates (i.e. *Strombidium* sp.) may also feed on large phytoplankters by engulfing and creeping along chains of *Skeletonema*, sucking soft substances from the cells, and leaving semi-digested and damaged chains behind (Fig. 4.8.2).

4.9 Zooplankton

M. Montú, A.K. Duarte, and I.M. Gloeden

Species Composition and Distribution

Zooplankton in the estuary is composed of holoplanktonic organisms (with their entire lives as plankton) and meroplankton (temporarily planktonic). Copepods are typical holoplankters, whilst barnacles and molluscs are typical meroplankters, spending their early lives floating in the estuary until they find a suitable substrate to metamorphose into attached forms. Other meroplanktonic organisms, like the pink shrimp, the blue crab, and some snails metamorphose into mobile forms.

The species composition of zooplankton (Appendix) in the estuary is directly related to local hydrographic conditions. Saltwater penetration and freshwater flushing introduce marine and freshwater species, respectively. In general, circulation patterns created by bottom characteristics and the influence of winds tend to concentrate species of marine origin along the eastern and those of freshwater origin along the western margin of the estuary. The estuarine salinity regime not only influences the distribution of species with different origins but also species diversity which augments with salinity increases. In order to describe patterns of species introduction and distribution, the hydrographic estuarine conditions can be divided into periods of inflowing seawater with high salinity, periods of salt and freshwater mixing, and periods of freshwater dominance during pronounced flushing (Fig. 4.9.1; Montú 1980; Castello 1985).

After saltwater inflow, species of marine origin, principally *Acartia tonsa*, *Euterpina acutifrons*, *Oncaea conifera*, *Sagitta tenuis*, larvae of cirripeds (*Balanus improvisus*) and of echinoderms dominate the zooplankton community. The larvae of most meroplanktonic species (polychaetes, molluscs, crustaceans) of adjacent coastal waters are introduced during this period and may reach the upper limits of the estuary (Montú 1980), though juveniles of the mysids are generally restricted to the lower estuary or to the inlet channel. During flushing periods, larval densities are drastically reduced and freshwater species like the copepods *Notodiaptomus carteri*, *Notodiaptomus incompositus*, *Mesocyclops annulatus*, and a large number of planktonic (*Diaphano-soma sarsi*, *Eubosmina tubicen*) and of small pleustonic cladocerans (*Simosa*, *Alona*, *Chydorus*, *Pleuroxus*, *Biapertura*, *Macrothrix*, *Eurycercus*, *Campto-cercus*, *Kurzia*, *Pseudosida*) become dominant. These cladocerans are regularly found in the shoals and often associate with sub- and intertidal vegetation (Montú 1980; Montú and Gloeden 1986). Periods of fresh-saltwater mixing promote the introduction of species of both marine (i.e. larvae of cirripeds, *Paracalanus parvus*,

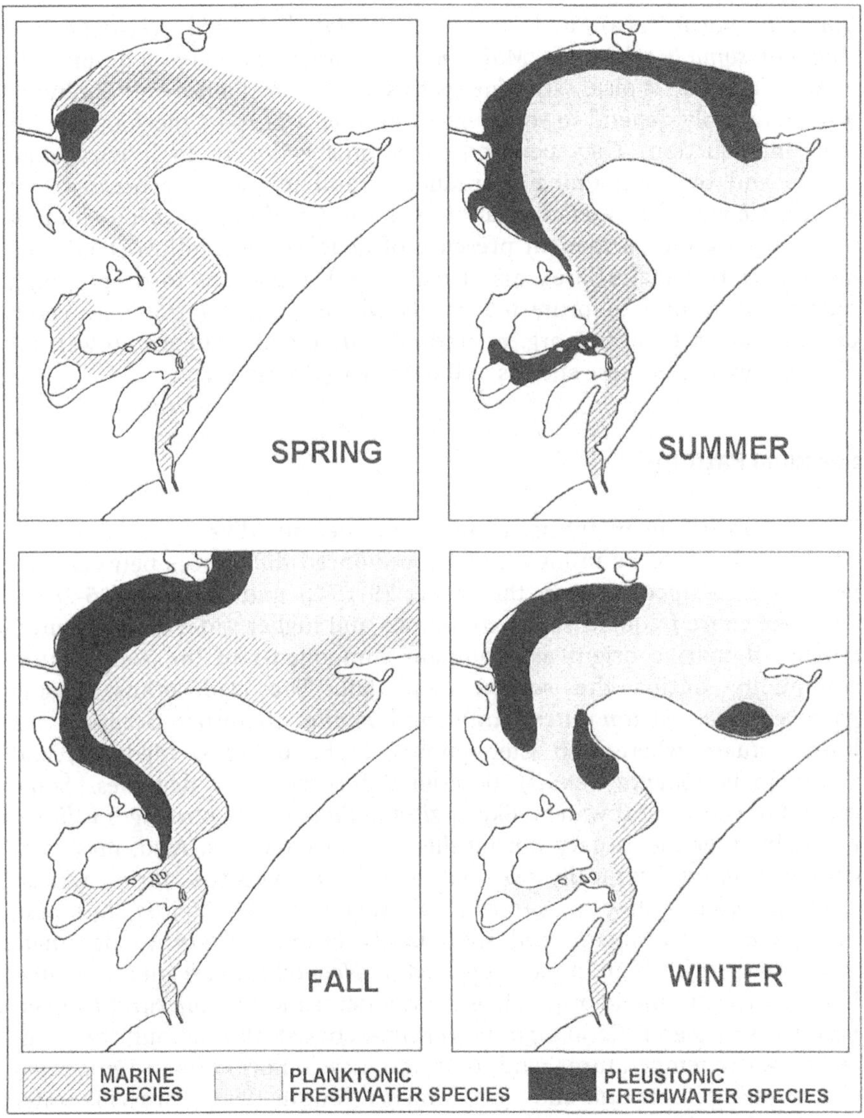

Fig. 4.9.1. Distribution of zooplankton in surface waters of the Patos Lagoon estuary. Distribution of pleustonic freshwater species does not imply an exclusion of other species

Euterpina acutifrons) and freshwater (*Notodiaptomus incompositus*, *Diaphanosoma brachyurum*, *Moina micrura*) origin as well as a dominance of *Acartia tonsa*. Typical estuarine species, like *Acartia tonsa* and *Euterpina acutifrons*, only depend to some extent on tidal transport mechanisms for their introduction. They persist in the estuary due to their euryhaline nature and by maintaining reproduction and migration processes that coordinate with the water circulation patterns. Continuous reproduction of these species, the permanent presence of nauplii/ copepodites stages, the abundance of females in excess of males, and migrations to deeper, high salinity strata and to protected shoals where *Acartia tonsa* may reach densities of up to 40 016 org. m^{-3} indicate successful strategies to warrant the permanence of these species in the estuary (Duarte 1986).

Seasonal Patterns

Seasonal variations in species composition, density, diversity, and growth cycles are likely to be influenced by pronounced differences between surface water temperatures in the winter (9-15^{0}C) and summer (25-28^{0}C). Owing to more frequent saline conditions and higher water temperatures, species of marine origin are common components of the zooplankton community during the summer and fall. The cladocerans *Pleopys polyphemoides*, *Podon intermedius*, and *Penilia avirostris* occur in the lower estuary where also Ctenophores, Siphonophores, Medusae, and Chaetognats (*Sagitta tenuis*) occasionally reach high densities. Some species of subtropical waters, like *Sagitta enflata* and *Centropages velificatus*, only enter the estuary during these warmer months. Also, mysids of coastal origin are typically seasonal, with lower densities during winter. These species follow a successional sequence. *Mysidopsis tortonesi*, *Metamysidopsis elongata*, and sporadically *Promysis atlantica* dominate from the spring to fall and are replaced by *Neomysis americana* in winter. Some species of limnic origin, like the cladoceran *Moina micrura*, *Diaphanosoma sarsi*, and *Ceriodaphnia cornuta* coexist throughout the year. Spring water temperatures tend to promote cyclomorphosis in *Ceriodaphnia*, resulting in *cornuta* and *rigaudi* forms (Montú 1980). Estuarine water temperatures also seem to influence the reproduction and the size structure of adult *Acartia tonsa*. Although reproduction occurs throughout the year, reduced reproduction and lack of resting eggs during colder month result in the dominance of larger individuals in winter (Duarte 1986). Also, the diversity, density, and biomass of zooplankton follow seasonal patterns. Larval densities tend to increase in the spring and summer when elevated numbers (12 473 org. m^{-3}) of *Balanus improvisus* larvae may competitively exclude other zooplankton species. The highest total zooplankton densities (40 016 org. m^{-3}) and biomass (0.2-11 ml m^{-3}) occur in the summer, whilst intermediate densities are typical of the spring (15 526 org. m^{-3}) and fall (14 190 org. m^{-3}) and the lowest densities occur during the winter (6251 org. m^{-3}; Montú 1980).

At sites with prolonged water residence, reduced circulation, and elevated levels of domestic and industrial effluents, estuarine pollutants appear to cause changes in the zooplankton community structure and anatomical anomalies. Under these conditions, species diversity and density are reduced and 14.29% of individuals of *Acartia tonsa* display malformations, expressed by adult females with legs of copepoditos stages IV and V, intersexes (females with legs of male copepodite stages IV and V), and ruptures of chitin between segments with extrusion of tissue. Anomalies are likely to be a consequence of exposure to pollutants during several life-cycles, suggesting that *Acartia tonsa* might be a suitable indicator for deteriorating environmental conditions in estuaries (Montú and Gloeden 1982).

4.10 Benthic Invertebrates

C. E. Bemvenuti

Species Composition and Diversity

The benthic community consists of animals that live in the sediments as infauna by burrowing or forming permanent or semi-permanent tubes, extending just above the sediment surface; animals that live on the sediment surface as either mobile or sedentary epifauna; and animals that form specialized live-bottom communities. As common to estuarine-lagoonal environments, euryhaline marine and estuarine, rather than limnic, species are the dominant components of the benthic macrofauna in the Patos Lagoon estuary (Appendix). Typical estuarine invertebrates are represented by about 15 species, whilst only 3 limnic species (*Tanais stanfordi*, *Palaemonetes argentinus*, *Heleobia parchapei*) occur in the estuary during periods of pronounced freshwater influence (Capítoli et al. 1978). The polychaetes *Laeonereis acuta*, *Nephtys fluviatilis*, *Heteromastus similis*, the tanaid *Kalliapseudes schubartii*, and the pelecipod *Erodona mactroides*, all components of the infauna, show the highest frequency of occurrence and dominance in mudflats (Bemvenuti 1983, 1987a, 1992), seagrass beds (Asmus 1984), and open water environments.

Epifaunal organisms, like the gastropod *Heleobia australis* and the crabs *Chasmagnathus granulata* and *Metasesarma rubripes* occur at high densities in surface layers of bottom sediments and in marginal salt marshes (Capítoli et al. 1978; D'Incao et al. 1992). The euryhaline and highly mobile decapods *Callinectes sapidus* and *Cyrtograpsus angulatus* use the marginal bays during the summer as growing sites, but migrate to deeper waters during the lower fall temperatures. The pink-shrimp *Penaeus paulensis* is commercially the most important decapod in the estuary. From the spring to early summer, larvae of *Penaeus paulensis* enter the estuary from the ocean. The shallow and protected habitats of estuarine shoals provide ideal conditions for their development into juveniles and sub-adults. By the fall, most of the females have matured and migrate back to the ocean to reproduce (D'Incao et al. 1990; D'Incao 1991). Polychaetes (i.e. *Sigambra grubii, Onuphis setosa, Magelona riojai, Hemipodus olivieri*) and peracarid crustaceans, represented by isopods (i.e *Synidotea marplatensis*), cumaceans (i.e. *Dyastilis sympterigiae*), and amphipods (i.e. *Bathyporeiapus bisetosus*), are major marine macrofaunal components in the estuary inlet.

The benthic macrofaunal community in the estuary is composed of only 38 species (Capítoli et al. 1978; Bemvenuti et al. 1992), most of which are r-strategists with pronounced seasonal and interannual variations in abundance. Low species diversity and high species abundance tend to be indicative of stressful estuarine conditions (McLusky 1981). Since choked estuarine characteristics and unpredictable wind and precipitation patterns promote unstable bottom substrates, modify hydrodynamics, and cause a general absence of conservative salinity gradients in the Patos Lagoon estuary, they may account for the low diversity of the macrobenthic fauna (Bemvenuti et al. 1992).

Vertical Distribution Patterns and Abundance

The upper zone of marshes, which fringe the estuary, are dominated by insects, but isopods and the amphipod *Orchestia platensis* also occur. The omnivorous crab *Chasmagnathus granulata* occurs at high densities in burrows in the lower marshes (Capítoli et al. 1978), whilst the preferred habitat of the crab *Metasesarma rubripes* is a small fissure at the base of *Spartina alterniflora* along eroded edges of the lower marsh. The density of infaunal marsh species, like *Orchestia platensis* (3023 ind. m^{-2}) and juvenile *Laeonereis acuta*, is low in comparison with subtidal environments (Capítoli et al. 1977). Extensive unvegetated mudflats are characteristically inhabited by the deposit feeder *Laeonereis acuta*, whose range extends from the intertidal to about 1 m depth and reaches densities of 5127 ind. m^{-2} and a biomass of 28.26 g m^{-2} (Bemvenuti 1987a, 1992). In spite of its wide depth distribution and pronounced seasonality, the gastropod *Heleobia australis* is most abundant (45 616 ind. m^{-2}) and achieves the highest biomass (246 g m^{-2}) in shallow waters (Capítoli et al. 1978). During periods of low density, *Heleobia australis* is replaced by the tube-dweller *Kalliapseudes schubartii* (12 808 ind. m^{-2}; Bemvenuti 1983), whilst *Erodona mactroides* becomes abundant (20 300 ind. m^{-2}, 216 g m^{-2}) after intense recruitment during the summer (Bemvenuti et al. 1978). Although shallow vegetated areas of the estuary, dominated by *Ruppia maritima,* may not be essential for the establishment of the benthic macrofauna, they nevertheless provide favourable conditions by either offering protection to young individuals of *Kalliapseudes* and *Erodona* or leafy substrate for juveniles of *Penaeus paulensis* and *Heleobia.* Since the biomass of these species is positively related to the *Ruppia* biomass, shallow vegetated areas of the estuary may have an important habitat function (Asmus 1984). In addition, deeper open waters (2-6 m) are dominated by the deposit feeder *Heleobia australis,* the suspension feeders *Erodona mactroides* and *Kalliapseudes schubartii,* and the carnivore or detritus feeder *Nephtys fluviatilis.* The highest mean densities of *Heleobia* (12 927 ind. m^{-2}) and mean biomass of *Erodona* (281 g m^{-2}) are found in the mid and upper reaches of the estuary, respectively (Bemvenuti et al. 1978). Benthic populations in deep channel environments are poor and typically composed of deposit feeders, though only *Heleobia australis* occurs in higher densities (Capítoli et al. 1978; Bemvenuti et al. 1992).

Recolonization Processes and Predation

A complex combination of abiotic and biotic factors controls recolonization processes and influences the resilience of soft-bottom substrates in the estuary. Furthermore, resilience processes to macrofaunal colonization are not limited to the species level, but also affect the size categories of invading individuals as a result of their changing niche requirements during development (Giller 1984). After defaunation, the soft bottom substrates in the Patos Lagoon estuary appear to be initially recolonized through recruitment of *Laeonereis acuta* larvae and invasion by adults of *Nephtys fluviatilis.* During intense predation by macropredators, these species give way to lateral migration by large individuals of *Heteromastus similis.* Owing to their limited dispersion, *Kalliapseudes schubartii,* large individuals of *Laeonereis,* and epifaunal peracarids will rarely participate in the first stages of colonization. Further colonization depends on habitat conditions, the predation pressure on initial invaders, and their reproduction increment. The available spaces might subsequently be occupied through recruitment by *Heteromastus* and *Nephtys* larvae (Bemvenuti 1992).

Exposure makes epifaunal peracarids, like isopods, amphipods, and tanaidaceans, susceptible to predation (Nelson 1979a; Virnstein et al. 1984), and they may therefore serve as indicators of predation pressure (Bemvenuti 1987a,b). The highest densities of peracarids occur in estuarine shoals where dense algal mats offer natural shelter; the abundance of crustaceans is associated with a reduced number of predators and protection and food offered by macroalgae during the fall and winter (Bemvenuti 1987b). Experimental field studies suggest that macropredators like fish and decapods keep the density of sedentary epi-and infauna, an important component of their diet (Araújo 1984; Asmus 1984; Bemvenuti 1990), below the carrying capacity of the estuary. However, both macropredators and infauna reach their highest abundance during the summer and early fall on vegetated (Asmus 1984) and unvegetated (Bemvenuti 1983, 1987a,b, 1992) subtidal bottoms. High infaunal densities are partially maintained through escape mechanisms (i.e. burrowing) from predation, where success of escape is a function of the size and mobility of the organisms (Virnstein 1979), but they also depend on species-specific utilization of shallow estuarine habitats and on efficient reproductive strategies, which permit the rapid replacement of depleted populations (Bemvenuti 1992). Trophic interactions and modifications of the soft-bottom community in the estuary are best represented by an intermediary trophic level (Fig. 4.10.1; *sensu* Ambrose 1984a). At this level an infaunal polychaete (*Nephtys fluviatilis*) and small crabs regulate the abundance of non-predatory infauna of specific sediment strata through predation and bioturbation (Bemvenuti 1992, 1994).

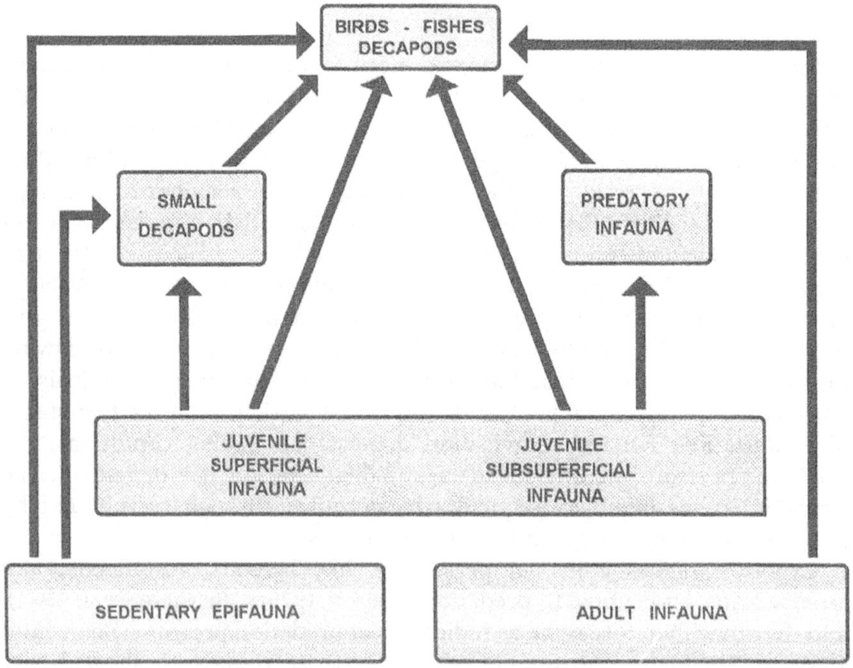

Fig. 4.10.1. Conceptual diagram of predation on the soft-bottom community in shallow estuarine shoals

4.11 Physiological Adaptations of Invertebrates and Fish

E.A. Santos and A. Bianchini

Among benthic estuarine invertebrates decapods play an important economic and ecological role. The shrimp *Penaeus paulensis* represents one of the principal estuarine fishery resources, and the crab *Chasmagnathus granulata* decisively interacts with the energy transfer in submersed and emersed estuarine habitats (D'Incao et al. 1992). Although the respiratory system of *Chasmagnathus granulata* is still highly dependent on the aquatic environment, the species is able to exploit the upper intertidal region effectively (Santos and Colares 1986; Santos et al. 1987). Aquatic decapods have acquired ecological and evolutionary advantages to withstand prolonged exposure to the atmosphere through morphological and physiological adaptations. Metabolic adaptations, which may involve the crustacean hyperglycaemic hormone (CHH), enable *Chasmagnathus granulata* to survive in the atmosphere (Santos et al. 1988; Santos and Stefanello 1991; Nery et al. 1993; Schmitt and Santos 1993b), and ion regulation processes during emersion appear to be strongly water-dependent (Schmitt and Santos 1993a). In general, the metabolism of *Chasmagnathus granulata* differs from that of typical aquatic decapods, like *Callinectes sapidus* (Santos and Colares 1990), being characterized by metabolic depression and dependence on lipids as the energy source under stressful conditions. In order to fulfill their vital functions, each life stage of *Penaeus paulensis* and *Chasmagnathus granulata* has physiologically adapted to the steep and highly unpredictable estuarine salinity gradients and to occasional high, localized concentrations of nitrogen compounds and possibly heavy metals (Seeliger and Knak 1982a,b; Baumgarten and Niencheski 1990a; Almeida et al. 1993) and pesticides. Also, some economically important molluscs, like *Mesodesma mactroides* and the abundant *Donax hanleyanus*, as well as the isopod *Excirolana armata* must tolerate salinity variations in shallow coastal nearshore waters.

Salinity and Temperature

The ability to cope with osmotic stress in media with highly unpredictable and fluctuating salinity regimes is clearly an essential prerequisit for the successful settlement of *Penaeus paulensis* and *Chasmagnathus granulata* populations in the Patos Lagoon estuary. *Chasmagnathus granulata* is an euryhaline species with wide (0-40) salinity tolerance, owing to a combination of efficient hyper-and hypo-osmoregulation of its extracellular fluid (Fig. 4.11.1), ionic (Na^+, K^+, Ca^{2+}, Cl^-) haemolymphatic regulation, intracellular fluid regulation, and cell volume control (Bromberg 1992). Although energy cost studies have emphasized the relative importance of carbohydrate and lipids for osmoregulatory processes (Fig. 4.11.2; Santos and Nery 1987; Nery and Santos 1993), the redistribution of energy otherwise available for somatic growth or reproduction to physiological mechanisms employed during salinity adaptation is likely to have a significant ecological consequence. Salinity also controls the responses of *Chasmagnathus granulata* to pH gradients. Under high salinity and low pH conditions the osmotic equilibrium is disturbed, but pH values above 4.0 are well tolerated (Miranda 1994), and the survival chances of *Penaeus paulensis* larvae increase with pH.

In addition, *Mesodesma mactroides* (15-35), *Donax hanleyanus* (10-35) (Souza-Santos 1991), and *Excirolana armata* (7-70) display considerable salinity tolerance. *Mesodesma mactroides* is a Na^+, K^+, and Ca^{2+} conformer, and salinity tolerance depends entirely on intracellular fluid regulation. However, the volume adjustment of the branchial and muscle cells appears

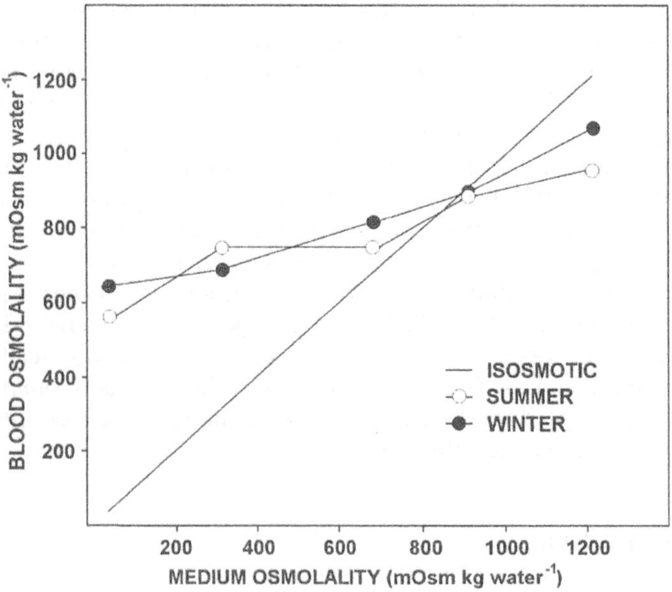

Fig. 4.11.1. Blood osmotic concentration in *Chasmagnathus granulata* in winter and summer, after 15 days of exposure to different salinities following a 30-day acclimation period at a salinity of 20

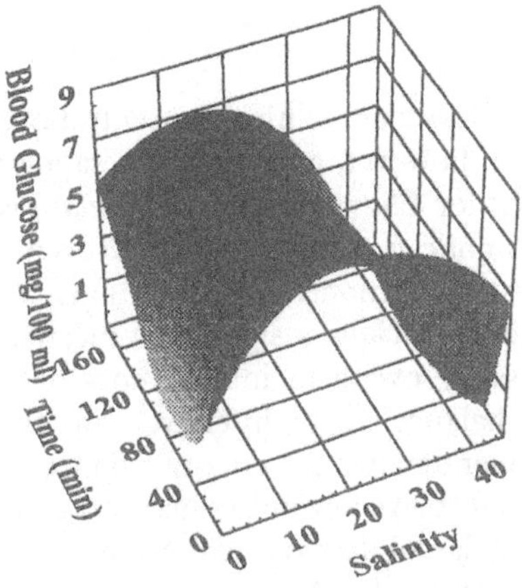

Fig. 4.11.2. Blood glucose levels of *Chasmagnathus granulata* exposed to different salinities as a function of time, following a 15-day acclimation period at a salinity of 30

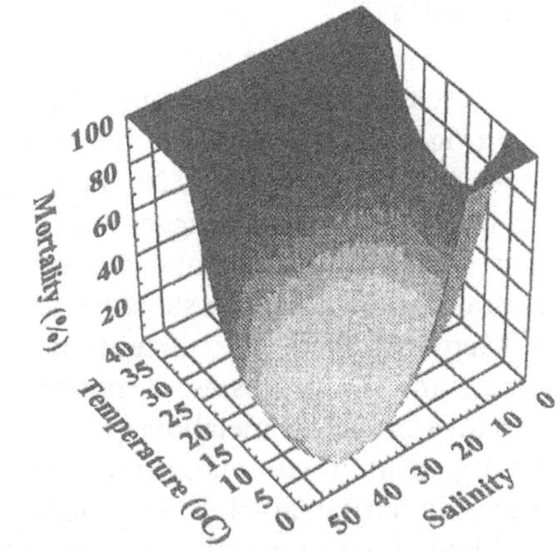

Fig. 4.11.3. Mortality of *Chasmagnathus granulata* (96 h) after exposure to different salinity and temperature combinations

to be regulated by intracellular variations of both inorganic ions and organic effectors, which are independent of carbohydrate energy sources. The metabolic responses of *Donax hanleyanus* and *Excirolana armata* vary with salinity (Souza-Santos 1991). At low salinities (<10), oxygen consumption is reduced in both species. Under these conditions, metabolic depression may decrease the energy expenditure of *Donax hanleyanus*. At higher salinities (approx. 20), protein and carbohydrate metabolism is stimulated without a concomitant increase in food uptake, which causes a pronounced decline in production. An increase of oxygen consumption at higher salinities (30) is likely to be associated with osmoregulatory processes and a concomitant reduction of protein metabolism.

Both field (D'Incao et al. 1992) and laboratory (Miranda 1994) data indicate the preference of *Chasmagnathus granulata* for temperatures between 20 and 25°C. At lower temperatures mortality increases with lower salinity (<20; Fig. 4.11.3), possibly as a result of changes in ion (i.e. Cl^-, Ca^{2+}) and osmoregulatory mechanisms (Miranda 1994). Similar to other land and semi-terrestrial crabs, *Chasmagnathus granulata* remains in burrows and becomes inactive at low temperatures. A relation between high temperatures and reproductive activities is indicated by high recruitment and the presence of ovigerous females throughout the summer (D'Incao et al. 1992). Local populations of *Penaeus paulensis* are eurythermal, with larvae tolerating temperatures between 8 and 38°C, though the best development occurs at intermediate temperatures (Boff and Marchiori 1984).

Pollutants

Since crustaceans are ammoniotelic, they tend to tolerate high concentrations of nitrogen compounds. Ammonia tolerance, however, appears to be salinity dependent. At salinities of 20, the LC_{50} (96 h) for *Chasmagnathus granulata* is 525 mg N-NH_3 l^{-1}. A significant increase of osmotic concentration and Cl^- occurs after 96 h of hyperosmotic shock with rising ammonia concentrations. In contrast, at low salinities sublethal ammonia levels cause a significant increase of haemolymphatic K^+ concentrations, possibly owing to inter-ference with ionic transport processes. A potassium imbalance might interfere with the oxidative branchial metabolism and consequently alter the energy balance of the organism. In contrast, *Penaeus paulensis* displays low ammonia tolerance, especially during larval development. Healthy larval development seems to be inhibited above ammonia concentrations of 2.8 mg l^{-1} (Ostrensky Neto 1991), and long-term exposure to low concentrations (0.14 mg N-NH_3 l^{-1}) may reduce growth up to 43%. The ammonia tolerance of *Excirolana armata* is similar to that of the other crustaceans, toxicity being partially related to the species water balance and osmoregulation.

Both *Penaeus paulensis* and *Chasmagnathus granulata* show significant responses to potential estuarine pollutants like copper ($CuSO_4$) and pesticides (methyl parathion), respectively. An LC_{50} (96 h) for *Penaeus paulensis* post-larvae at 1.33 ppm Cu and a decline of wet weight increase (42.69%) at 0.22 ppm are likely to be either a function of enzyme inhibition for protein synthesis and disturbance of osmoregulatory processes or an indirect effect of copper leading to reduced feeding. In *Chasmagnathus granulata* pesticide toxicity (LC_{50}-96 h) increases with temperature, possibly owing to hypoxia and higher oxidative enzymatic de-sulphurization of the compound (Monserrat and Bianchini 1995).

Adaptations in Fishes

Some teleost fish also display physiological adaptation to environmental gradients associated with estuaries and coastal waters. In general, juveniles of the flatfish *Paralichthys orbignyanus* are euryhaline (0-45) over a wide range of temperatures. Lower oxygen consumption (0.1 mg g^{-1} h^{-1}) is associated with low salinity and temperature (12°C; Wasielesky 1994). Juveniles of *Paralichthys orbignyanus* and *Mugil platanus* are also resistant (LC_{50} 96 h) to elevated ammonia (0.67 and 0.90 mg l^{-1}, respectively) and nitrite (30.57 and 80.97 mg l^{-1}, respectively) concentrations. Higher salinity increases the ammonium and nitrite tolerance in *Paralichthys orbignyanus* and ammonium tolerance in *Mugil platanus* (Wasielesky 1994; Bianchini et al. 1996). *Paralichthys orbignyanus* has also low tolerance to pH stress. An increase of the ventilatory rate at pH 5.2 to 4.7 and 100% mortality at pH 4.0 is probably related to an acid-base disequilibrium, owing to plasmatic HCO_3^- depletion (Wasielesky 1994).

4.12 Ichthyoplankton

C. Sinque and J. H. Muelbert

Species Composition and Abundance

The Patos Lagoon estuary plays an important role in the early life history of many local fish species. The estuary serves as a nursery ground for coastal and estuarine species. A variety of habitats, which provide an abundant food supply and protection against predators, make this estuary a suitable environment for the development of eggs and larvae (Weiss and Krug 1977; Weiss and Souza 1977a,b; Castello and Krug 1978; Weiss 1981; Muelbert 1986; Pereira 1986; Muelbert and Weiss 1991). The dynamic nature of the estuary also contributes to the sporadic presence of many oceanic species in the ichthyoplankton.

Eggs and larvae of at least 29 fish species are found in the estuarine and adjacent coastal waters (Appendix). These species belong to different ecological categories (Chao et al. 1982), and their presence and abundance in early stages of development reflect the degree to which they utilize the estuary. Estuarine resident fish (those who complete their entire life-cycle in the estuary) are represented by the eggs and larvae of the flatfish *Achirus garmani* and by larvae of Atherinidae, *Syngnathus folletti*, Blenniidae, Gobiidae, and Gobiesocidae. These species are not that abundant (0.4-3.3%) and are usually found in shallow inland estuarine regions.

A large number of marine species depend on the estuary for their development, such as those with an abundant number of larvae, e.g. the Clupeidae *Brevoortia pectinata*, the Engraulididae *Lycengraulis grossidens* (=*L. olidus*), and the Sciaenidae *Micropogonias furnieri*. They are present in egg and larval stages and together account for 88% of the eggs and 66% of the larvae found in the estuary. A less abundant group, which only opportunistically uses the estuary as a nursery ground, is present only in the larval stage. It includes many commercially important species, like the Sciaenidae *Macrodon ancylodon*, *Menticirrhus* spp., and *Paralonchurus brasiliensis*, and the flatfishes *Paralichthys* sp. and *Symphurus jenynsi*. In addition, *Trichiurus lepturus*, present in both egg and larval stages, can be included in this group. Freshwater species, mainly the Siluriforme *Parapimelodus nigribarbis*, use the estuary opportunistically during periods of intense freshwater runoff.

A last group of species, composed of larvae of marine origin, is occasionally caught in more saline waters. These species only occur in the estuary under conditions of salt water intrusion. Included in this group are *Anchoa marinii*, *Ophyctus gomesi*, *Urophycis brasiliensis*, *Porichthys porosissimus*, *Epinephelus* sp., *Prionotus punctatus*, *Cynoscion* spp., *Umbrina canosai*, and *Peprilus paru*.

Spatial Distribution

The degree of transport and the location of spawning are the key factors controlling the spatial distribution of ichthyoplankton. With the exception of the estuarine species *Achirus garmanii*, the eggs of all other fish species in the estuary are of marine origin (Weiss 1981; Chao et al. 1982), and thus eggs are generally more abundant closer to the mouth of the estuary. The distribution within the estuary reflects spawning location and utilization of the estuary as a nursery ground. Species such as *Engraulis anchoita* and *Anchoa marinii* that spawn further offshore and do not depend on the estuary for their development, are confined in their distribution to the estuary inlet. Species like *Micropogonias furnieri* and *Achirus garmani* which use the estuary during their development lay eggs distributed throughout the estuary.

This pattern of horizontal distribution is also observed for larvae. The larvae of oceanic species (*Anchoa marinii, Porichthys porosissimus, Prionotus punctatus, Synagrops sp., Parona signata, Cynoscion striatus, Umbrina canosai, Mugil spp., Trichiurus lepturus, Peprilus paru, Symphurus jenynsi*) occur in the estuary during periods of strong saltwater intrusion (Muelbert and Weiss 1991), whilst larvae of coastal water *species (Brevoortia pectinata, Lycengraulis grossidens (=L. olidus), Macrodon ancylodon, Menticirrhus spp., Micropogonias furnieri, Paralonchurus brasiliensis*) use the entire estuary as a nursery area. Atherinidae, *Syngnathus folletti*, Blenniidae, Gobiidae, and *Achirus garmani* reside and are uniformly distributed in the Patos Lagoon estuary. The freshwater species *Parapimelodus nigribarbis* exhibits a decrease in abundance towards the ocean.

The vertical distribution of fish eggs is influenced by their buoyancy and by the salinity structure of the water column. In the Patos Lagoon estuary the maximum number of eggs is related to salinities between 25 and 30 (Weiss 1981). Since these values generally correspond to surface waters near the estuarine inlet and upper estuarine bottom waters, they induce a vertical distribution pattern of the eggs and may reflect an ideal density for egg buoyancy (Muelbert 1986). Although most fish larvae in the estuary show a uniform vertical distribution, some groups of larvae seem to present different patterns. Less saline surface waters favour *Brevoortia pectinata, Lycengraulis grossidens, Parapimelodus nigribarbis*, and Atherinidae, while *Micropogonias furnieri* and *Trichiurus lepturus* are frequently found in more saline bottom waters. During anomalous periods of freshwater run-off, this pattern may change, and *Parapimelodus nigribarbis* may become more abundant at the bottom than near the surface (Muelbert and Weiss 1991). Based on the available information on the spatial distribution of ichthyoplankton, Weiss (1981) has proposed a general model for the dynamics of fish eggs and larvae in the estuary of the Patos Lagoon (Fig. 4.12.1).

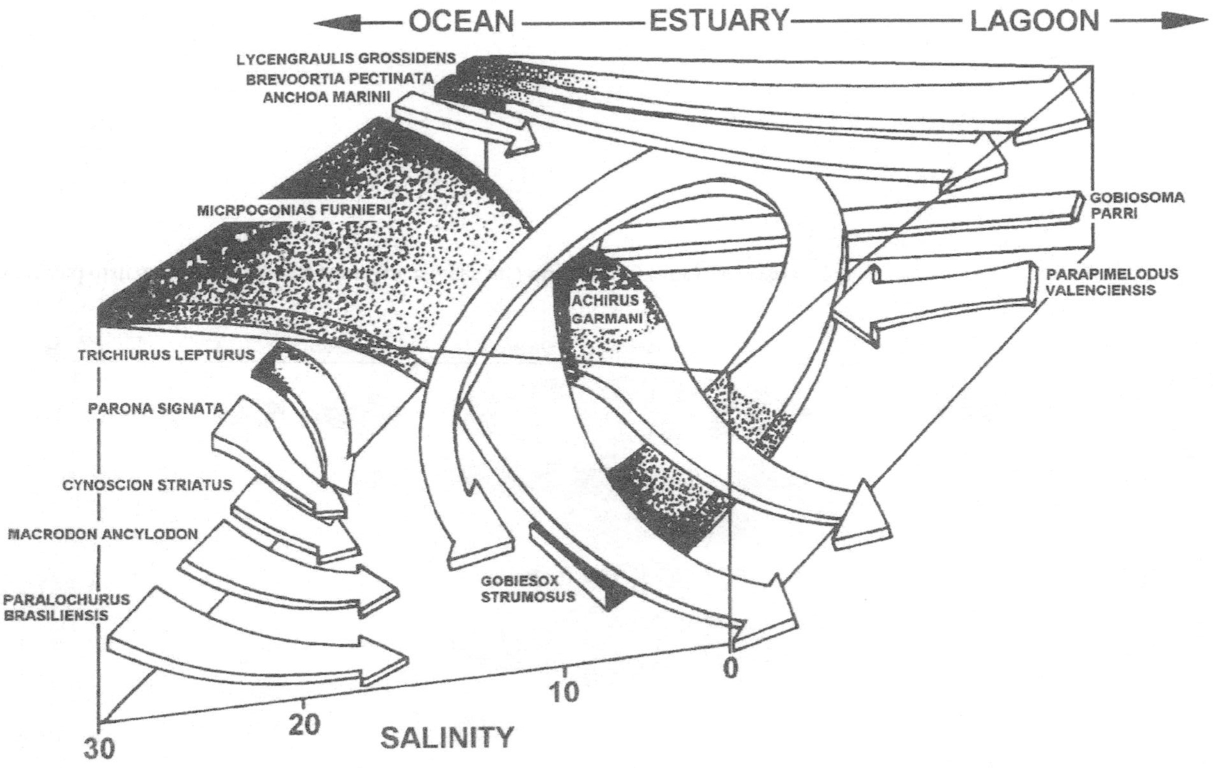

Fig. 4.12.1. General model summarizing the different early life-history stages of fish in the Patos Lagoon estuary. The width of the *arrows* represents the relative abundance of different stages of development (egg stages: *shaded*; larvae: *clear*. Modified after Weiss 1981)

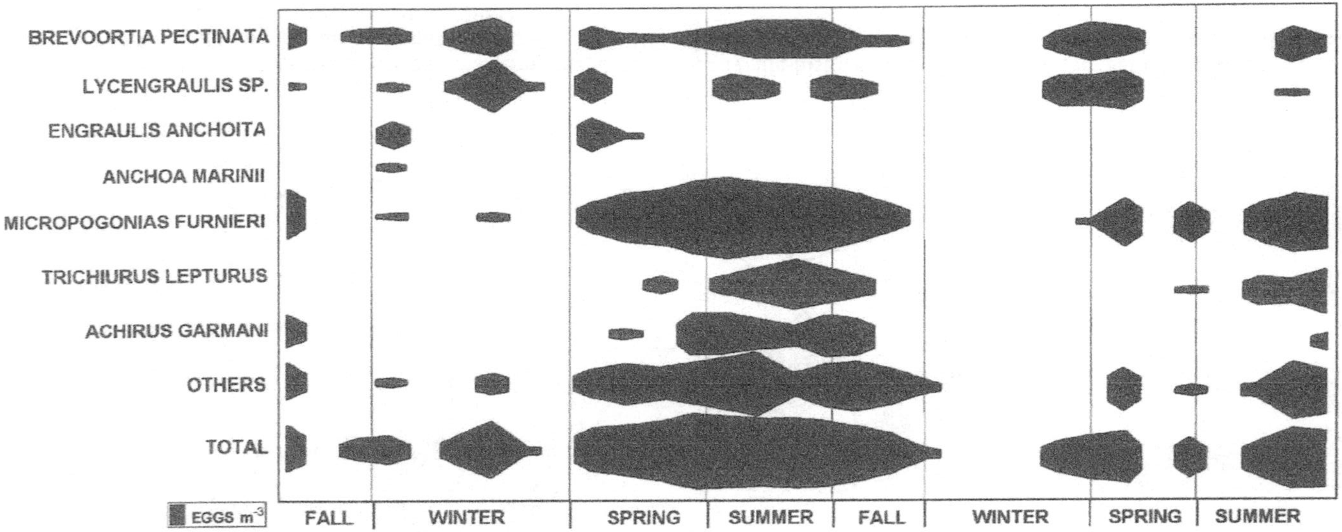

Fig. 4.12.2. Seasonal distribution of fish egg abundance (eggs m^{-3}) during a 2-year cycle (1981-1983)

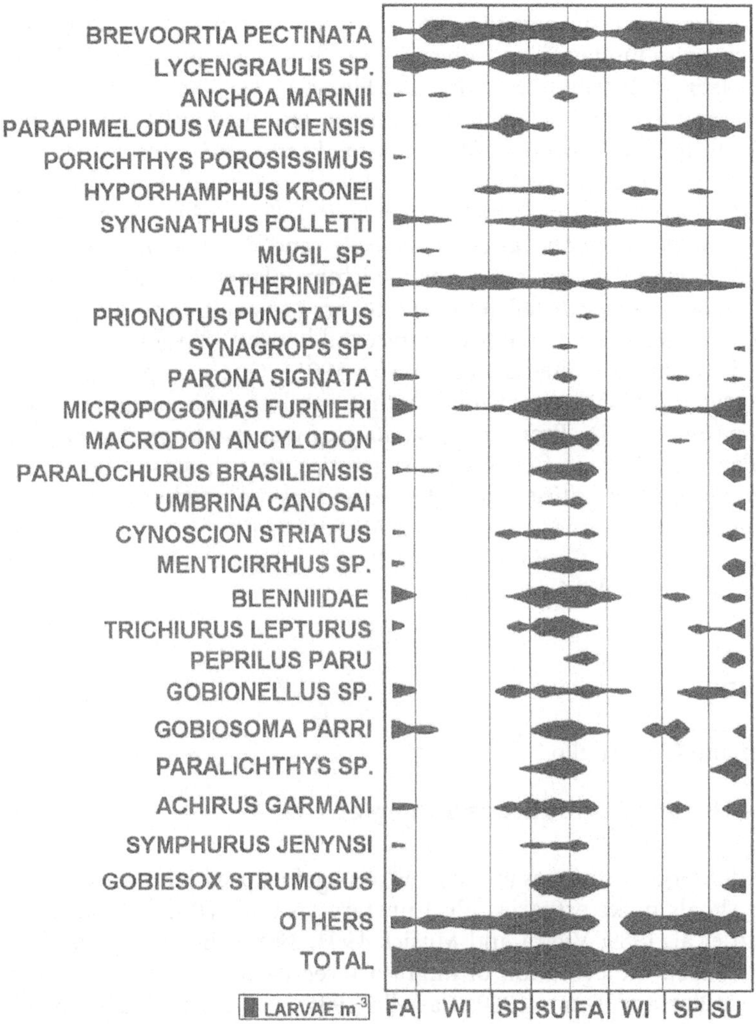

Fig. 4.12.3. Seasonal distribution of larval abundance (larvae m^{-3}) during a 2-year cycle (1981-1983)

Temporal Distribution

Temperature rather than salinity seems to control the seasonal distribution of spawning and of ichthyoplankton in the estuary. In general, the highest abundance of eggs (Fig. 4.12.2) and larvae (Fig. 4.12.3) occurs during the summer. With the exception of *Engraulis anchoita*, the eggs of all identified

species are present in the spring (18%) and summer (80%). During the fall and winter egg abundance and diversity are low. Eggs and larvae follow similar seasonal distribution patterns (Fig. 4.12.3). The highest larval abundance and diversity occur in the summer, and the lowest larval abundance in the fall, with *Brevoortia pectinata*, *Lycengraulis grossidens*, and Atherinidae as the dominant taxa, and *Achirus garmani*, *Parapimelodus nigribarbis*, *Hyporhamphus kronei*, *Syngnathus folletti*, *Mugil* spp., *Micropogonias furnieri*, *Paralonchurus brasiliensis*, and *Gobiosoma parri* also present. Although in winter and spring the ichthyoplankton concentration is generally higher than in the fall, fewer species are present in the estuary.
Fish larvae in the Patos Lagoon estuary exhibit two different patterns of seasonal distribution (Muelbert and Weiss 1991). One group of larvae is present throughout the year, while another exhibits a discontinuous seasonal cycle (Fig. 4.12.3). *Brevoortia pectinata*, *Lycengraulis grossidens*, and Atherinidae compose the first group, suggesting that these species reproduce throughout the entire year. The second group predominates from early spring to late summer and includes *Micropogonias furnieri*, *Macrodon ancylodon*, Blenniidae, *Gobionellus* spp., *Peprilus paru*, *Trichiurus lepturus*, *Paralichthys* sp., and *Achirus garmani*.

4.13 Fish Fauna

J.P. Vieira and J.P. Castello

Estuarine Use and Reproductive Strategies

The abundant food resources and protection against predation provided by estuarine shoals make estuaries ideal nursery grounds (Potter et al. 1983; Loneragan et al. 1989; Vieira and Musick 1993, 1994) and may have decisively influenced the evolution of migratory recruitment patterns of juvenile fish (Miller and Dunn 1980; Castello 1986; Vieira and Musick 1993, 1994). About two-thirds of the commercial fishery in the USA is based on estuarine-related species (McHugh 1966, 1985), and similar estimates apply in southern Brazil (Chao et al. 1986). Several commercially important species utilize the Patos Lagoon estuary as their principal nursery during part of their life-history, while others may use shelf areas under the influence of estuary runoff. Therefore, only a limited number of species is entirely estuarine-dependent, but most species are facultative or opportunistic migrators, as has been observed elsewhere (Day et al. 1989).

The ichthyofauna of the Patos Lagoon estuary is composed of approximately 110 species (Appendix; Chao et al. 1982, 1985; Vieira 1991a), but only a few of these are abundant or frequent. Depending on whether species complete their entire life-cycle in the estuary (estuarine residents *sensu* McHugh 1967) or whether during specific life-cycle phases marine and freshwater species migrate into the estuary (estuarine-dependent *sensu* Blaber et al. 1989; estuarine-opportunists *sensu* Lenanton and Potter 1987), seven categories of local estuarine use by fish can be distinguished.

Estuarine residents are represented by *Odontesthes argentinensis*, *Atherinella (=Xenomelaniris) brasiliensis*, *Jenynsia lineata*, *Ramnogaster arcuata*, and several species of the Blenniidae, Gobiidae, and Poecilidae which are largely restricted to estuarine shoals, whilst *Achirus garmanii* and *Genidens genidens* are common to deeper waters. The marine species *Mugil platanus*, *Paralichthys orbygnianus*, and the sciaenids *Pogonias cromis* and *Micropogonias furnieri* spawn in the sea or in waters proximate to the estuary, but their post-larval, juvenile, and sub-adult life-cycle stages seem to be obligatory users of estuarine shoals and deeper waters as nursery grounds. Early life-cycle stages of another group of marine species (*Macrodon ancylodon*, *Menticirrhus americanus*, *Paralonchurus brasiliensis*, *Symphurus jenynsi*, *Porichthys porosissimus*, *Urophycis brasiliensis*, *Lycengraulis grossidens*, *Anchoa marinii*, *Trichiurus lepturus*, *Pomatomus saltatrix*) use deeper estuarine waters only facultatively or opportunistically as a nursery. However, under favourable conditions sub-adults may persist in the estuary throughout the year. Yet another considerable number of marine species (like *Cynoscion guatucupa*, *Umbrina canosai*, *Ctenosciaena gracilicirrhus*, *Prionotus punctatus*, *Peprilus paru*, *Lagocephalus laevigatus*, and *Trachinotus marginatus*) occasionally penetrate the estuary under favourable conditions as juveniles or sub-adults. *Netuma barba* and *Netuma planifrons* are anadromic species which spend the major part of their life-cycle in the sea but migrate through the estuary to reproduce in limnic and pre-limnic zones. The juvenile stages of these species also use the estuary as a nursery and feeding ground. Additionally, juveniles and sub-adults of the freshwater species *Parapimelodus nigribarbis* and *Pimelodus maculatus* use the estuary opportunistically. Especially during periods of intense freshwater flushing, their eggs, larvae, and juveniles spread throughout the entire estuarine area. Finally, a large group of freshwater (chiclids and charachiforms) and tropical marine species (Carangidae, Serranidae, Pomacentridae, Gerreidae, Balistidae), coastal elasmobranchs, and anguiliforms are occasional visitors to the estuary.

Specific Diversity and Space-Time Segregation

Most of the species which compose the fish assemblage in the Patos Lagoon estuary are of marine origin. In general, the type of fish encountered and migration patterns of their life-cycle phases are similar to those observed in estuaries elsewhere (Vieira and Musick 1994). Only a few species dominate the assemblage and are commonly caught by one or more fishing methods throughout the year (Chao et al. 1982, 1985). Of the approximately 110 species (Chao et al. 1985), 22 are caught by all fishing methods employed (i.e. bottom trawl, mid-water trawl, beach seine). The bottom trawl method nets 81 species. *Micropogonias furnieri*, *Netuma barba*, and *Genidens genidens* alone total more than 50% of the catch, whilst seven sciaenids, two ariids, and one batrachoid represent 90% in number and weight. The mid-water trawl method catches 43 species, all of which are also found in bottom trawls. Although the proportion of engraulidids and *Trichiurus lepturus* increases in mid-water trawls, *Micropogonias furnieri* continues to be an important member. The species (n = 61) composition of beach seine catches differs from the other two methods. Atherinidae, Jenynsiidae = Anablepidae, and juveniles of Mugilidae represent more than 70% of the catch in numbers, though *Micropogonias furnieri* is still an important contributor.

Based on the fishing methods employed and on the spatial and temporal distribution of species, the estuarine ichthyofauna can be divided into deeper water, shallow water, and pelagic associations (Vieira and Musick 1994). Depth gradients further separate the deeper water association into a group below 4 m depth and another between 2 and 4 m. The major components of the deeper water estuarine association are bottom-oriented epibenthic and demersal fish, like *Micropogonias furnieri*, *Netuma barba*,

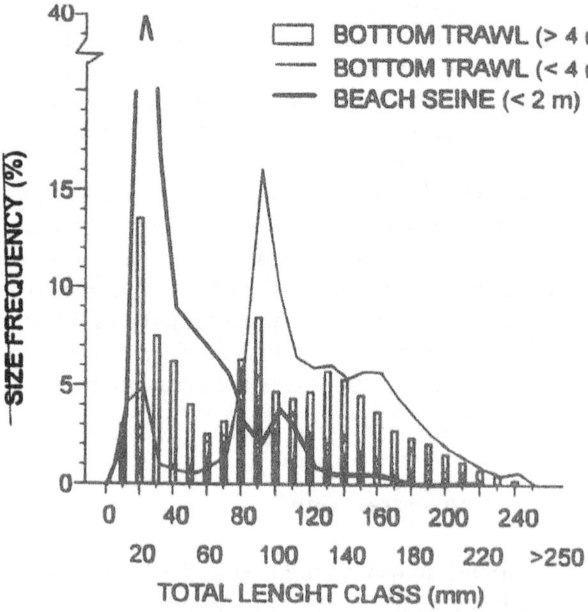

Fig. 4.13.1. Size frequency distribution (10-mm size class) of all species caught by bottom trawls and beach seines at different depths in the Patos Lagoon estuary

Genidens genidens, and several species of Sciaenidae (*Macrodon ancylodon, Menticirrhus americanus, Paralonchurus brasiliensis, Cynoscion guatucupa, Umbrina canosai*), which comprise an important portion of the southern Brazilian fishing industry. Species collected by bottom trawls in channels and adjacent areas (>4 m depth) are largely represented by post-larvae, young-of-the-year, and sub-adults between 10 and 250 mm total length (TL) (Fig. 4.13.1). Several young-of-the-year sciaenid species and *Porichthys porosissimus*, which uses the estuary as a nursery ground, may reach high abundance in the summer. However, throughout the year juveniles and sub-adults of *Micropogonias furnieri*, *Netuma* spp. and sub-adults of *Paralonchurus brasiliensis, Macrodon ancylodon*, and *Menticirrhus americanus*, which use the Patos Lagoon as a feeding ground, represent the dominant component in channels and adjacent areas (Fig. 4.13.2a). With the exception of *Netuma barba*, all dominant species are represented by individuals of less than 70 mm TL, and several sciaenids (e.g. *Ctenosciaena gracilicirrhus, Umbrina canosai, Cynoscion guatucupa*) are restricted to this size class. *Netuma barba* and *Anchoa marinii* dominate the size class between 70-90 mm TL, whilst larger (>90 mm TL) individuals are mostly represented by the sciaenids (*Micropogonias furnieri, Paralonchurus brasiliensis, Macrodon ancylodon, Menticirrhus americanus*), and Ariidae (*Netuma barba, Genidens genidens*). Species collected by bottom trawls in open estuarine waters (2-4 m depth) range from 10 and 250 mm TL (Fig. 4.13.1; 4.13.2b), and *Netuma barba* and *Anchoa marinii* dominate the size class between 70 and 90 mm TL. However, individuals below 70 mm TL are less abundant and are principally represented by *Micropogonias furnieri*. The larger size class (>100 mm TL) is mainly composed of *Micropogonias furnieri, Netuma barba*, and *Genidens genidens*.

Fish species of the shallow water association (<2 m) caught by beach seines (littoral fishes *sensu* Day et al. 1989) are typically dominated by small estuarine resident fish (Atherinidae, Jenynsiidae = Anablepidae) and juveniles of estuarine-dependent marine species of the Mugilidae and *Micropogonias furnieri* (5 to <200 mm TL; Figs. 4.13.1; 4.13.2c). Together, these species make up 90% in number of beach seine catches and display three clearly defined annual abundance peaks (Chao et al. 1985). Young individuals of three species of Mugilidae (<50 mm TL) compose 42% (*M. platanus* 34.6%) of the catch. *Mugil curema* and *Mugil gaimardianus* occur during summer/fall, whilst *Mugil platanus* is present throughout the year, with a peak abundance from late winter to early summer. However, although absent from bottom trawl or beach seine samples, large monospecific schools of mature adults of *Mugil platanus* are abundant during the late summer and fall (Vieira and Scalabrin 1991). Both juveniles and adults of *Mugil platanus* feed on diatoms, bacteria, and associated inorganic matter (Vieira 1991b). Species of the Atherinidae are present throughout the year and total about 30% of the catch. The peak abundance

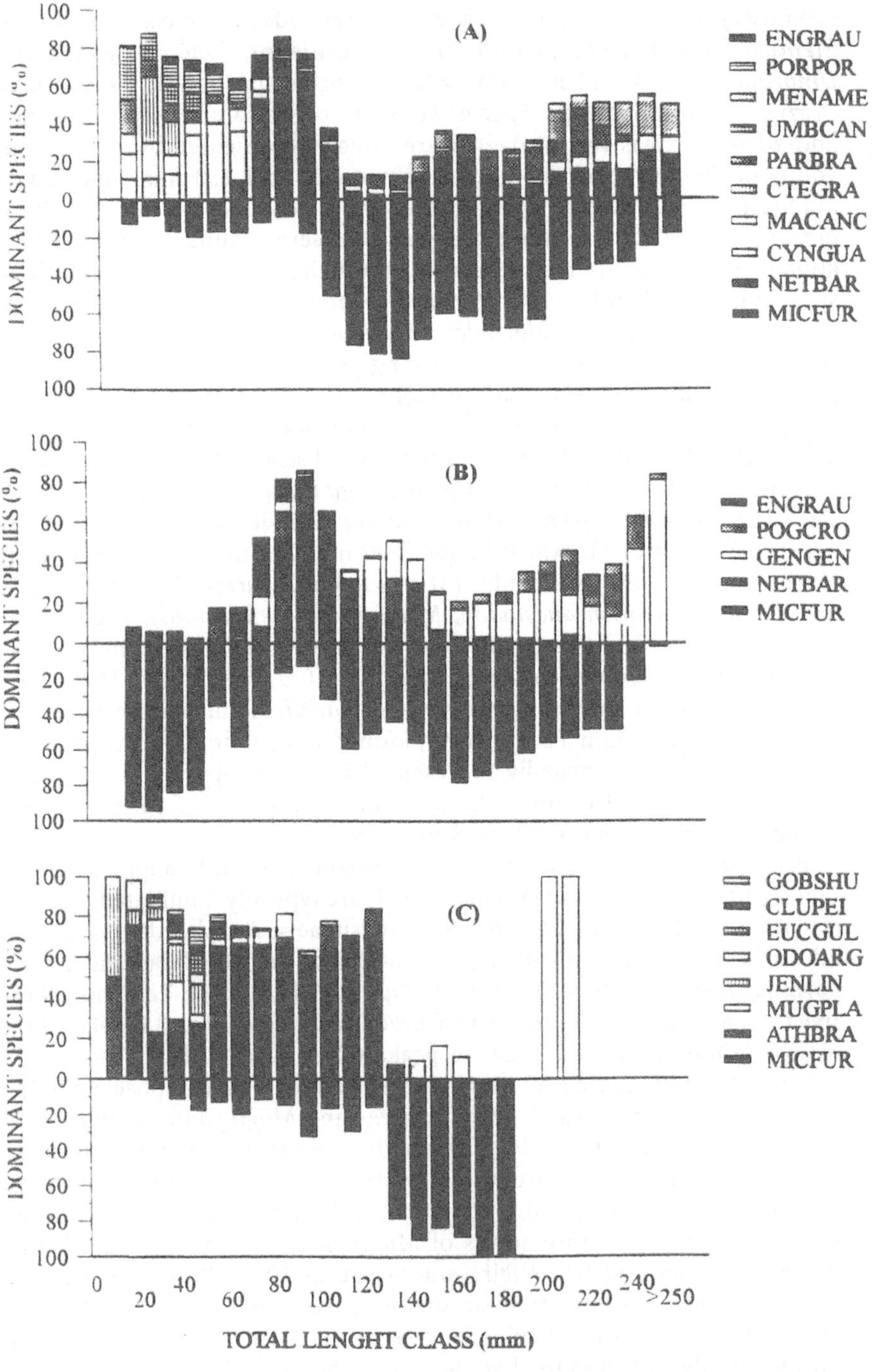
(A)
(B)
(C)
ENGRAU
PORPOR
MENAME
UMBCAN
PARBRA
CTEGRA
MACANC
CYNGUA
NETBAR
MICFUR
POGCRO
GENGEN
GOBSHU
CLUPEI
EUCGUL
ODOARG
JENLIN
MUGPLA
ATHBRA
DOMINANT SPECIES (%)
100
80
60
40
20
0
20
40
60
80
100
0
20
40
60
80
100
120
140
160
180
200
220
240
>250
TOTAL LENGHT CLASS (mm)

Fig. 4.13.2.

of *Atherinella brasiliensis* (22.6%) and *Odontesthes argentinensis* (7.7%) occurs during the fall/winter and winter/spring, respectively, the first always preceding the second by 1 or 2 months. The size classes of these species rarely coincide in time and space (Bemvenuti 1984), probably owing to similar feeding habits. Juveniles (<70 mm TL) of both species are plankton feeders, with a preference for phytoplankton followed by zooplankton, whilst adults graze upon benthic organisms (Bemvenuti 1990). The two abundance peaks of *Jenynsia lineata* (20.9%) coincide with the species' reproductive cycle between the spring and fall (Chao et al. 1985). This small omnivorous species appears to lack ontogenetic changes in diet preference and grazes principally upon benthic vegetation (Betito 1984).

Species of the pelagic association swim freely through the entire water column and are sampled by mid-water trawls, although they are also frequently caught by bottom trawls and beach seines. The engraulidids (<120 mm LT) *Lycengraulis grossidens, Anchoa marinii, Engraulis anchoita*, and juveniles of *Peprilus paru* and *Selene setapinis* total 40% of mid-water trawl catches. Young individuals of *Trichiurus lepturus* also form an important component of the pelagic association. Plankton-feeding juveniles of the clupeids *Brevoortia pectinata, Ramnogaster arcuata*, and *Platanichthys platana* are abundant in somewhat shallower areas. Engraulidids are plankton feeders and partition the available resources through differential occupation of space during the year. Juveniles of *Engraulis anchoita* enter the estuary during the fall and winter, but *Anchoa marinii* and *Lycengraulis grossidens* are present throughout the year. Salinity distributions in the estuary are largely responsible for the vertical and horizontal segregation of the two latter species (Buckup 1984). Salinity also appears to segregate different age cohorts of *Lycengraulis grossidens*, with larvae preferring oligohaline waters for their development (Castello and Krug 1978). Top predators, such as Pomatomidae, Trichiuridae, and several Sciaenidae, are often abundant members of the pelagic association in large estuaries (Vieira and Musick 1994). Since these species are scarce or absent from the pelagic association in the Patos Lagoon estuary, it is likely that the spatial and temporal distribution of the local ichthyofauna is controlled by environmental factors and competition for available food resources, rather than by predation.

Fig. 4.13.2. Contribution of dominant species at different total length classes in **A** bottom trawls at more than 4 m depth; **B** bottom trawls at less than 4 m depth; and **C** shallow water (<2 m) beach seine catches. **ATHBRA** *Atherinella brasiliensis*; **CLUPEI** *Brevoortia pectinata, Ramnogaster arcuata, Platanichthys platana*; **CTEGRA** *Ctenosciaena gracilicirrhus*; **ENGRAU** *Anchoa marinii, Lycengraulis grossidens*; **EUCGUL** *Eucinostomus gula*; **GENGEN** *Genidens genidens*; **GOBSHU** *Gobionellus shufeldti*; **JENLIN** *Jenynsia lineata*; **MACANC** *Macrodon ancylodon*; **MENAME** *Menticirrhus americanus*; **MICFUR** *Micropogonias furnieri*; **MUGPLA** *Mugil platanus*; **NETBAR** *Netuma barba*; **ODOARG** *Odontesthes argentinensis*; **PARBRA** *Paralonchurus brasiliensis*; **POGCRO** *Pogonias cromis*; **PORPOR** *Porichthys porosissimus*; **CYNGUA** *Cynoscion guatucupa*; **UMBCAN** *Umbrina canosai*

4.14 Bird Fauna

C.M. Vooren

The open waters and protected shallow bays of the estuary and the wetlands and beaches at their margins comprise habitats which in the last century contained a bird fauna so abundant and diverse as to draw the attention of visiting scientists and explorers (Appendix). Von Ihering (1887) states that highly abundant breeding birds such as the black-necked swan (*Cygnus melancorhyphus*), the roseate spoonbill (*Ajaia ajaja*), gulls and terns of the genera *Larus* and *Sterna*, carão (*Araunus guarauna*), grebes of the genus *Podiceps*, and several species of Anatidae and Threskiornitidae were exploited commercially for feathers, eggs, and meat, and that already in those days a great decline in their abundance was noted, especially of the black-necked swan. In spite of all this, the bird fauna of the myxohaline waters and shores of the Patos Lagoon estuary has never been studied systematically. Some general features may be deduced from the studies on estuarine and marine birds in general (Escalante 1970, 1983; Barattini and Escalante 1971), from the reviews by Belton (1984, 1985) and Sick (1984), from the study on the shorebirds and coastal marine birds of Cassino Beach (Vooren and Chiaradia 1990), and from unpublished personal observations in the estuary. Thirteen species of birds are common or otherwise of interest, and are discussed below.

The area supports six species of piscivorous birds. The bigua cormorant *Phalacrocorax olivaceus*, one of the most common birds in the estuary, fishes by diving in shallow water. No breeding places of this bird are known in or near the area. Trudeau's tern (*Sterna trudeaui*), common tern (*Sterna hirundo*), and black skimmer (*Rynchops niger)* prey on fishes at the water surface. *Sterna trudeaui* probably breeds in the nearby inland marshes, although Belton (1984) was unable to confirm this, while *Sterna hirundo* is a summer visitor from North America (Vooren and Chiaradia 1990). Both these terns feed mostly at sea but fish in the estuary when the sea is rough. *Sterna hirundo* roosts in numbers of up to 5000 birds on a beach in the estuarine inlet. *Rynchops niger* roosts in flocks of up to 300 birds on sand banks in the same area. This species does not breed in the area, and it is not known from where the birds come. The birds fish mostly at night (Escalante 1970), probably in the estuary. The snowy egret (*Egretta thula*) and the green heron (*Butorides striatus*) breed in nearby marshes and fish habitually in the shallow water of estuarine margins and creeks. *Cygnus melancorhyphus* breeds in marshes south of the estuary (Taim ecological reserve), and groups are occasionally seen in the estuary. The species is herbivorous and may graze on beds of *Ruppia maritima*, but this remains to be confirmed. The latter is also true for the red-gartered coot (*Fulica armillata*), which is common in the estuary, probably breeds in the area, and is mostly herbivorous (Escalante 1983). The brown-hooded gull (*Larus maculipennis*) breeds in nearby inland marshes. This gull scavenges in large numbers in the fishing harbours and at sewage outlets in the estuary, and roosts at night in protected bays of islands. The kelp gull (*Larus dominicanus*) is also a scavenger but feeds mostly on the seashore and only enters the lower reaches of the estuary to roost in large numbers on the beaches. The species breeds on the coast of Uruguay and is a non-breeding visitor to this area. The muddy shore and banks of the inner reaches of lower estuarine bays are used by waders. The hudsonian godwit (*Limosa haemastica*) from North America, the rufous-chested dotterel (*Charadrius modestus*) from Patagonia, and the common stilt (*Himantopus himantopus*) from nearby inland breeding areas have been seen there. These and other waders may also use such areas in other bays of the estuary. The small islands of the estuary are important as roosting places for gulls and herons. *Rynchops niger* roost in large numbers on the sandy peninsula of the inlet. Adjacent salt marshes and the sandy beaches which border them are important roosting places for *Sterna trudeaui* and *Larus dominicanus* and are at high tide and rough weather a refuge for sandpipers and plovers from the seashore. These critical areas need to be studied in detail so that measures for their conservation can be taken.

4.15 Marine Mammals

M.C. Pinedo

Species Composition and Movements

The bottlenose dolphin (*Tursiops truncatus;* Pinedo et al. 1992), previously referred to as *Tursiops gephyreus* (Castello and Pinedo 1977a; Pinedo 1982, 1986), is common in the Patos Lagoon estuary. Between 31 (Möller et al. 1994) and approximately 100 individuals (generally in groups of 3-4) enter the estuary throughout the year (Castello and Pinedo 1977a; Pinedo 1982), though they are more frequent in May and November (Pinedo, unpubl.). Since the species tolerates low salinities (2-3; Castello and Pinedo 1977a), specimens may reach the limits of the upper estuary near the São Gonçalo Channel. Individuals of the sea lion (*Otaria flavescens*) only occasionally enter the estuary, but in contrast to the bottlenose dolphin, specimens of *Otaria flavescens* may occasionally reach the upper limits of the lagoon near Porto Alegre, probably in search of food (Pinedo 1990). Although *Pontoporia blainvillei* has been reported in the estuary (von Ihering 1892; Cabrera and Yepes 1940; Carvalho 1961, 1975) the species has not been seen since 1976.

Feeding and Behaviour

Various types of behaviour, like leaps out of the water, tail beating of the surface, belly exposure, or vertical body positioning are displayed by *Tursiops truncatus* (Castello and Pinedo 1977a). Rapid displacements and occasional leaps always precede the capture of fish. The most common prey is *Micropogonias furnieri*, followed by *Trichiurus lepturus* and *Trachinotus* spp. These species are first bitten and, after floating on the surface, ingested (Pinedo 1982). Their abundance is likely to influence the occurrence of dolphins in the estuary (Castello and Pinedo 1977a). A cooperative fishing method between bottlenose dolphins and fishermen, using a circular nylon throw-net, is practised at 28^0 29' S (Pryor et al. 1990) but has not been observed in the Patos Lagoon. On rare occasions *Tursiops truncatus* may be caught in gill nets set for shrimp fishery in the estuary (Castello and Pinedo 1977a).

5 Energy Flow and Habitats in the Patos Lagoon Estuary

5.1 Primary Production Cycles

U. Seeliger, C.S.B. Costa, and P.C. Abreu

The functional primary producers in the Patos Lagoon estuary include emersed and submersed macrophytes, benthic and floating macroalgae, cyanobacteria, and epibenthic, epiphytic, and planktonic microalgae. All supply food for estuarine and coastal consumers and interact with estuarine chemical cycles; some perform critical habitat functions during different life stages of inumerous other organisms (Kennish 1986; Day et al. 1989). In the Patos Lagoon estuary, the various primary producers display different production patterns, and the rates of production are likely to vary in response to biotic and abiotic conditions; thus, the availability of live organic matter and detritus may fluctuate considerably within and between years.

	1966	1967	1968	1969	1970	1971	1972	1973	1974	1975
JANUARY	100	68	88	66	91	150	121	173	171	66
FEBRUARY	100	66	50	85	66	133	73	141	101	91
MARCH	135	51	136	60	93	70	91	45	70	75
APRIL	65	33	50	30	41	28	46	81	20	66
MAY	45	106	36	101	116	85	65	86	116	83
JUNE	125	178	41	68	133	118	146	83	80	60
JULY	168	108	66	70	96	81	158	208	75	75
AUGUST	125	146	28	81	105	82	208	73	103	158
SEPTEMBER	100	136	110	66	30	98	126	58	110	110
OCTOBER	95	111	66	40	91	35	113	70	70	51
NOVEMBER	55	50	93	75	35	37	75	18	66	83
DECEMBER	135	35	100	50	105	116	70	66	85	60
TOTAL	1248	1088	864	792	1002	1033	1292	1102	1067	978

Fig. 5.1.1. Minima and maxima of mean monthly precipitaton between 1966-1975 in the Patos-Mirim Lagoon watershed

In most estuaries physico-chemical factors vary predictably on an hourly to seasonal scale because they are largely controlled by regular tidal and seasonal flushing cycles (Wolfe and Kjerfve 1986). In contrast, the choked lagoon characteristics and negligible tidal amplitude make physico-chemical parameters in the Patos Lagoon estuary largely dependent on wind and rainfall. Precipitation patterns also tend to vary from year to year (Fig. 5.1.1; Castello and Möller 1978), owing to the influence of El Niño and anti-El Niño phenomena on the climate in the southwestern Atlantic (Arntz and Fahrbach 1991; IAI 1992). As a consequence, the unpredictable short-term variations of physico-chemical factors with their irregular amplitude and timing, their reduced seasonal oscillation, and their pronounced inter-annual variability become important abiotic forcing factors in the Patos Lagoon estuary. Further-more, the physiographic heterogeneity of the estuary with deeper open waters and protected, semi-open, and open shallow marginal bays induces different site-specific environmental characteristics. Common to all sites are seasonal cycles of surface water temperature and incident light (Coutinho 1982; Cafruni 1983) and generally reduced and highly unstable underwater light conditions, owing to frequent resuspension of fine sediments by winds (Costa et al. 1988a; Moreno 1994). Differences between sites are largely due to hydrology, degree of wind exposure, and sediment characteristics.

Phytoplankton biomass and production in the estuary, with a mean concentration of 0.07 g C m^{-2} (0.03-7.0 g C m^{-2}; Abreu, unpubl.) and production rate of 0.072 g C m^{-2} h^{-1} (0.004-0.320 g C m^{-2} h^{-1}; Abreu et al. 1994a), are largely determined by the contribution of flagellates (<20 μm). Peak biomass occurs in the spring and summer, while values close to zero are common in the winter and fall when heterotrophic processes prevail. Seasonal variations of phytoplankton biomass and production are principally related to insolation and temperature. The lowest primary production values tend to coincide with water temperatures below 20^{0}C and mean water column irradiance below the critical light intensity (Riley 1967) of 40 gcal cm^{-2} day^{-1} (Abreu et al. 1994a). Furthermore, especially during the summer, the phytoplankton biomass is influenced by zooplankton grazing. Although grazing will reduce the biomass, it seems to stimulate phytoplankton productivity per unit of biomass, owing to the increase in available light and higher nutrient concentrations (Abreu et al. 1994b).

In contrast to tide-controlled estuaries, both the short-and long-term phytoplankton dynamics in the Patos Lagoon estuary are directly related to hydrological and meteorological events. Since short-term chlorophyll a variability is a function of the residence time of water in the estuary, which is controlled in turn by wind direction and velocity, high chlorophyll a biomass arises during runoff after prolonged periods of salt water penetration (Abreu 1987). Long-term interannual chlorophyll a oscillations depend on the precipitation patterns in the watershed and estuarine region. Phytoplankton biomass and production peaks are closely related to the input of high nitrogenous estuarine nutrient concentrations after elevated rainfall. Since large-scale El Niño events augment precipitation in southern Brazil, they are likely to promote increased estuarine phytoplankton biomass and production (Abreu et al. 1994a; Proença et al. 1994; Bergesch et al. 1995)).

The considerable biomass (6-22 g C m^{-2}) of benthic microphytes is dominated by pennate diatoms. After resuspension in the water column during high incident light, the photosynthesis of benthic cells is inhibited, suggesting their adaptation to low light conditions (Abreu 1992). The variability of mean light intensities seems to be the key controlling factor for these organisms and might explain why phytobenthic cell peaks occur during the winter, contrary to maximum phytoplankton biomass in the spring and summer (Bergesch et al. 1995).

Benthic macroalgae are responsible for most of the biomass production in subtidal areas during the winter and early spring. The highest mean monthly biomass of the dominant *Enteromorpha* sp. (34.3 g C m^{-2}), *Rhizoclonium riparium* (308 g C m^{-2}), *Ulothrix flacca* (2.2 g C m^{-2}), and *Vaucheria longicaulis* (70.4 g C m^{-2}) is linked to regional light and temperature regimes and to the site-specific availability, permanence, and exposure of suitable substrate (Coutinho and Seeliger 1986). Throughout the year cyanobacteria represent the largest single component of the subtidal flora, with dominance of Lyngbya confervoides in the summer (peak = 55.7 g C m^{-2}) and *Microcoleus chthomoplastes* in the fall (peak = 19.8 g C m^{-2}; Coutinho 1982; Coutinho and Seeliger 1986). Although the primary productivity values of cyanobacteria in saltmarsh environments are lacking, carbon and nitrogen fixation levels are probably high (Yunes et al. 1990).

Despite the local occurrence of *Zannichellia palustris* during prolonged periods of low salinity (Cafruni et al. 1978; Cafruni 1983; Moreno 1994), *Ruppia maritima* is responsible for most of the submersed spermatophyte biomass formation. Rather than being a stable feature of the benthos, *Ruppia maritima* may exhibit either perennial growth or annual growth cycles with massive die-off during the fall, though approximately seasonal cycles of biomass formation are most common (Cafruni 1983; Asmus et al. 1984; Moreno 1994). Biomass peaks may reach 82.8 g C m^{-2}, but the average value in the estuary is approximately 10.8 g C m^{-2}, though it may vary greatly from site to site (Cafruni 1983; Moreno 1994). Data on *Ruppia* productivity are still lacking, but preliminary estimates indicate a net annual production of 39.6-43.2 g C m^{-2} $year^{-1}$ (Seeliger, unpubl.). *Ruppia maritima* also serves as the principal semi-permanent substrate for benthic micro- and macroalgae in the estuary (Ferreira and Seeliger 1985). Since abiotic conditions, the availability of colonizer species, size and density of *Ruppia* beds, and age and turnover of leaves influence epiphyte biomass, production patterns are likely to be highly variable. Epiphyte biomass (113.1 g C m^{-2})

grows towards the summer and fall and represents between 50 and 70% of the total *Ruppia maritima* biomass (Coutinho 1982; Seeliger, unpubl.), though occasionally it may exceed 100%.

The species-specific seasonal growth cycles (Fig. 5.1.2) of dominant marsh species with either C_3 *(Scirpus maritimus)* or C_4 (*Spartina alterniflora, Spartina densiflora*) carbon metabolism are likely to be a function of different air temperature optima (Rozema et al. 1991). Lower air temperatures (<15°C) favour rapid vegetative growth of *Scirpus maritimus* during the early spring, followed by peak live aerial biomass at the end of spring and maximum dead and total aerial biomass at the end of summer (Costa, in press). In contrast, the highest vegetative growth of leaves and shoots of *Spartina alterniflora* (Cunha 1994) and *Spartina densiflora* (Silva et al. 1993) coincides with the end of spring when temperatures surpass 20°C, whilst peak dead and total aerial biomass occur in the early fall and during fall-winter, respectively. In addition, interannual temperature differences may decisively affect primary production of the three dominant marsh species (Fig. 5.1.2). A mean temperature decrease of 2.8°C during the growing season results in 11% and 33% reduction of aboveground and total *Spartina alterniflora* primary production, respectively (Cunha 1994). Mean net aboveground production ranges from 287.7 to 808.4 g C m^{-2} $year^{-1}$ and is

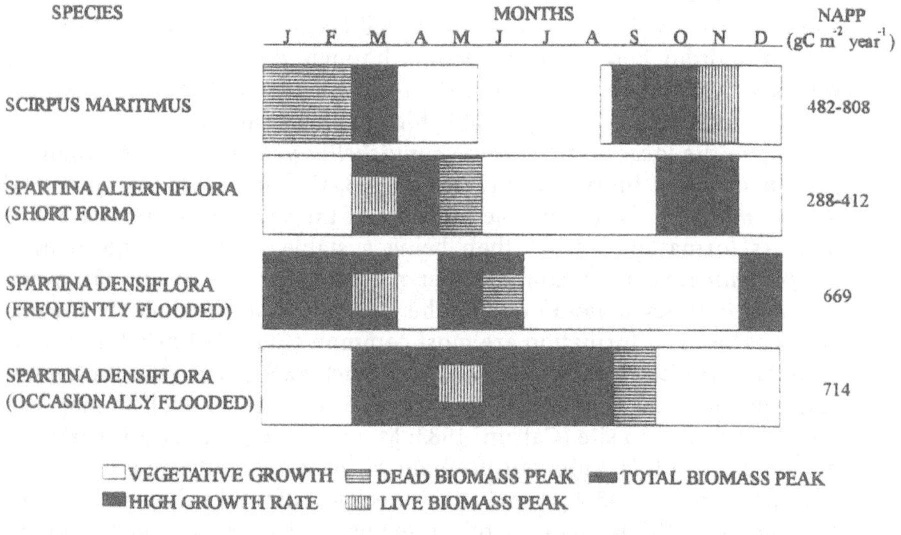

Fig. 5.1.2. Aerial growth and biomass cycles and net aerial primary production (NAPP) estimates based on dry weight to carbon conversion (0.43) of dominant salt marsh plants in the Patos Lagoon estuary

comparable to highly productive, tide-controlled *Spartina alterniflora* marshes elsewhere (Turner 1976; Long and Mason 1983; Adam 1990). Net belowground production of 727.9-2095.4 g C m^{-2} $year^{-1}$ and 1510.2-2282.9 g C m^{-2} $year^{-1}$ of short *Spartina alterniflora* (Cunha 1994) and *Scirpus maritimus* (Costa, in press), respectively, represent more than 70% of their total net primary production.

Conservative estimates of net primary production based on maximum and minimum biomass difference (Roberts et al. 1993) suggest that the dominant salt marsh plants, macroalgae, and cyanobacteria together are responsible for as much as 86% of autochthonous carbon addition to the estuary, though much of the annual marsh production may be recycled in situ, owing to irregular flooding events (Costa, unpubl). If the annual biomass turnover rates

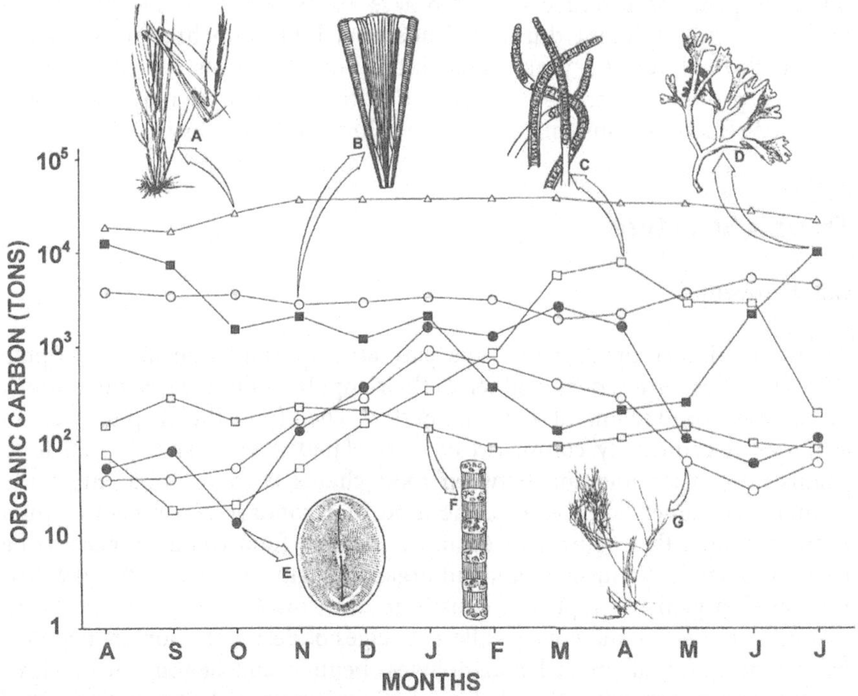

Fig. 5.1.3. Total monthly organic carbon contribution by marsh plants (A), epibenthic microalgae (B), cyanobacteria (C), benthic macroalgae (D), epiphytic microalgae (E), phytoplankton (F), and *Ruppia maritima* (G) in the Patos Lagoon estuary

reported for salt marsh plants (Schubauer and Hopkinson 1984; Costa, in press), seagrasses (Hillman et al. 1989), and benthic macroalgae (Lüning 1990) are applied, net annual carbon contributions increase even more. In particular, the net annual contributions of phytoplankton will be enhanced by turnover rates of up to 50 times (Abreu et al. 1994a).

Carbon contributions to the Patos Lagoon estuary are the result of both year-round supply by some producer groups (i.e. salt marshes) and sequential biomass pulses by others (i.e. macroalgae, *Ruppia maritima*, and epiphytes; Fig. 5.1.3). The apparent seasonal programming of carbon fixation is one of the functional processes that sustains high estuarine production (Rojas-Galaviz et al. 1992). Some producer groups (i.e. *Ruppia maritima* and salt marsh plants) will also enhance biomass generation by other primary producers via substrate formation, reduction in hydrodynamics, and sediment-water column modifications. The combined influence of different primary producers on biomass formation and decomposition processes with different duration leads to a high autochthonous carbon supply throughout the year. Large interannual changes of total annual primary production and subsequent detrital production may have pronounced effects on consumers and the ecology of the estuary as a whole.

5.2 Trophic Structure

C.E. Bemvenuti

Different primary producer groups play an important role in the trophic structure of estuaries, owing either to their supply of diverse grazing options or to their production of detrital matter. Therefore, the trophic web in estuaries is principally composed of interrelated grazing and detrital food chains. The discrimination between food chains is more apparent at the primary consumer level because the food of secondary consumers is often derived from either source. In estuaries, detrital food chains appear to be more important (Kennish 1990), and organic detritus particles with attached microbial organisms typically constitute the lowest trophic level. Large quantities of detrital material at different stages of decomposition are supplied by emersed and submersed macrophytes, benthic and floating macroalgae, cyanobacteria, and epibenthic and epiphytic microalgae. Additionally, part of the net phytoplankton production is incorporated into the bottom detritus pool. Although energy contributions by the different primary producers and detrital fractions to the various animal groups have not yet been quantified in the Patos Lagoon estuary, studies on food chain partitioning indicate their differential role in different estuarine habitats (Fig. 5.2.1).

Organic detritus particles represent a large portion of the diet of macrobenthic invertebrates and the meiobenthos, and thus these organisms are likely to provide the major link of energy flow in an oligohaline estuary. As in estuaries elsewhere (Day et al. 1989), the meiobenthic community in soft-bottom substrates of the Patos Lagoon estuary is dominated by nematodes and ostracods, followed by harpacticoid copepods and turbelarians. Some nematode species feed on detritus and microbial organisms in deeper anaerobic layers and therefore facilitate the energy flow to upper aerobic sediments. Most meiobenthic organisms, however, feed on detrital particles in superficial sediments and may be consequently ingested by deposit feeders (Coull and Bell 1979) or selectively preyed upon by epibenthic crustaceans and fishes (Castel 1992). Because meiobenthic organisms are very abundant and have high population turnover rates, they assume an important function at specific trophic levels in the estuarine detritus food chain (Day et al. 1989; Kennish 1990; Castel 1992).

Macrobenthic invertebrates display diverse feeding habits (Bemvenuti 1983, 1992; Asmus 1984), but detritus appears to be an obligate food item for most species. Infaunal deposit feeders of the first consumer level, like *Laeonereis acuta,* ingest non-selectively detritus particles of surface layers, whilst the burrower *Heteromastus similis* feeds on organic matter in deeper (<15 cm) galleries (Bemvenuti 1988). Similarly, the burrowing suspension feeders *Tagelus plebeius* and *Kalliapseudes schubartii* consume detritus and

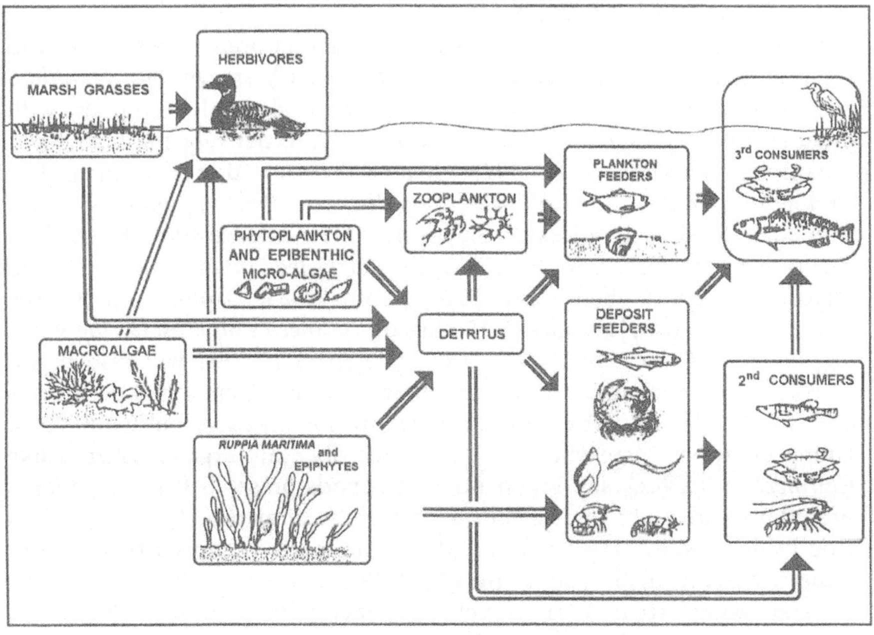

Fig. 5.2.1. Conceptual diagram of trophic relationships in the Patos Lagoon estuary

epibenthic microalgae in soft sediments (Capítoli et al. 1978). Predation of the infaunal polychaete *Nephtys fluviatilis* on *Heteromastus similis* represents an important intermediate link between the non-predatory infauna and epifaunal macropredators, which aids in the transfer of energy to higher trophic levels (Bemvenuti 1994). Infaunal predators frequently regulate the abundance of other non-predatory infaunal organisms, leading to the establishment of multiple trophic levels in soft-bottom sediments (Ambrose 1984a,b; Commito and Schrader 1985). In the Patos Lagoon estuary, however, low infaunal diversity is likely to be responsible for abbreviated food chains in soft-bottom substrates (Bemvenuti 1994).

The first consumer level of the epifauna is dominated by *Heleobia australis*, which feeds on microbes in superficial sediments and on submersed macrophyte leaves. *Tanais stanfordii* largely ingests epibenthic diatoms, but the diet of the amphipod *Melita mangrovi* also includes macroalgae and copepods (Asmus 1984). The infaunal *Kalliapseudes schubartii* (Araújo 1984; Asmus 1984; Bemvenuti 1990) and epifaunal isopods, amphipods, and tanaidaceans (Reise 1985) are an important food source for secondary consumers. In particular, the abundance of epifaunal organisms, owing to their exposure, may provide a measure of predation pressure by fish and decapods in the estuary (Bemvenuti 1987b). Among the fishes, primary consumers are represented by deposit-feeding species of the genus *Mugil* which ingest large quantities of epiphytes and detrital matter (Vieira 1985; Fig. 5.2.1).

Several highly motile decapods with diverse diet preferences are able to act at different consumer levels. The marsh crabs *Chasmagnathus granulata* and *Metasesarma rubripes* either share detrital food resources with infaunal and sedentary epifaunal species or alternatively prey on these organisms (Capítoli et al. 1977; Asmus 1984). Other decapods, like *Cyrtograpsus angulatus*, *Callinectes sapidus*, and *Penaeus paulensis*, may shift their food preferences as they grow. After intense recruitment into shallow vegetated bottoms, their early life stages feed on detrital matter and attached microbes (Bemvenuti 1983; Asmus 1984; Capítoli and Ortega 1993). As they grow, juveniles (5-15 mm) of *Callinectes sapidus*, *Cyrtograpsus angulatus*, and the small crab *Rhithropanopeus harrissi* will also ingest ostracods and nematodes of the meiofauna, as well as epifaunal peracarids and small infaunal organisms (Bemvenuti 1992). Larger individuals of young decapods increasingly prey on macrobenthic organisms and, like *Callinectes sapidus*, may ingest other decapods or small fish. In general, pre-adults and adults of decapods are typically omnivorous and opportunistic feeders, which exploit different trophic levels (Asmus 1984) as food items become available (Bemvenuti 1983, 1992).

Open water estuarine areas, with a deeper (2-6 m) water column and strong hydrodynamic activity, are comprised of soft bottoms with low organic matter concentrations. Under these conditions, the dominance of the suspension feeders *Erodona mactroides* and *Balanus improvisus* suggests an increasing importance of the grazing food chain. Owing to its high abundance and wide distribution, *Erodona mactroides* is the most important infaunal phytoplankton grazer (Bemvenuti et al. 1992), whilst fish of the clupeid and engraulidid families are prominent zooplankton feeders (Fig. 5.2.1).

In limnic areas of the Patos Lagoon, *Erodona mactroides* occurs in all size classes and represents the greatest infaunal biomass (Bemvenuti 1987a). The larger individuals are the major diet items of adult *Callinectes sapidus* and larger fish, like *Micropogonias furnieri, Netuma barba, Netuma planifrons,* and *Genidens genidens* (Araújo 1984). The larvae and juveniles of *Erodona mactroides* assume an important trophic role in the shallow estuarine shoals after intense recruitment in the summer (Bemvenuti et al. 1978). Here they form, together with *Balanus improvisus,* the principal food source for fish and decapods which use the shoals as nursery grounds (Asmus 1984).

In conclusion, the typical environmental stresses of oligohaline estuarine systems (McClusky 1981; Day et al. 1989) are especially enhanced in the Patos Lagoon estuary (Capítoli et al. 1978; Castello 1985; Bemvenuti et al. 1992) and may have contributed to the development of a soft-bottom community with wide trophic niches and abbreviated food chains.

5.3 Irregularly Flooded Marginal Marshes

C.S.B. Costa

Spatial Heterogeneity

Much of the extensive salt marsh surface (40 km^2) surrounding the Patos Lagoon estuary (Fig. 5.3.1), particularly that of the southern margin and of the small islands, displays a considerable variety of physiography, including small pools, a dense web of drainage channels, and marginal levees (Fig. 5.3.2). Owing to the choked characteristics of the estuary, reduced tidal oscillation, and highly variable seasonal and annual precipitation, the marshes are irregularly flooded by waters of varying salinity. Euryhaline conditions tend to prevail in the summer which, during periods of extreme water deficit, may occasionally approach hypersaline conditions. Oligo-and mesohaline waters are most commonly found in the winter and spring, when elevated precipitation causes intense runoff (Castello and Möller 1978; Costa et al. 1988a; Vieira and Rangel 1988). In consequence, the salinity regimes of the surface and interstitial water in the marshes are seasonally predictable (Fig. 5.3.3). The interstitial water of the more frequently inundated lower marshes tends to be more saline than that of the rarely flooded upper marshes (Bastos et al. 1993a; Silva et al. 1993), where salts are rapidly leached from increasingly sandy sediments, owing to

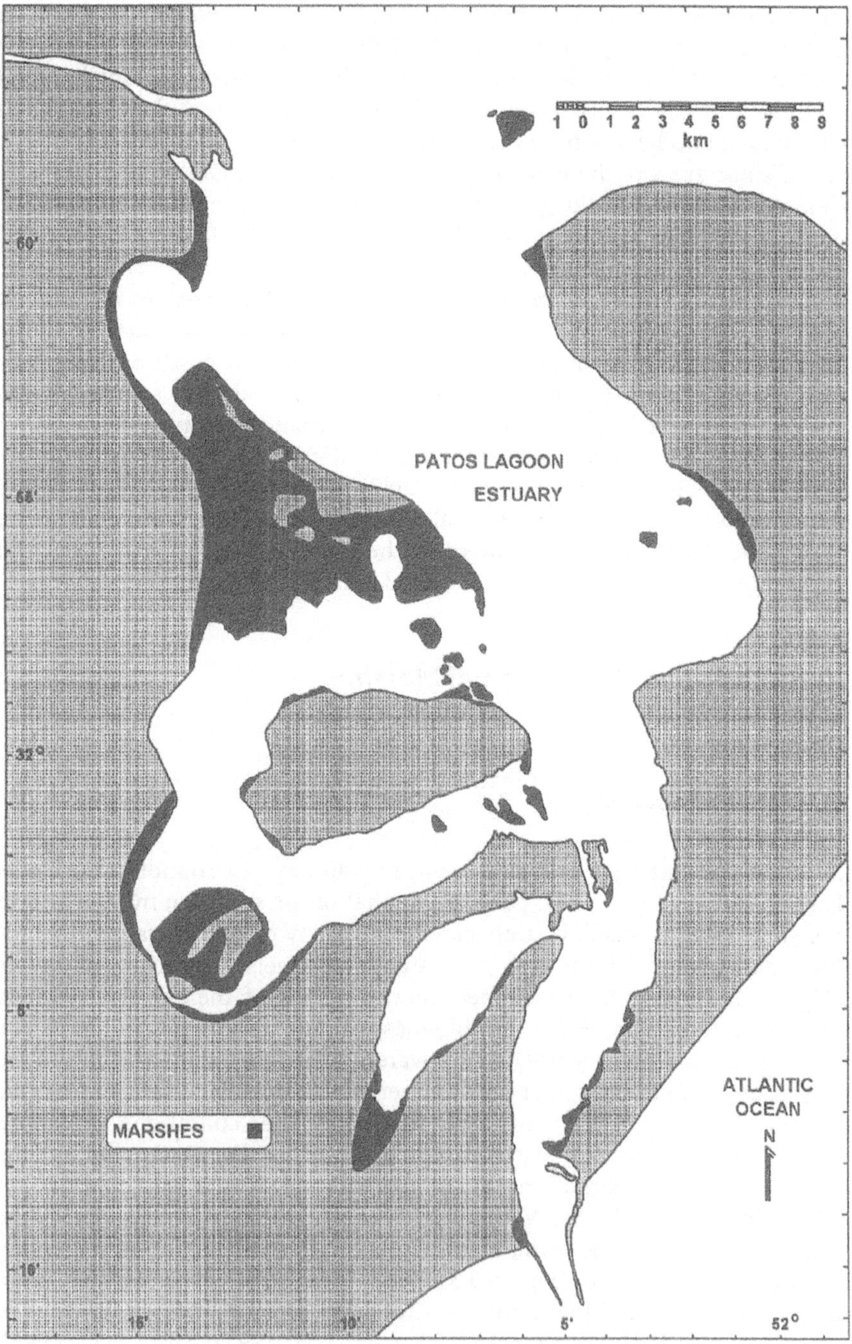

Fig. 5.3.1. Distribution of marsh habitats in the Patos Lagoon estuary

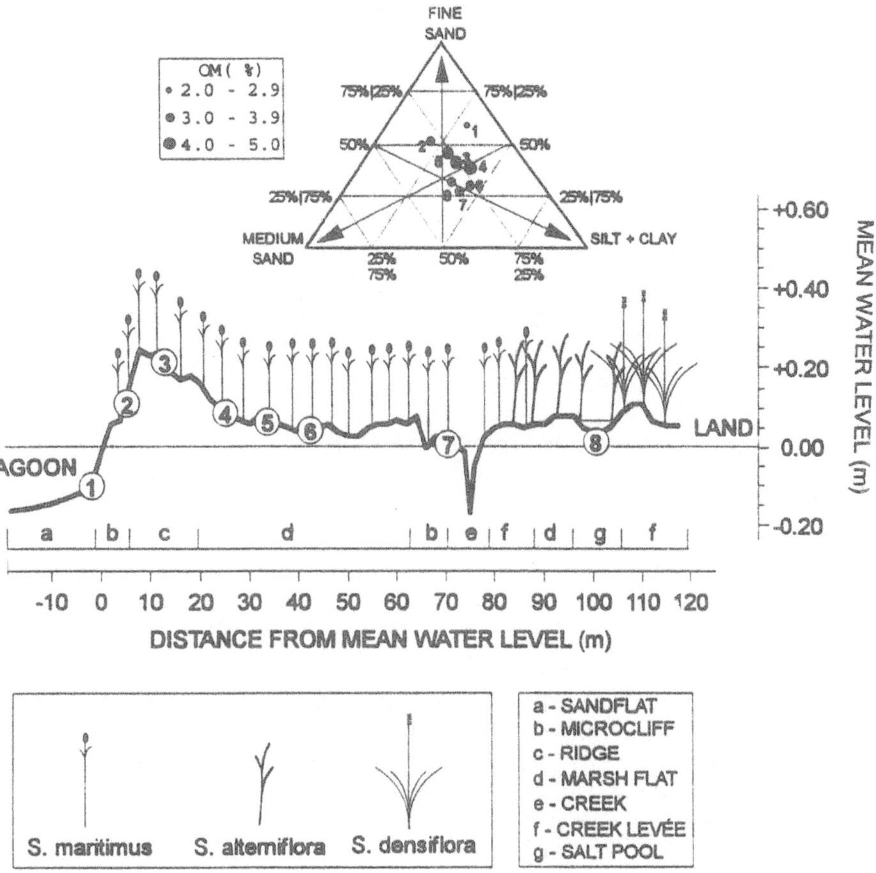

Fig. 5.3.2. Physiographic features of a typical Patos Lagoon estuary salt marsh with percent organic matter (*OM*) content and texture of sediments

high mean annual precipitation (1350 mm; Vieira and Rangel 1988). Even in marshes without pronounced physiographic diversity, the vertical and horizontal influence of different flooding patterns on marsh sediment generates environmental heterogeneity. Some of the dominant species (i.e. *Spartina densiflora*) act as important geomorphological agents. They favour accretion of the marsh surface and are able to modify drainage patterns and local physiography (Figueroa et al. 1985) because their canopy promotes deposition of fine sediments from the water column and their dense root system resists erosion of marsh creek sediments. In conjunction, these factors open new niches and provide for pronounced spatial heterogeneity with a marked influence on diversity, abundance, and distribution of plants and animals.

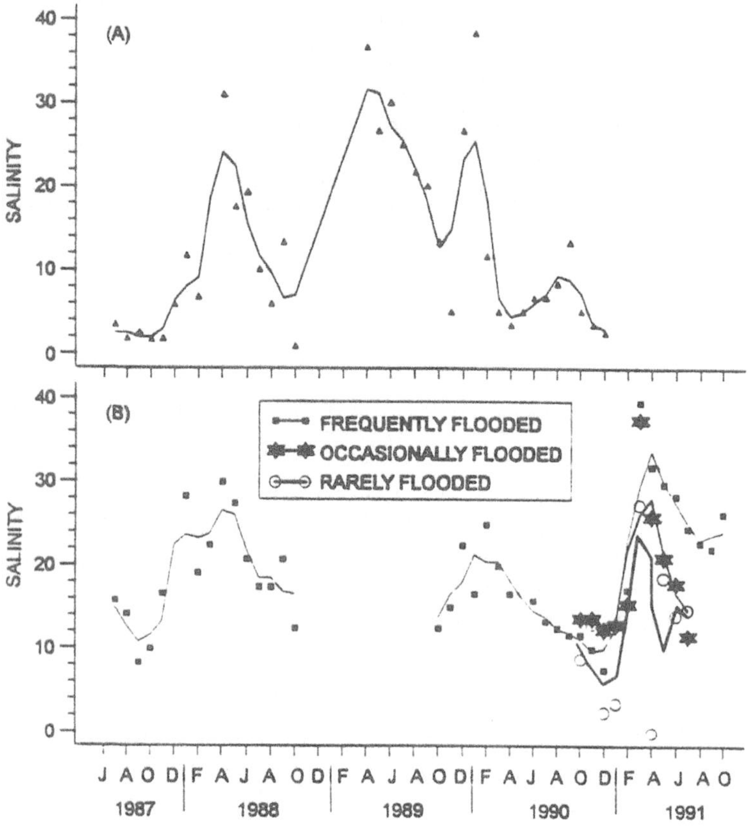

Fig. 5.3.3. Salinity of estuarine surface water (**A**) and of interstitial water in the rooting zone at different salt marsh levels (**B**) at the Patos Lagoon inlet. *Curves* are based on a simple moving average of three observations

Habitat Function

The fauna of the estuarine marshes is composed of species of aquatic and terrestrial origin. Since large extensions of the upper marshes are infrequently and unpredictably flooded, terrestrial organisms tend to be more common. However, aquatic organisms have received more attention, especially communities dominated by the crabs *Chasmagnathus granulata* (Capítoli et al. 1977, 1978; D'Incao et al. 1992) and *Metasesarma rubripes* (Capítoli et al. 1977). *Chasmagnathus granulata* is a burrowing omnivore of the lower marshes (D'Incao et al. 1990, 1992) with larval dependence on nearshore coastal waters. Although the smaller *Metasesarma rubripes* is a herbivore and, rather than burrowing, hides among the roots and shoots of

Spartina alterniflora, Spartina densiflora, Scirpus maritimus, and *Salicornia gaudichaudiana*, it competes for space and food with *Chasmagnathus granulata* (Capítoli et al. 1977). Both species act upon the fragmentation and remobilization of the belowground macrophyte biomass, thus decisively influencing the recycling of organic matter in the marshes.

Owing to the origin of sediments and local hydrodynamics, substrate characteristics vary among and within estuarine marshes. Different silt-clay ratios in the lower (3.1-72.4%) and upper (0.6-10.5%) marshes are likely to lead to changes of infaunal composition. Especially in creeks and pools, nematodes and annelids are common inhabitants of superficial sediments, some being detritivorous (e.g. *Heteromastus similis, Laeonereis acuta*) and others predators (e.g. *Nephtys fluviatilis*). Since the plant biomass and organic matter concentrations are considerably higher in marshes (up to 20%) than in estuarine sediments (1-3.5%; Martins 1966; Moreno 1994), epi- and infaunal estuarine organisms are attracted to the marshes during flooding. Under these conditions, the gastropod *Heleobia australis* frequently feeds on epiphytic and epipsamic microalgae in the lower marshes. Detritivorous aquatic (*Bathyporeiapus bisetosus*) and terrestrial (*Balloniscus* spp.) isopods, amphipods (*Orchestia platensis*), spiders, and insects are abundant throughout the marsh. The abundance of deposit feeders in particular is likely to correlate with sites of organic matter accumulation in depressions and areas with dense vegetation cover. Most of these organisms are an important food source for juvenile fishes (*Odontesthes argentinensis, Atherinella= Xenomelaniris brasiliensis, Micropogonias furnieri, Mugil platanus*; Bemvenuti 1990; Vazzoler 1991; Vieira and Scalabrin 1991) and birds which rest in the estuary (see Vooren Sect. 4.14).

Organic Matter Contribution

Salt marshes have been described as producer systems that subsidize estuarine and nearshore waters (Odum 1971). Their ability to fix carbon at remarkable rates remains unchallenged, but since marshes display a large degree of individuality, the ecological fate of the fixed organic matter appears to be highly variable (Haines 1979; Dixon 1980). In regions with a rapid sea-level rise, marshes may accumulate plant litter in the sediments. Estuaries with elevated river flow are likely to export large amounts of net primary marsh production during runoff, whilst small semi-protected marshes may recycle large portions of fixed organic matter. In consequence, the fate of organic matter production by the physiographically diverse and irregularly flooded marshes of the Patos Lagoon estuary is likely to be manifold. In occasionally (25% of the time) flooded marshes, up to 20.8% of the annual marsh detritus pool (2290 g m^{-2}) may be exported, whilst 25.5 and 31% are more likely during autolysis/leaching and microbial decay, respectively (Costa et al., unpubl.).

5.4 Unvegetated Intertidal Flats and Subtidal Bottoms

C.E. Bemvenuti

Habitat Characteristics

The majority of the Patos Lagoon estuary is comprised of unvegetated soft bottoms (300 km^2); their environmental characteristics and the structure of the associated benthic community change along depth gradients. Shallow soft-bottom habitats are typically represented by intertidal flats between sandy shorelines or marshes of marginal bays and both vegetated and unvegetated shallow subtidal zones. Since estuarine water levels are largely dependent on wind and precipitation, rather than on tides (Kantin 1983), the intertidal flats are irregularly inundated and may be exposed during prolonged periods in the summer. The larger part of the estuarine shoals is dominated by shallow (<1.5 m) subtidal soft bottoms, either with or without the occurrence of *Ruppia maritima* beds, but with epibenthic microalgal growth and occasional aggregations of macroalgae. Deeper (<6 m) subtidal soft bottoms are void of plant growth and extend to the edge of the natural and artificial estuarine channels. The channel bottoms (6-11 m) are predominantly muddy, with the exception of the estuarine inlet (15 m), where sediments become increasingly sandy.

Macrobenthic Community

The motile epibenthic organisms of subtidal soft bottoms in the estuary are largely decapods (Capítoli et al. 1978; Bemvenuti 1987b). Shallower areas are dominated by the small crab *Rhithropanopeus harrissii* and juveniles of *Cyrtograpsus angulatus*, *Callinectes sapidus*, and *Penaeus paulensis* (Fig. 5.4.1). Between the end of spring and beginning of fall, juveniles use shallow subtidal shoals as nursery grounds and together with juvenile fish (Sciaenidae, Bothidae, Jenynsiidae, Atherinidae) exert a severe predation pressure on the macrobenthic community (Bemvenuti 1987b). At the end of the fall, as is typical for decapods of temperate regions (Reise 1985), *Penaeus paulensis*, *Callinectes sapidus*, and *Cyrtograpsus angulatus* migrate to deeper subtidal areas.

The sedentary epifauna of soft bottoms is principally represented by the widely distributed gastropod *Heleobia australis* (Hydrobiidae) and by peracarid crustaceans, with a dominance of the tanaid *Tanais stanfordi*, the amphipods *Melita mangrovi*, *Ampithoe ramondi*, *Leptocheirus* sp., and the isopods *Dies fluminensis* and *Munna peterseni* (Fig. 5.4.1). Small sedentary organisms, like peracarids, are often under severe predation pressure (Nelson 1979b; Virnstein et al. 1984; Reise 1985), and their survival depends on the structural diversity of the habitat, which is often provided by submersed

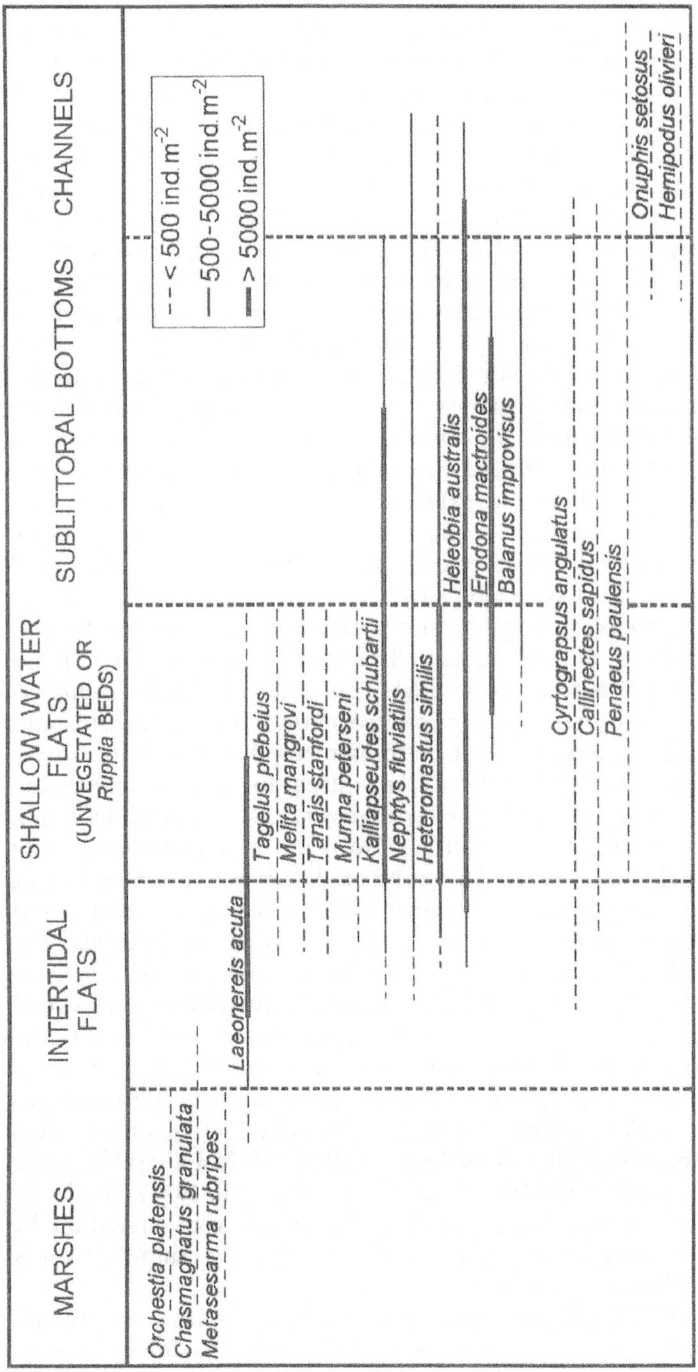

Fig. 5.4.1. Zonation of principal macrobenthic organisms in soft-bottom estuarine habitats

vegetation (Orth et al. 1984; Summerson and Peterson 1984). As a consequence, the densities of epifaunal peracarids, which are typically restricted to the lower reaches of intertidal flats and shallow subtidal soft bottoms in the estuary, grow as a result of macroalgal aggregations in shallow subtidal areas, which supply protection against predation (Bemvenuti 1987b).

The distribution of *Heleobia australis* ranges from tidal flats to the soft bottoms of deep channels (Capítoli et al. 1978). Densities may exceed 40 000 ind. m^{-2}, though pronounced spatial and temporal changes in density are common (Bemvenuti et al. 1978, 1992; Lana 1986) since gastropods of the Hydrobiidae are generally subject to intense predation (Reise 1985). In shallow inter-and subtidal areas, elevated densities are associated with the occurrence of filamentous algae (Bemvenuti et al. 1978; Orth et al. 1984; Bemvenuti 1987b). Drastic density fluctuations can also be attributed to low salinities, which cause osmoregulatory stress (Chomenko and Schaffer 1984; Bemvenuti et al. 1992). Under this condition, *Heleobia australis* moves from the epistratum to the water surface and, utilizing the water surface tension, migrates to regions with higher salinity, like deeper subtidal areas and channels. The small number of competitors and the access to an ample trophic and spatial niche of the epistratum result in elevated densities of *Heleobia australis* in the deeper subtidal areas.

Infaunal polychaetes, pelecipods, and peracarids are characteristic and abundant inhabitants of environmentally stressful intertidal flats (Reise 1985), though they also occur in subtidal areas. The polychaete *Laeonereis acuta* is common to estuaries in the southwestern Atlantic (Orensanz and Gianuca 1974) and is found in soft bottoms of the Patos Lagoon estuary down to about 1 m depth (Fig. 5.4.1). As is usual for infaunal burrowing species (Holland et al. 1980), adults of *Laeonereis acuta* escape predation in deeper (approx. 20 cm) sediment strata. However, in the superficial substrate their juvenile stages are under strong predation pressure, which leads to high mortality and controls the density of the population (Bemvenuti 1992). Dense recruitment, either in shallow areas of the intertidal flat (Bemvenuti et al. 1978; Capítoli et al. 1978) where preying fish and decapods have limited access or during the winter (Asmus 1984) when several predators abandon estuarine shoals (Bemvenuti 1987b), counteracts the otherwise severe impact of predation and guarantees a larger number of adult individuals. The polychaetes *Nephtys fluviatilis* and *Heteromastus similis* have a wider vertical distribution (Capítoli et al. 1978) than *Laeonereis acuta*, with higher densities in shallow subtidal areas but lower densities in the deeper subtidal (>2 m) (Bemvenuti et al. 1978). Similar to *Laeonereis acuta*, *Heteromastus similis* seeks refuge from predation by burrowing into deeper sediment strata of shallow subtidal zones (Bemvenuti 1987b).

Individuals of the burrowing pelecipod *Tagelus plebeius* characteristically form patches in silty and clayey areas along the lower edge of intertidal flats, whilst the isopod *Pseudosphaeroma mourei* frequently occupies increasingly sandy sediments of more unstable, shallow, subtidal zones (Capítoli et al. 1978). The distribution of the tube-dwelling tanaiid *Kalliapseudes schubartii*, which lives in deeper (15 cm) strata of sandy mud bottoms, ranges from the lower reaches of intertidal flats to the margin of channels at about 6 m depth (Fig. 5.4.1). In spite of intense predation (Bemvenuti 1983, 1987b, 1992) by fish and decapods (Araújo 1984; Asmus 1984; Bemvenuti 1990), this peracarid is the dominant species (>10 000 ind. m^{-2}) in shallow subtidal areas. Apart from seeking shelter against predators in sub-surface sediment strata, *Kalliapseudes schubartii* is a typical r-strategist (*sensu* McCall 1977), with intense reproductive activity, marsupial spawning, and recruitment, which maintains elevated population densities (Bemvenuti 1987b, 1992) in shallow subtidal areas. Lower densities (1000 ind. m^{-2}) in deeper subtidal zones (Bemvenuti et al. 1978) may partly reflect the substrate and water column instability of these areas (Capítoli et al. 1978; Kantin 1983). The largest macrobenthic biomass is produced by the pelecipod *Erodona mactroides*. In the upper estuary, where all size classes are present, larger individuals (up to 30 mm length) at lower densities (461 ind. m^{-2}) form an average biomass of 281 g m^{-2}. In the lower estuary, the high mortality of *Erodona mactroides* during the first year results in individuals of reduced size (< 13 mm), and despite high densities (3722 ind. m^{-2}), the mean biomass rarely exceeds 105 g m^{-2} (Bemvenuti et al. 1978). Especially at the end of the summer, high mortalities of *Erodona mactroides* appear to be related to elevated deposition of fine particles which may obstruct the filtration system of the organism. The density and size of *Erodona mactroides* also influence in part the estuarine distribution of the cirriped *Balanus improvisus* which, in the absence of any other hard substrate, tends to attach to the shells of *Erodona*. Since small individuals of *Erodona mactroides* are unsuitable as substrate, the density of *Balanus improvisus* tends to be higher (719 ind. m^{-2}) in the upper estuary.

The abundance and diversity of the macrobenthic community drastically decrease between the lower reaches of deeper (>6 m) subtidal areas and channel bottoms (Bemvenuti 1987a), though in the entrance channel diversity may temporarily increase during periods of inflowing seawater, owing to the addition of peracarids and small polychaetes (Fig. 5.4.1; Bemvenuti et al. 1992).

Habitat Function

The high production by different primary producer groups and the continuous supply of detrital matter make inter-and subtidal bottoms ideal nurseries for invertebrates and fish. The benthic communities of the intertidal flats, shallow and deeper subtidal zones, and channel bottoms are interdependent and linked by continuous feedback processes. Although intertidal flats generally lack any conspicuous vegetation cover and, owing to the dominance of burrowing species, appear to be faunistically poor, they are areas of elevated secondary production with high rates of energy transfer between infaunal organisms and highly motile predators (Reise 1985). In shallow subtidal zones, the deposit feeders represent an important link between large amounts of detritus and higher trophic levels. The deeper subtidal areas are dominated by suspension and deposit feeders, which largely depend on allochthonous detritus and, in the absence of any protective vegetation, are exposed to intense predation. These areas are therefore important feeding sites for fish and decapods (Summerson and Peterson 1984) and serve as a refuge for decapods during the colder months (Reise 1985). Even the otherwise inhospitable deep channels constitute an important migratory route for many estuarine organisms, and the large water column favours the presence of pelagic species.

5.5 Seagrass Meadows

U. Seeliger

The size and physiographic heterogeneity of the Patos Lagoon estuary provides for diverse and extensive habitats. Rather than being isolated, habitats interface along abiotic gradients via detrital matter transport and migration of organisms. About 120 km^2 of *Ruppia maritima* beds in the Patos Lagoon estuary (Fig. 5.5.1) serve as a fishery habitat because they provide substrate and shelter and offer nursery and feeding grounds for a diverse assemblage of organisms, which may have adapted their life-cycles accordingly. The leaves, stems, roots, and rhizomes of *Ruppia* form a structurally complex habitat of calm water and stable substrate. The abundant plant and detrital matter supports dense populations of motile and sessile organisms, and the high infaunal and epifaunal abundance attracts predators. Additionally, drift algae may form an associated habitat of potential importance as a nursery.

Seagrass-Epiphyte Complex

In the absence of consolidated sediments, the leaves and stems of *Ruppia maritima* provide the principal semi-permanent substrate for many micro- and macroalgae in the estuary, although none of the species appears to be host-specific. As the leaves of *Ruppia* emerge from the sediments, epiphyte colonization commences with a dense cover of the pennate diatom *Cocconeis placentula*. Its flat morphology and copious mucus production enable this species to cling firmly to the leaves, thereby becoming a monospecific pioneer stage. A secondary, still selective colonization by *Synedra fasciculata* takes place over the living *Cocconeis* cells and associated detritus.

The formation of dense *Synedra* populations in the median region of leaves favours attachment of *Amphora*, *Nitzschia*, and naviculoides. During the final stage of colonization, the apical region of the leaves is completely covered by a diverse assemblage of diatoms and macroalgae (*Enteromorpha* sp., *Rhizoclonium riparium*, *Achrochaetium flexuosum;* Ferreira and Seeliger 1985). Since the shading effect of epiphytes is likely to decrease the lifespan of individual leaves (and possibly of the entire population), epiphytic community turnover must be relatively rapid.

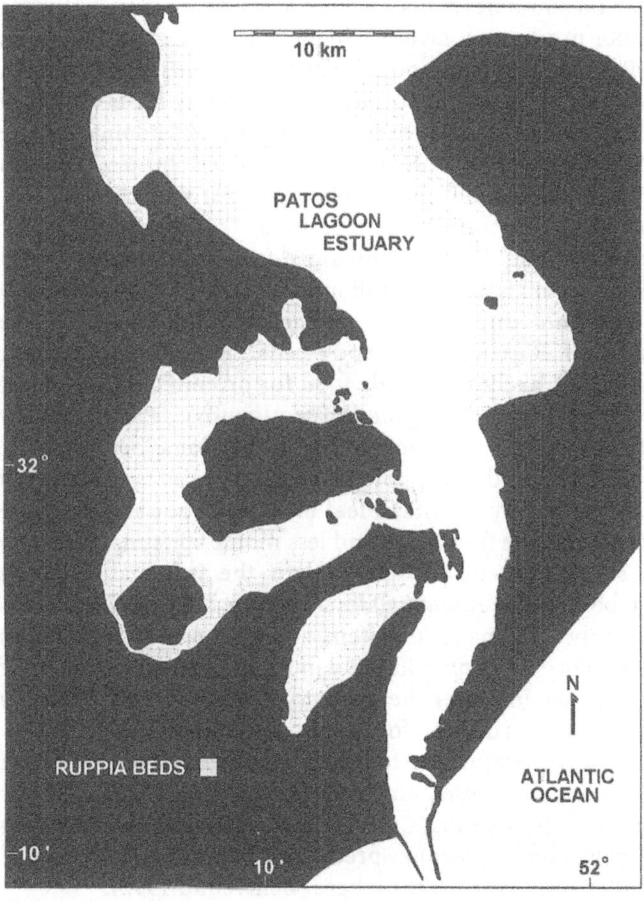

Fig. 5.5.1. Distribution of *Ruppia maritima* beds in the Patos Lagoon estuary

Faunal Relations

Polychaetes, molluscs, amphipods, the crabs *Callinectes sapidus* and *Cyrtograpus angulatus* (Bemvenuti 1987a,b; Asmus 1989), the pink shrimp *Penaeus paulensis*, and juvenile and sub-adult fish live either on and between the leaves or in burrows between rhizomes of *Ruppia* (Asmus 1984); thus, vegetated areas tend to exhibit higher faunal abundance than adjacent unvegetated bottoms. Since the size distribution of particles and the organic matter contents of sediments do not vary (Moreno 1994), differences in animal densities can be directly attributed to macrophyte biomass. As the *Ruppia* meadow develops, increasing plant density, epiphyte colonization, and litter production favour a higher abundance and biomass of consumer populations. The gastropod *Heleobia australis*, for example, which normally lives on the sediment surface, preferentially feeds on dense bacterial films on *Ruppia* leaves, and reaches its highest abundance during peak plant growth (Asmus 1989). Many of the invertebrates feed on epifauna attached to the leaves, but they probably ingest a mixture of epiphytic algae concurrently. In consequence, most of the invertebrates are opportunistic omnivores, like *Cyrtograpus angulatus*, which occasionally show a preference for plant tissue (Capítoli and Ortega 1993). Also waterfowl, like *Fulica armillata* and *Cygnus melancorhyphus*, appear to be important herbivores of *Ruppia* tissue and seeds. The abundant plant detritus in the meadows is likely to provide food for juvenile and sub-adult fish (*Odontesthes bonariensis*, *Atherinella brasiliensis*). Predator-prey relationships in the beds are influenced by interactions between the behavioural, physiological, and morphological characteristics of animals and variations in shoot density, biomass, leaf area, and root and rhizome depth of the *Ruppia* plants. The free-living and less motile epifaunal *Heleobia australis* is more susceptible to predation than the infaunal *Erodona mactroides* and the tube-dwelling *Kalliapseudes schubartii*. While *Kalliapseudes schubartii* appears to be the main diet item of the predator *Callinectes sapidus* during the absence of *Ruppia* in the winter, the protective cover of plants during the summer impedes the predator's motility and *Heleobia australis* may become the principal food source (Asmus 1989). Patchy *Ruppia* beds may provide favourable foraging areas for fish, like *Mugil platanus*, *Odontesthes argentinensis*, and *Atherinella brasiliensis*, and for the highly mobile *Callinectes sapidus*, *Cyrtograpsus angulatus*, and *Penaeus paulensis*, which may exert a strong predation pressure over several macrobenthic juveniles seeking protection here (Bemvenuti 1992).

Nursery Grounds

One of the most important roles of *Ruppia* beds in the estuarine ecosystem is that of a nursery ground in which postlarval stages of many invertebrate and some fish species concentrate and develop (Asmus 1984). The shrimp *Penaeus paulensis* and the blue crab *Callinectes sapidus* are an important local fishery resource, and both are estuarine dependent, since they spawn in coastal water and their post-larvae or megalopi are carried back into the estuary during the spring, where they find protection in the *Ruppia* beds. Although shrimps and crabs are found throughout the estuary, many continue to forage in the *Ruppia* beds. Furthermore, *Ruppia* beds may also be important nursery grounds for numerous species of commercial fish, and for small forage species that serve as food for the larger carnivores. At some stage in their life-cycle, several species of the families Sciaenidae, Mugilidae, Atherinidae, and Syngnathidae are associated with the beds. While all large shallow estuarine areas could potentially be occupied by *Ruppia*, their size may naturally vary as much as 40% between years (Moreno 1994), and growth may commence at any time between the early spring and early summer and cease between the late summer and early winter. In consequence, periods of high faunal recruitment into the estuary may not coincide with years or periods of optimal *Ruppia* bed size and permanence, which may impose habitat limitations for estuarine fisheries stock (D'Incao 1991).

Energy Contribution and Export

Most *Ruppia* tissue only becomes an available food source as detrital matter. During the growth period, leaves are continuously shed and tend to accumulate within the beds, where they undergo rapid decomposition and serve as food for detritivore organisms (Bemvenuti 1987a, 1992). At the end of the growth period, whole *Ruppia* plants and the remaining detritus are removed by winds and water currents and accumulate in deeper subtidal areas and channels. During periods of flushing, the detritus along with recently detached green leaves and uprooted plants is discharged into coastal waters and transported by NE winds and longshore currents up to 50 km southward along the beach (Seeliger, unpubl.). Therefore, in addition to being an important food source in the estuary, the export of live and dead plant material and detrital matter represents an important trophic link and energy source for organisms in adjacent and distant coastal nearshore waters.

5.6 Rubble Structures and Hard Substrates

R.R. Capítoli

Colonizing Organisms

In spite of the overwhelming dominance of unconsolidated sediments in the Patos Lagoon estuary, a variety of artificial structures (approx. 0.18 km^2) provide for hard sub-and intertidal substrates. Among the different types inside the estuary are revetments, piers of port installations, and wooden pilings which contribute considerably to the total area. The 4-5 km long jetties of the Patos Lagoon inlet (Fig. 5.6.1) represent the largest continuous extension of permanent substrate.

Permanent substrates in the estuary and the channel side of the jetties are under the direct influence of turbid oligo-and mesohaline estuarine waters with pronounced salinity variations and reduced tidal oscillation. These environmental conditions account for an encrusting community of low floristic and faunistic diversity, dominated by a few tolerant species (Fig. 5.6.1), like *Balanus improvisus*, as has been observed in estuaries elsewhere (Cory 1967; Dean and Bellis 1975). *Balanus improvisus* larvae settle on permanent substrates throughout the year, but their densities are lowest in the winter. Winter water temperatures of 12°C retard the growth of recently attached larvae, sexual maturation, and hatching of new larvae, and thus the abundance of planktonic larvae in the winter and spring tends to be low (Montú 1980; Capítoli 1983; Baumgarten and Niencheski 1990b). Intense colonization occurs in the summer and fall together with larvae of *Stylochus* sp., *Obelia* sp., and *Boccardia hamata*. However, owing to crowding and high predation pressure, principally by *Stylochus* sp., the initially high abundance of *Balanus improvisus* is rapidly reduced (Capítoli 1983). The dead shells of the cirripeds supply new spatial niches which are occupied by the polychaete *Boccardia hamata*, the amphipod *Amphithoe ramondi*, and the tanaiid *Tanais stanfordi*. As a result of strong physical and biological disturbances within the estuary and, similar to the model for „unstable environments“ proposed by Wood (1987), the community development appears to lack seasonal progression and biotic succession. Turbid estuarine waters also limit algal colonization to the upper 15 cm (*Lyngbia confervoides*, *Ulvaria oxysperma*, *Polysiphonia* sp., *Enteromorpha* sp.) or 50 cm (*Melosira moniliformis*, *Biddulphia* sp.) of permanent substrates (Coutinho 1982).

Environmental conditions and species composition in the estuary and inlet channel differ from those on the seaward side of the northern and southern jetties, each of which display specific habitat conditions with distinct benthic communities. The seaward side of the southern jetty is

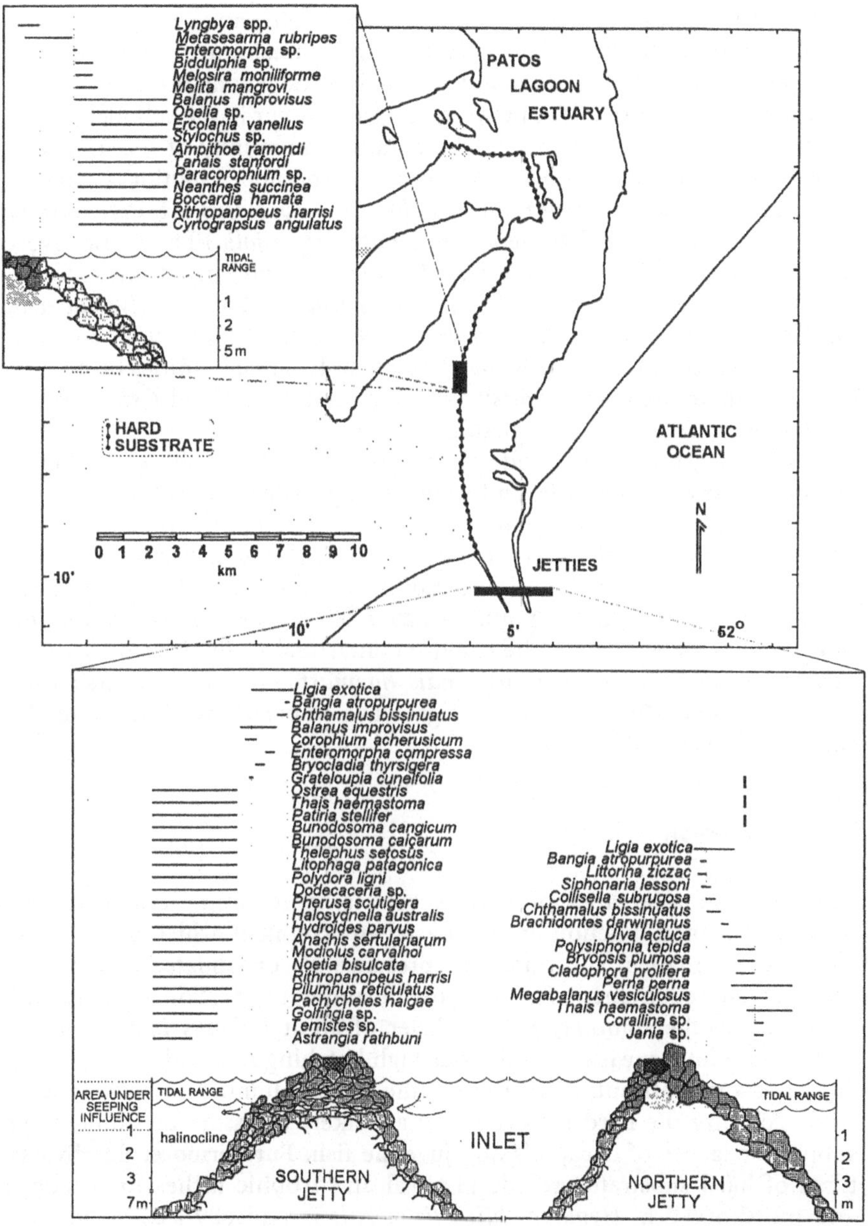

Fig. 5.6.1. Vertical distribution of principal macro-invertebrates and algae on rocky substrate in the Patos Lagoon estuary

protected from swell and wave-overwash during NE winds. Since estuarine runoff flows southward, the generally turbid freshwater tends to seep through the gaps of the jetty, and thus the intertidal zone is dominated by low salinities, whilst salinities between 25 and 30 occur in the subtidal (>1 m; Coutinho 1982). In consequence, the benthic community of the intertidal (cyanophytes, diatoms, *Bangia atropurpurea*) and upper subtidal (<50 cm) zones (*Chthamalus bissinuatus*, *Balanus improvisus*, *Enteromorpha compressa*, *Bryocladia thyrsigera*, *Grateloupia cuneifolia*) has a low diversity. During prolonged drought periods, when seawater penetrates into the estuary, *Brachydontes darwinianus*, *Perna perna*, and algae (*Ulva lactuca*, *Cladophora flexuosa*, *Gelidium crinale*, *Gymnogongrus griffthsii*) colonize the intertidal zone temporarily. More stable polyhaline salinities below 1 m depth promote increased diversity and the establishment of *Ostrea equestris* and associated species (Fig. 5.6.1).

In contrast, the seaward side of the northern jetty is characteristically exposed to swell and waves, and water transparency and salinity (30) are high (Coutinho 1982). The colonization of typical marine organisms follows a clear pattern of zonation (Fig. 5.6.1). *Littorina ziczac* occupies the upper and *Siphonaria lessoni* and *Collisella subrugosa*, the lower spray zone, followed by a sparsely populated zone of *Chthamalus bissinuatus* and *Brachidontes darwinianus*. The extensive intertidal is dominated by *Perna perna*, *Megabalanus vesiculosus*, *Thais haemastoma*, and a diverse algal flora (Coutinho 1982). Owing to high wave impact, observations of the subtidal zone are still lacking.

Habitat Function

Although the total area of hard substrate is reduced, it assumes an important habitat function in this coastal environment, which otherwise is devoid of permanent substrate. The presence of consolidated substrate is an essential requirement for the growth and reproduction of cirripeds. Especially in the summer, a dense layer of adult *Balanus improvisus* on rocks and pilings means a continuous high hatching rate and consequently an important contribution of larvae to the water column. These will successively colonize the hard substrate and are likely to assure an ample food supply for larvae of decapods and juvenile fish. Furthermore, the diverse types of hard substrate provide physical and trophic niches for resident and transitory fish. *Hypleurochilus fissicornis* is typically associated with hard substrates in the estuary and the inlet channel, where it feeds on soft parts of *Balanus improvisus* and amphipods (Capítoli 1983). Other species are likely to visit the jetties, where they supply the resources for sport fishing activities. Finally, the jetties have biogeographic significance, because they provide settlement conditions for northern tropical and southern cold temperate sedentary organisms, which are otherwise unable to bridge the 400 km long barrier of unconsolidated sandy substrate between 30° and 34° S (Coutinho and Seeliger 1986).

6 Coastal and Marine Environments and Their Biota

6.1
Geomorphological Setting

L.J. Calliari

Continental Margin

The southern Brazilian continental margin is characteristic of low intensity tectonic evolution processes (Villwock 1987), which levelled the basement, formed thick sedimentary layers, and led to the development of an extensive coastal plain (Fig. 6.1.1). The gently sloping continental shelf (2 m km^{-1}) between Santa Marta Grande Cape (28° 40' S) and Chuí (34° 40' S) has a width of 100-180 km (Martins et al. 1972) and reaches its largest extension between 31° and 33° 30' S. Remnant valleys of ancient fluvial channels typically cut the shelf in this region. Characteristic morphological features of the shelf include marine terraces (Koswmann et al. 1977) and beachrock outcrops (Calliari and Abreu 1984), whilst linear sand ridges typically occur on the inner shelf (Figueiredo 1975). The transition between shelf and upper slope starts at 160-190 m depths at a rate of 20 m km^{-1}. The considerable width of the slope (up to 250 km near Cabo Santa Marta) as well as the gradual drop to approximately 3000 m depth suggest the dominance of depositional processes during its formation. At the central part of the Pelotas Basin, the slope resembles the shape of a cone, which extends from the outer shelf to about 4000 m depth. A well-developed continental rise with a maximum width of 400 km (30° S) extends between 2700 and 4000 m depth (Zembruscki et al. 1972). An important feature of the oceanic basin is the Rio Grande Rise, which extends in an east-west direction between the continental rise and the flanks of the Mid-Atlantic Ridge and separates the ocean basins of Brazil and Argentina (Palma 1984).

Since the early Pleistocene (1.8 million B.P.) sea-level changes have been responsible for the sequential deposition of several barrier islands over the coastal plain. However, only sea-level changes after the Wisconsin glaciation in the late Pleistocene (18 000 B.P.) have decisively influenced the sedimentary characteristics of shelf and slope. At that time, sea-levels were 120-130 m lower and approximately corresponded with the shelf break

U. Seeliger et al.
Subtropical Convergence Environments

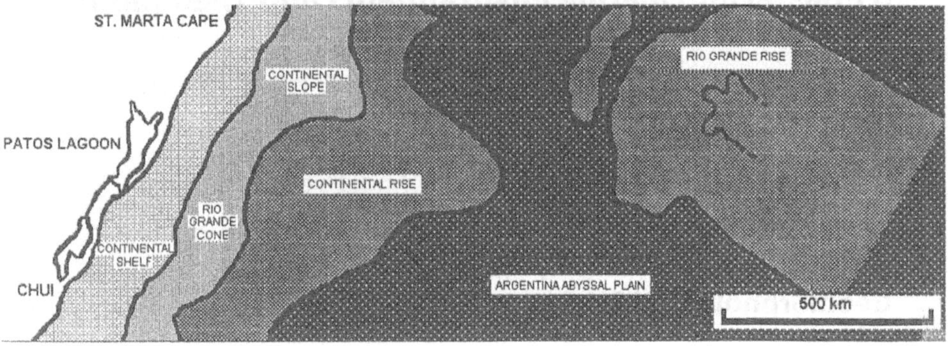

Fig. 6.1.1. Main physiographic features of the continental margin between 29 and 35° S

(Corrêa 1990). The subsequent phases of rising sealevel during the Holocene transgression stabilized around 100-110, 60-70, 32-45, and 10-25 m and imprinted erosive and depositional features of old shorelines over the shelf (Koswmann et al. 1977; Corrêa 1990). At the maximum of Holocene transgression (5000 B.P.), the sea was 4 m above present levels and inundated the older Patos-Mirim Lagoon system. The following regression phase, interrupted by small trangressions, formed a sequence of beach ridges and depressions with small lagoonal systems. At the end of the regression phase (2000 B.P.), the sea-level was 1-2 m below the present level, followed by a gradual increase to its current height (Villwock and Tommazelli 1989).

The sedimentation patterns of the shelf and upper slope are predominantly terrigenous and, based on the sand/silt/clay contents, define eight distinct facies (Corrêa 1987; Fig. 6.1.2). The inner and outer relict sand facies of the shelf, composed of medium and fine quartz sand similar to that of today's beaches and dunes, were deposited during the Pleistocene regression and reworked during the Holocene transgression (Martins et al. 1972). Relict clayey silt and silty clay facies dominate over the middle and outer shelf. Continental runoff during the Wisconsin transgression phase was probably responsible for their deposition. Palimpsest sediments are represented by transitional facies, composed of similar proportions of terrigenous sediments and mixtures of biodetritus with sand or mud. The deposition of recent silty clay and clayey silt facies was a result of sediment transport from the La Plata River and the Patos Lagoon.

Surface carbonate deposits of shell fragments (Martins et al. 1972; Corrêa 1983) take up approximately 1000 km^2 of the shelf area, and coarse sand deposits exist near beachrock outcrops south of the Patos Lagoon inlet (Calliari and Abreu 1984; Calliari et al. 1994). Glauconitic sands are common sediments of the outer shelf and upper slope, and phosphate concretions (up to 16% P_2O_5) have been dredged from the upper slope (Klein, unpubl.).

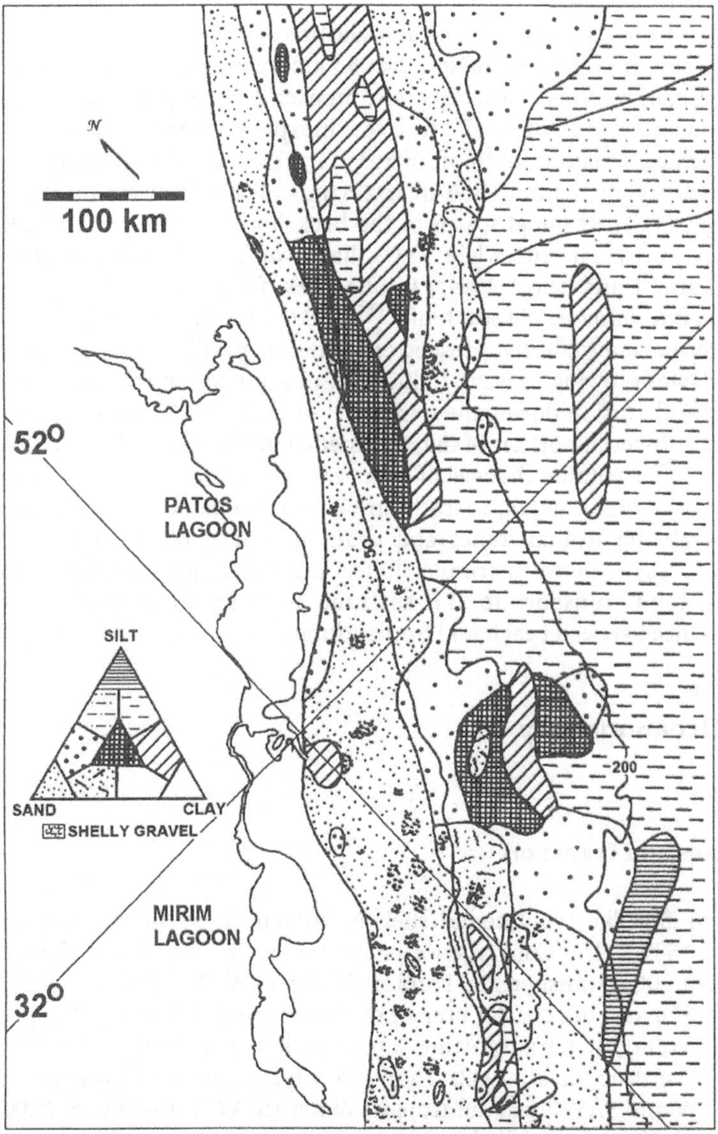

Fig. 6.1.2. Superficial sediments of the continental shelf and upper slope. (Modified after Corrêa 1987)

Beaches

Except for sandstone, basalt, and volcano-clastic rock formation near 29° S, which offer a certain degree of shelter, the 653 km long coastline between 29° and 35° S is totally exposed. Beaches are composed of fine quartz sand, and most (95%) have a low gradient (2°) with few inexpressive beach cusps. Only a 45 km stretch of the shore south of Albardão lighthouse (33° S) is composed of coarse shell fragments and fine quartz sands. These beaches display a higher gradient (4°), erosional scarps at the backshore, and frequently well-developed cusps (Calliari and Klein 1993). Based on morphodynamic characteristics (Wright and Short 1984), the beaches are considered generally intermediate, but immediately south of 32° and 33° S the beaches are dissipative and intermediate/reflective, respectively. Beach profiles demonstrate a seasonal behaviour, owing to wave-induced sand accretion in the late spring and summer and erosion during the remainder of the year. During cold front passages, the sub-aerial portion of the intermediate/reflective and dissipative beaches may suffer volume changes of 40 m^3 m^{-1} and 4.9 m^3 m^{-1}, respectively. During periods of high rainfall, a large number of ephemeral creeks (approx. 1.5 km^{-1}) break the frontal dunes and drain landward depressions, with concomitant erosion between frontal dunes and the inner foreshore. The considerable sand transport to the surfzone is likely to determine the sediment balance between beach and frontal dunes.

6.2 Physical Oceanography

C.A.E. Garcia

Water Masses and Transport

The southern Brazilian continental shelf is influenced by the Subtropical Convergence (STC). The Convergence is formed by the southward flowing Brazil Current (BC), transporting Tropical Water (TW; T >20°C, S >36.00) (Emilsson 1961), and the coastal branch of the northward flowing Malvinas (Falkland) Current (MC), transporting Subantarctic Water (SAW; T 4-15°C, S 33.70-34.15; Sverdrup et al. 1942; Thomsen 1962). The convergence and mixture of TW and SAW form Subtropical Water (STW; T 10-20°C, S 35.00-36.00; Emilsson 1961), also called South Atlantic Central Water (Sverdrup et al. 1942), which flows northward in deeper layers in a counter-clockwise gyre and may reach the continental shelf (Miranda 1972; Hubold 1980a). In contrast to oceanic zones, water transport over the continental shelf is largely a consequence of the impact of offshore currents on the shelf's topographic features, though the wind regime also plays an important role in circulation. Long-term current velocity changes follow seasonal patterns and spatial distribution of wind regimes along the continental margin (Luedemann 1971, 1978; Pereira 1977), whilst short term changes of current velocities are associated with atmospheric conditions as a result of cold front passages in 2-10 day intervals.

Owing to the influence of the Atlantic gyre, the southward flowing BC becomes deeper and wider between 24° and 31° S. The convex curvature of the bathymetry at 24° S and the smooth topography of the shelf cause instability of the flow and oscillations of the front along the shelf break (Garfield 1990). The geostrophic circulation and the transport of the water mass associated with BC vary (Cacciari 1986), but the flow is generally directed southward. However, although TW and STW masses of the upper water column flow south, water masses below the thermocline move in opposite directions. Antartic Intermediate Water (AIW) below the thermocline flows north (Deacon 1933; Wüst 1935), whilst deep North Atlantic Water (DNAW) flows south. It is likely that the BC extends vertically down to the interface between DNAW and northward flowing Antartic Bottom Water (ABW). Due to the anti-cyclonic bottom circulation, some of the AIW may be transported southward along the slope (Reid et al. 1977; Piola and Gordon 1989). Computations of the southward transport of the BC at 32-33° S (Evans et al. 1983; Stramma 1989) suggest that during most of the year tropical water is transported over the shelf (Garfield 1990).

Two types of upwelling processes are common along the southern Brazilian coast. Coastal upwelling is more frequent during the spring and summer, owing to the strong influence of NE winds, which provoke a divergence of superficial waters and cause advection of STW over the shelf between 28° and 32° S (Miranda 1972; Hubold 1980a). Upwelling at the shelf break is more common in the winter and spring. The BC and the coastal branch of the MC form frontal vortices of cyclonic circulation which cause upwelling of bottom water, principally STW (Hubold 1980a,b).

Thermohaline Structure

The seasonal migration of the Convergence (STC) causes the advection of TW, SAW, and STW, with contrasting thermohaline characteristics over the shelf (Olson et al. 1988; Gordon 1989). Furthermore, the thermohaline properties of the coastal water (CW) vary between seasons and years as a function of freshwater runoff from the Patos Lagoon and the La Plata River (Miranda 1972; Miranda et al. 1973; Tseng 1976; Castello and Möller 1977,1978; Miranda and Castro Filho 1979; Godoi 1983; Cacciari 1986; Ciotti 1990; Fillmann 1990). In consequence, the seasonal and interannual variabilities of the thermohaline properties of the shelf waters are pronounced, with a predominance of TW and SAW during the summer and winter, respectively (Miranda 1972).

The centers of warm anticyclones and cold cyclones directly control the depth of the thermocline (100-150 m) of the oceanic region (Silva 1981; Godoi 1983). In contrast, over the shelf the vertical thermohaline structure varies with season, continental runoff, and direction and velocity of the winds. During the summer, the shelf (>50 m depth) and the slope are covered by the tropical surface water of the BC. Owing to reduced continental freshwater runoff, the vertical thermohaline structure is moderately stratified, and the thermocline occurs at a depth of about 40 m. During the winter, the BC flows along the outer shelf, and the advection of the northward flowing, cold, and less saline MC water mass results in a strong horizontal thermohaline gradient. Horizontal current sheer may cause upwelling at the front between the two water masses over the shelf break. Under these conditions, the shelf waters become weakly stratified (Lima 1992). High freshwater discharge by the Patos Lagoon and the La Plata River (Castello and Möller 1978) and less frequent south winds correct this situation during the spring. Coastal waters over the shelf with less than 50 m depth become strongly stratified, with vertical sigma-t gradients of 2.5 m^{-1} (Brandhorst and Castello 1971; Podestá 1990; Lima 1992).

Shelf water surface temperatures (Miranda 1972; Castello and Möller 1977; Hubold 1980a,b) are largely a consequence of the different water masses. The seasonal variation of mean surface temperature, with 23^{0}C in February and 12.5^{0}C in August, is most pronounced over the southern shelf. The presence of tropical water over the shelf during the summer causes high monthly mean surface temperatures with a low standard deviation, whilst high monthly surface temperature deviations are typical for the spatial heterogeneity of water masses in the winter (Lima 1992; Belém, unpubl.).

6.3 Chemical Characteristics

L.F. Niencheski and G. Fillmann

The chemical characteristics of shelf and slope water largely depend on continental runoff and on the influence of the different water masses at each season (Magliocca 1971, 1973; Castello and Möller 1977; Hartmann et al. 1980; Hubold 1980a,b; Kantin 1982, 1983; Kantin et al. 1982; Magliocca et al. 1982; Dohms 1983; Carreto et al. 1986; Fillmann 1990). Sediment-water column exchange, assimilation, remineralization, and enrichment processes define the chemical characteristics of each water mass.

The Brazil Current (BC) is characterized by oligotrophic Tropical Water (TW) which reaches the southern Brazilian shelf most of the year. Low average dissolved nutrient concentrations (phosphate 0.31 µM, nitrite 0.09 µM, nitrate 1.43 µM, ammonium 0.78 µM, silicate 4.83 µM) and low mean total suspended matter levels (4.45-5.96 mg l^{-1}) of their water indicate a lack of enrichment from continental runoff. In contrast, the Subantartic Water (SAW) mass represents an important source of nutrients for shelf waters. Especially during the winter, the addition of high dissolved nitrate, ammonium, phosphate, and silicate, with maximum concentrations of 11.1 μM, 6.6 μM, 1.1 μM, and 10 μM, respectively, fuels the primary production over the shelf (Fillmann 1990; Ciotti 1990; Sales Dias, unpubl.; Fig. 6.3.1).

Continental runoff from the La Plata River and the Patos Lagoon influences the chemical characteristics of coastal water (CW) over the shelf. In consequence, both short-term and seasonal changes in volume and composition of runoff cause high variability of chemical parameters in CW. Dissolved mean oxygen concentrations (8.18 mg l^{-1}) and saturation levels (96.06%) are higher than in the other water masses, though extremely low values (3.06 mg l^{-1}; 38.15%) may occur during the spring, and the pH (7.78 - 8.80) may also fluctuate considerably. Average total suspended matter (>16 mg l^{-1}) and silicate (23.10 μM) are highest in CW. The generally low nitrate (2.61 μM) concentrations and N:P ratios indicate that nitrogen may limit primary production here (Fig. 6.3.1).

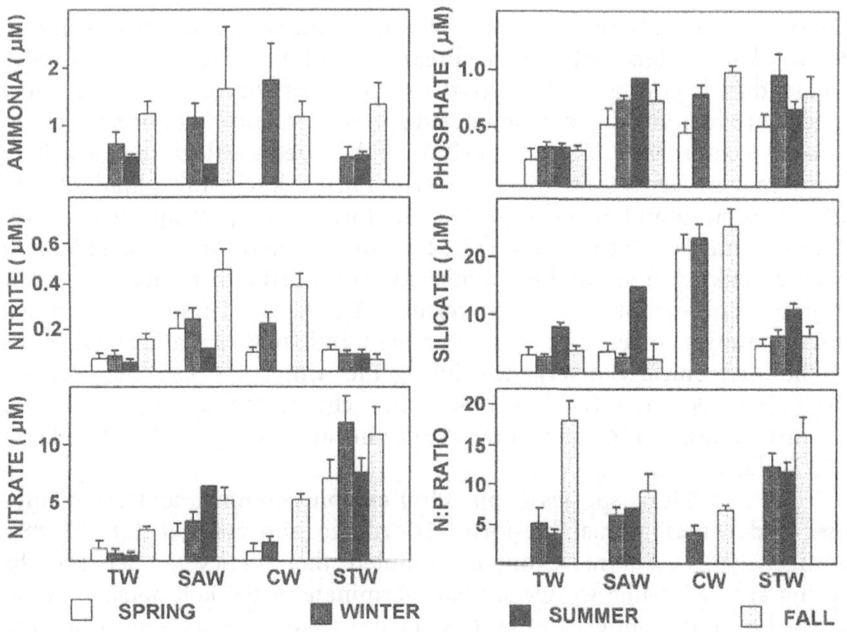

Fig. 6.3.1. Mean seasonal nutrient concentrations and N:P ratios (CV=95%) of different water masses over the continental shelf between 29 and 35° S

Subtropical Water (STW) is a major source of dissolved nitrate, with a mean concentration of 9.29 μM and a range between 0.22 and 26.28 μM. This water is also rich in dissolved inorganic phosphate (mean 0.73 μM, maximum 1.81 μM), while mean nitrite (0.08 μM) and ammonium (0.74 μM) concentrations are low. The suspended matter concentration (4 mg l^{-1}) of the deep STW is low. The oxidation of organic matter during transport leads to low dissolved oxygen (mean 6.6 mg l^{-1}) and oxygen saturation levels (approx. 79%), which also results in increased nitrate values. The influence of STW on the continental shelf strongly depends on the direction of dominant winds since the great depth of these waters obstructs input processes. Nutrients rarely reach the euphotic zone (Fillmann 1990) and only cause elevated productivity and chlorophyll *a* levels during coastal and shelf break upwelling (Hubold 1980a; Ciotti 1990).

6.4 Coastal Foredune Flora

U. Seeliger

Floristic Composition and Affinities

Coastal foredunes between 32° and 34° S are exclusively colonized by herbaceous plants, whilst arboreous climax vegetation is restricted to older inland dunes. Owing to the recent formation of the coastal plain, most species are migrants from neighbouring provinces, and endemic species are rare (Waechter 1985). The composition of the foredune flora changes along latitudinal gradients (Legrand 1959; Reitz 1961; Eskuche 1973; Pfadenhauer 1980; Cordazzo and Seeliger 1987). The floristic composition of the foredunes of Brazil's extreme south (30-34° S) and northern Argentina (37-39° S) is much alike, but shows little similarity with northern tropical (28° S) or southern cold-temperate (42° S) regions (Fig. 6.4.1). These distributional discontinuities correspond to the northern and southern latitudinal limits of the Subtropical Convergence during the winter and summer, respectively (Legeckis and Gordon 1982), thus characterizing this coast as a warm-temperate transition zone (Coutinho and Seeliger 1986; Cordazzo and Seeliger 1988a).

The 71 foredune species (Appendix) exhibit perennial, as well as summer and winter annual life-forms (Cordazzo and Seeliger 1987, 1988a; Seeliger 1992a). Summer annuals complete their life-cycles between the spring and fall, while winter annuals germinate in the fall, persist vegetatively during the winter, and set flowers and seeds the next year. All perennials exhibit drastically reduced growth vigour during the winter (Cordazzo and Seeliger 1988b), owing either to suboptimal environmental conditions or endogenously induced resting periods, as has been observed for *Hydro-*

cotyle bonariensis, *Gamochaeta americana*, and *Blutaparon portulacoides* in the Brazilian tropics (Hueck 1955; Ormond 1960). True dormancy only occurs in *Oenothera affinis*, *Pterocaulon purpurascens*, and *Achyrocline satureoides*, which overwinter through underground organs. Germination and flowering of all species appear to be largely controlled by local soil moisture conditions or regional temperature and light regimes, respectively (Cordazzo and Seeliger 1988b).

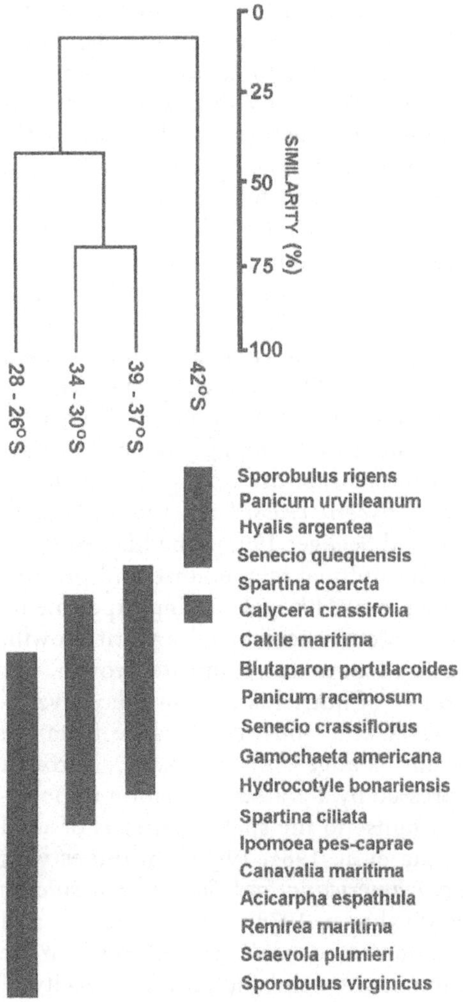

Fig. 6.4.1. Similarity and species composition of dominant coastal foredune vegetation between southern Argentina (42° S) and St. Catarina, Brazil (26° S)

Biology of Dominant Species

Their biomass and abundance classify *Blutaparon portulacoides*, *Panicum racemosum*, *Spartina ciliata*, *Hydrocotyle bonariensis*, *Andropogon arenarius*, and *Androtrichum trigynum* as foredune dominants. Their distribution and abundance are largely a function of substrate stability and water table distance (Cordazzo 1985; Cordazzo and Seeliger 1993). Sand accretion and erosion are particular controlling factors for the perennial rhizomatous *Blutaparon portulacoides*, which is common to the Atlantic ocean beaches between northern Brazil and Argentina (Hueck 1955; Pfadenhauer 1978). Succulent and frequently shed leaves, foliar salt excretion glands, and a reduced number of stomata enable this species to tolerate salt-stressed backshore conditions (Dillenburg et al. 1986). The clonal growth favours the occupation of the unstable substrate, and the leaves and rhizomes aid in the capture of windblown sand (Pfadenhauer 1980; Bernardi and Seeliger 1989). Although cyclic events like extreme storm tides drastically reduce the biomass and density of the *Blutaparon* population (Pfadenhauer, 1978), the seasonal recuperation (Cordazzo 1994) from rhizome fragments is rapid (Bernardi et al. 1987; Bernardi and Seeliger 1989; Cordazzo and Seeliger 1993).

Stands of the perennial rhizomatous grass *Panicum racemosum* extend from northern Brazil to Argentina (Smith et al. 1982). This species is dominant on foredunes with variable substrate stability and nitrate (<0.5 mg kg^{-1}), phosphate (0.2-1.2 mg kg^{-1}), and potassium (3-35 mg kg^{-1}) deficient sands (Costa 1987). Occasionally, *Panicum racemosum* occurs together with the strandline species *Cakile maritima*, *Calycera crassiflora*, and *Calystegia soldanella* (Seeliger 1992b; Cordazzo and Seeliger 1993). The highest flowering and growth response occur under optimal temperature and precipitation during spring in sand accreting areas. Although cutting experiments with *Panicum* tillers show that sand-accretion accompanies plant growth, the mechanical deposition of sand itself does not stimulate growth, but rather provides a substrate and a source of nutrients for continuous vertical growth of rhizomes and tillers. Over time and with distance from the sea, *Panicum racemosum* populations change from "invader", through "mature", to "regressive" stages, expressed by a gradual reduction in population density and tiller vigour, in response to the spatial patterns of sand deposition and salt spray input (Costa et al. 1984, 1991). On dunes with reduced sand accretion, *Gamochaeta americana* and *Senecio crassiflorus* may locally become dominant (Pfadenhauer 1978, 1980; Cordazzo and Seeliger 1987, 1988a). The distribution of *Senecio crassiflorus* can be explained by the species' germination success and the growth capacity of seedlings in response to sand accretion (Cordazzo and Souza 1993).

The distribution and biology of the other dominant species are largely controlled by the groundwater table distance and inundation stress (Cordazzo and Seeliger 1988a). In areas close to the groundwater table, with reduced sandflow and low levels of nutrients, *Spartina ciliata* assumes dominance. Although both *Spartina* and *Panicum* maintain their populations principally via clonal growth, each species adopts different colonization strategies (Costa et al. 1984, 1991). While rhizome fragments are the principal dispersal units of *Panicum*, a large number of salt-tolerant seeds permit *Spartina ciliata* to rapidly occupy disturbed sites, owing to high germination success and efficient seedling establishment (Cordazzo 1994). At drier sites, the number of live *Spartina* leaves/tillers relates to precipitation. High death rates occur during dry spring and summer, and high birth rates during the fall. At humid sites, birth and death peaks alternate, and *Spartina* tillers reach their maximum height and leaf number. These differences suggest that the density of leaves/tillers depends on the carrying capacity of each site (Costa and Seeliger 1988a).

The pronounced phenotype plasticity and the extensive and fast-growing integrated rhizome system, which compensates for nutrient limitation or competition (Costa and Seeliger 1988b, 1990), enable *Hydrocotyle bonariensis* to opportunistically exploit seasonally flooded humid areas (Costa 1987). Similar distribution patterns of *Hydrocotyle bonariensis* have been observed in dune systems elsewhere (Moreno-Casasola and Espejel 1986; Barbour et al. 1987). The local co-dominance of humid sand indicators, like *Paspalum vaginatum* (Pfadenhauer 1978, 1980; Pfadenhauer and Ramos 1979; Cordazzo and Seeliger 1988a, 1993; Costa et al. 1988b), confirms this patterns for the southern Brazilian coast. The leaf demography and formation of reproductive structures of *Hydrocotyle bonariensis* depend on the carrying capacity of each site as well as on resources in belowground organs. A high below/above ground biomass ratio indicates that *Hydrocotyle* depends principally on rhizome-propagated growth. Groundwater table fluctuations promote different leaf morphologies, with the rapid sprouting of small leaved and long stemmed plants after flooding (Costa and Seeliger 1988b, 1990).

Similar to *Hydrocotyle bonariensis*, areas with a seasonally fluctuating watertable are dominated by the perennial grass *Andropogon arenarius*. This species is endemic to stable dunes of southern Brazil and Uruguay and grows in association with *Spartina ciliata*, *Imperata brasiliensis*, *Androtrichum trigynum*, and *Hydrocotyle bonariensis* (Cordazzo 1985). Within distributional limits, the growth of *Andropogon* appears to be influenced by competition for nutrients and/or space with *Spartina ciliata*. Site-specific abiotic factors favour different strategies of spatial occupation. At dry sites, with desiccation stress and/or sand accretion or erosion, *Andropogon* seedlings rarely become established, and spring growth initiates from clonal budding of rhizomes, which results in dense clusters of plants. In contrast, wet areas promote seed germination and seedling survival, which leads to the random growth of single shoots with low density. Low aerial biomass during winter suggests a pre-adaption mechanism to seasonally flooded conditions (Costa 1987; Costa et al. 1988a).

Winter-inundated sites are typically dominated by the perennial Cyperaceae *Androtrichum trigynum*, which is endemic to southern Brazil and northern Argentina (Costa 1987; Costa et al. 1988b). Associates include *Phyla canensis*, *Bacopa monnieri*, *Pluchea sagittalis*, and *Paspalum vaginatum*, inundation-tolerant species of drier sites (*Andropogon arenarius*, *Hydrocotyle bonariensis*, *Spartina ciliata*), and marsh remnants, like *Juncus acutus* and *Typha domingensis* (Cordazzo and Seeliger 1987, 1993). The change of dominance between *Andropogon* and *Androtrichum* can be attributed to different growth and reproductive responses of these species to inundation stress. The aerial biomass of *Androtrichum* is positively related to the watertable distance, and plants are absent from areas with a deeper (>2 m) mean annual groundwater table. The rapid occupation of wet areas during spring seems to be related to an efficient uptake of nutrients which become available after drainage (Lacerda et al. 1986). A high proportion of monocarpic flowering shoots induces high mortality in the fall after flowering. In contrast, a low proportion of flowering shoots in dry areas results in low and constant mortality rates throughout the year (Costa et al. 1988b).

6.5 Coastal Foredune Fauna

N.M. Gianuca

As has been reported for dune systems elsewhere (van Heerdt and Morzer Bruyns 1960; Callan 1964; Caussanel 1970), insects are the dominant fauna in the extensive coastal foredunes of Brazil's extreme south and represent the most diverse group of organisms, though some vertebrate species are also important (Gianuca 1985, 1988; Appendix).

Insects

Ants, with nine species, are among the most conspicuous organisms throughout the year, although in the spring and summer elevated noon surface sand temperatures drastically reduce their activity on the surface. Despite an apparently low diversity of food sources, diet preferences vary between species. *Pheidole spininodis*, *Monomorium* sp., and two species of the genus *Conomyrma* are omnivorous, whilst *Camponotus punctulatus* and *Solenopsis geminata* are essentially herbivorous and carnivorous, respectively. The basic food of *Mycetophylax simplex* and *Brachymyrmex* sp. are fungi on dead organic matter and faecal pellets (Bicho, unpubl.).

The coleopterans (>40 species) are the most diverse group. Sand burrowing scarabs like *Thronistes rouxi*, *Ligyrus gianucai*, *Plectris bonariensis*, *Athyreus chalybeatus*, and *Psammodius* sp. are typical of the dunes. Their adult forms are largely restricted to the summer and only surface from the sandy sediments on wind-still nights with high temperature and humidity. Both larvae and adults of scarabaeids feed on detrital matter of roots and rhizomes of resident vegetation. In contrast, all species of carabids are predators. *Tetragonoderus variegatus* remains buried at night and preys during the day on collembolans and delphacids, whilst *Tetragonoderus undatus* surfaces and feeds at night. The carabids *Sphizogenius costiceps*, *Bembidion* sp., and *Scarites* spp. are usually restricted to slacks, where they prey on larvae and adults of other insects, preferentially the rove beetles *Bledius bonariensis*, *Bledius microcephalus*, *Bledius fernandezi*, the heterocerid *Efflagitatus freudei*, the hydrophilid *Paracymus rufocinctus*, and the orthopteran *Neotridactylus carbonelli*, which reaches high densities in the spring and summer. Sand microalgae (Garcia-Baptista 1993) and organic matter derived from the decomposition of plant litter make up the basic diet of these slack dwellers. Several coleopterans are associated with the aerial parts of dune plants. The anticid *Lagrioida nortoni* is abundant on the leaves and spikes of the grasses *Panicum racemosum* and *Spartina ciliata*. The weevil *Listroderes uruguayensis* lives between and feeds on plants of *Hydrocotyle bonariensis*. Caterpillars of the moth *Ecpantheria indecisa* graze upon the leaves of *Senecio crassiflorus* (Pereira 1957) as well as on *Blutaparon portulacoides*, which is also mined by larvae of the small fly *Hapopleudes vogti* (Agromyzidae). Other species associated with the vegetation of frontal dunes include the cicadas *Proarna uruguayensis* and *Proarna* sp. and the delphacid *Delphacodes kuscheli*. The scavenging mole cricket *Scapteriscus riograndensis* and the predator earwig *Labidura riparia* (Labiduridae) are common species throughout the year.

Some flies (i.e. *Eccritosia rubriventris*, Asilidae) and several wasps are well adapted to the sand dune environment owing to both their ability to burrow in loose sand and to locate and catch either flying or buried insects. The spider wasp *Anoplius bilunulatus* attacks the white spider *Moenkhausiana halophila* (Zoodaridae), which serves as food for the larvae in sub-surface galleries. The larvae of one of the largest species (*Campsomeris cineraria*, Scoliidae) parasitize the white grubs of the scarabs *Thronistes rouxi* and *Ligyrus gianucai*. *Eucoila* sp. (Cynipidae) attacks larvae of flies. Underground nests of *Tachytes ornatipes* (Larridae) are most common in the slacks, where adults paralyze and store the pigmy mole cricket *Neotridactylus carbonelli* to feed the larvae.

Vertebrates

The frontal dunes are the preferred habitat of several vertebrate species. At night, the sand toad *Bufo arenarum arenarum* (Bufonidae) and the sand frog *Pleurodema darwini* (Leptodactylidae) are the dominant predators of different insects and spiders. During daylight, the lizard *Liolaemus occipitalis* (Iguanidae) feeds opportunistically on the available insects. Small toads, frogs, and especially lizards are preyed upon by the hognose snake *Lystrophis dorbignyi* (Colubridae). Two rodents, the mouse *Calomys laucha* and the tuco-tuco *Ctenomys flamarioni* (Ctenomyidae), are common to the drier frontal dunes, where they ingest seeds and plant tissue of foredune vegetation. During most of the day, *Ctenomys flamarioni* hides in an extensive network of tunnels, and storage food and building material for the nest is almost always collected at night. The sheltered areas and pastures behind the foredunes are inhabited by species like the burrowing owl *Athene cunicularia* (Strigidae), the skunk *Conepathus chinga* (Mustelidae), the armadillo *Dasypus hybridus* (Dasypodidae), and the fox *Dusicyon gymnocercus* (Canidae), all of which undertake feeding incursions into the foredunes at night.

6.6 Bacterioplankton

P. C. Abreu

Both autotrophic picoplankton (<2 μm) and heterotrophic bacteria are important components of oligotrophic oceanic water masses because of their contribution to primary production (Platt and Li 1986) and their influence on remineralization processes and importance as trophic links (Azam et al. 1983), respectively. In spite of their relevance, little is yet known about the biology of bacterioplanktonic organisms and their ecology in coastal and oceanic waters of this region. In superficial shelf and slope waters (31^0 30' - 34^0 30' S) the number of free bacteria (length 0.5-1 μm; width 0.3-0.5 μm) oscillates between 0 and 1.35 x10^5 cells l^{-1}, whilst the number of bacteria (length 1-2 μm; width 0.3-0.5 μm) attached to particles varies between 0 and x5.25 10^5 cells l^{-1}. Free and attached bacteria are more abundant during the spring and distribution patterns generally follow those of phytoplankton biomass (chlorophyll *a*), being highest in coastal water and during the influence of subantartic water. The biomass of total free and attached bacteria and phytoplankton is of similar magnitude (Fig. 6.6.1). The present data indicate that bacteria efficiently metabolize a large part of the organic matter produced by phytoplankton, which is an important carbon source for higher trophic levels.

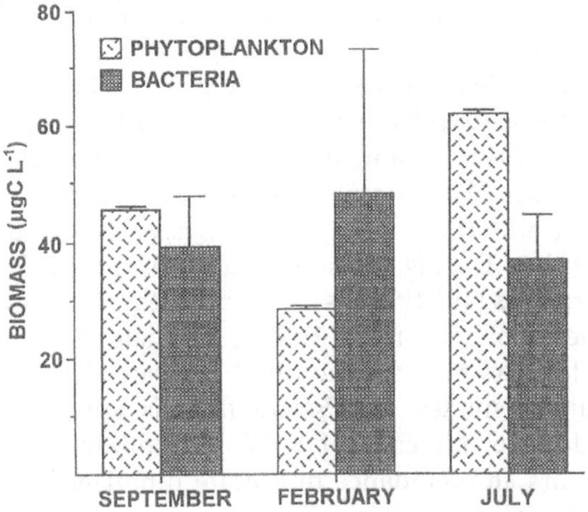

Fig. 6.6.1. Mean bacterial and phytoplankton biomass with standard deviation in southern Brazilian coastal waters

6.7 Phytoplankton

C. Odebrecht and V. M. T. Garcia

Taxonomic Composition, Cell Abundance, and Diversity

Descriptions of phytoplankton species of the southern Brazilian shelf were first done early this century (Cunha and Fonseca 1918; Peters 1932; Käsler 1932; Gemeinhardt 1932). These and subsequent studies showed that a large number of the diatom species belong to the genera *Chaetoceros*, *Coscinodiscus*, *Rhizosolenia*, and *Nitzschia* (= *Pseudonitzschia*), and that *Ceratium*, *Dinophysis*, *Prorocentrum*, *Scrippsiella*, *Noctiluca*, *Oxytoxum*, and Gymnodiniales are important dinoflagellates (Rosa 1979; Kremer and Rosa 1983; Dohms 1983; Ciotti 1990). Early estimates of total phytoplankton abundance (diatoms, naked and thecate dinoflagellates, nanoflagellates, coccolitophorids) at various depths (Hentschel 1932) reported concentrations of 10^4-10^5 cells l^{-1}, though recent studies have shown concentrations of 10^6 cells l^{-1} in coastal and offshore waters (Barth 1968; Ciotti 1990), and 10^7 -10^9 cells l^{-1} in the surfzone (Odebrecht et al. 1995a).

The convergence between the Tropical Water (TW) of the Brazil Current (BC) and Subantartic Water (SAW) of the Malvinas Current (MC), as well as the freshwater input from the La Plata River and the Patos Lagoon, all with distinct phytoplankton communities, results in high phytoplankton diversity over shelf and slope (Appendix). At the southern end of the shelf during the winter, the Shannon-Weaver index (3.4-4.6) for diatoms and silicoflagellates compares favourably to values of all phytoplankton taxa from highly diverse marine ecosystems elsewhere (Lange and Mostajo 1985). Of the 131 diatoms and silicoflagellates species identified by these authors, 45% are of neritic and 16% of oceanic origin. The dominant species, like *Rhizosolenia setigera, Thalassiosira rotula, Thalassionema nitzschioides, Dictyocha fibula*, and *Dictyocha speculum*, have a cosmopolitan and temperate distribution. During the late winter and spring elevated phytoplankton concentrations (10^5-10^6 cells l^{-1}) occur both nearshore and offshore. Under the influence of continental runoff, the diatom *Skeletonema* is abundant in coastal waters near the mouth of the Patos Lagoon estuary, while *Rhizosolenia* and pennates like *Pseudonitzschia* are abundant in offshore regions under the influence of SAW. The influence of Coastal Water (CW) also favours an abundance of thecate dinoflagellates (*Scrippsiella* cf. *trochoidea, Prorocentrum* spp., *Dinophysis* spp., *Ceratium* spp.) in the northern central shelf regions. The autotrophic ciliate *Mesodinium rubrum* is common along the entire southern Brazilian coast. Nanoflagellates largely dominate oceanic areas under the influence of TW but share dominance with diatoms and dinoflagellates over the shelf and in coastal regions in the winter and spring (Ciotti 1990; Castello et al. 1990). Under the strong influence of the BC in the summer, phytoplankton abundance is usually low, and warm water diatoms (*Palmeria hardmanniana, Pseudosolenia calcar-avis, Rhizosolenia robusta, Rhizosolenia imbricata, Proboscia indica, Chaetoceros decipiens, Chaetoceros peruvianus, Chaetoceros coarctatus*) extend into nearshore waters (Dohms 1983).

Chlorophyll *a* and Primary Production

High chlorophyll *a* concentrations and primary production rates occur off the coast of southern Brazil. Based on the data of six cruises (Table 6.7.1), the mean annual particulate primary production is approximately 160 g C m^{-2} $year^{-1}$. High productivity rates result mainly from northward transport of nutrient-rich SAW, continental runoff, and upwelling of Subtropical Water (SACW; Fig. 6.7.1). As a consequence, the distribution patterns of phytoplankton chlorophyll *a* concentrations and primary production rates are directly a function of the temporal and spatial variability of the different water masses. Large-scale meteorological processes (El Niño-Southern Oscillation) may cause pronounced interannual variations of hydrographic characteristics as a result of differences in the total annual precipitation. Therefore, chlorophyll *a* concentrations are higher in CW during El Niño years (warm periods in the Pacific Ocean) than in La Niña years (cold periods; Ciotti et al. 1995).

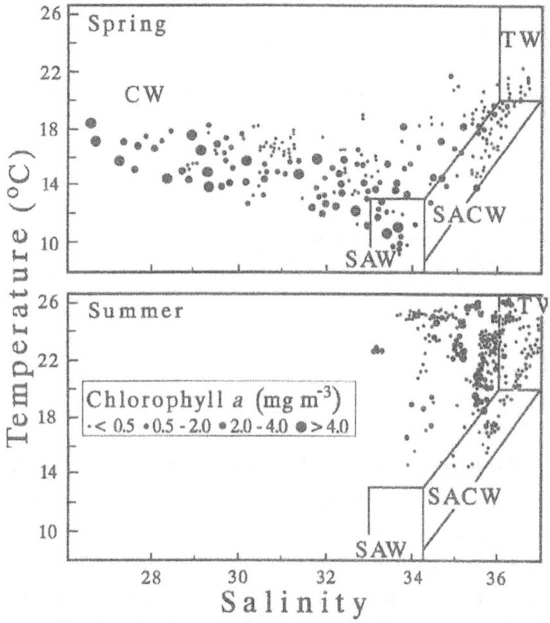

Fig. 6.7.1. T-S diagrams with water masses and chlorophyll *a* concentrations off southern Brazil. Subantarctic Water (*SAW*); Tropical Water (*TW*); Coastal Water (*CW*); and South Atlantic Central Water (*SACW = STW)*

In late winter and spring, the greatest chlorophyll *a* values and primary production rates are related to nutrient input from SAW and continental runoff, with sub-surface and surface nuclei of high biomass, respectively (Ciotti et al. 1995). In the summer, the low chlorophyll *a* concentrations and primary production rates are due to nutrient limitation during oligo-trophic TW influence. Furthermore, phytoplankton concentrations, production rates, and chlorophyll *a* concentration may be limited by light in the winter. Shelf break and coastal STW upwelling favour high biomass concentrations in slope and coastal waters, respectively, the extent of STW intrusion over the shelf being an important factor. Large-scale coastal upwelling events lead to high chlorophyll *a* values (up to 7 mg m^{-3}) in surface waters (Hubold 1980a). In contrast, incomplete upwelling and mixing result in sub-surface chlorophyll *a* maxima (approx. 3 mg m^{-3}), owing mainly to the growth of large diatoms with an increased cell chlorophyll *a* content, rather than to an increase in cell number (Odebrecht and Djurfeldt 1996).

Table 6.7.1 Particulate primary production rates (mg C m^{-2} h^{-1}) in different seasons off the coast of southern Brazil.

Season	mean	min.	max.
Winter[a]	12.3	8.5	17.2
Late winter[b]	78.6	44.8	176.7
Spring[c]	120.3	52.6	344.3
Summer[d]	27.3	9.0	56.7
Fall[c,e]	36.0	3.4	98.1

[a]Garcia (unpubl.). [b]Ciotti (1990). [c]Teixeira et al. (1973). [d]Odebrecht (unpubl.). [e]Gonzalez (1994).

Bloom Events

Under favourable climatic, hydrographic, and physico-chemical conditions, phytoplankton blooms are a common phenomenon along the southern Brazilian coast. Unidentified red-coloured surface patches have been observed since the beginning of the century (Cunha and Fonseca 1918), and high concentrations of the diatoms *Coscinodiscus*, *Rhizosolenia*, and *Nitzschia* (Barth 1968), of the cyanophyte algae *Anabaena* and *Trichodesmium*, and of the dinoflagellate *Ceratium massiliense* (Rosa 1979) occurred at several occasions in nearshore and coastal waters. Especially in the surfzone, environmental conditions appear to favour blooms of some phytoplankton species.

Large concentrated patches (10^9 cells l^{-1}) of the diatom *Asterionellopsis* (=*Asterionella*) *glacialis*, with chlorophyll *a* concentrations up to 1647 mg m^{-3}, are a common feature of the surf zone in southern Brazil. Patch formation is related to cycles of cold front passages every 6 to 10 days, which force coastal waters onshore, and diurnal cycles (Odebrecht et al. 1995a). Since mucus production of *Asterionellopsis glacialis* cells in unialgal cultures is stimulated by the addition of fine sediments, diurnal chlorophyll *a* oscillations in the surf zone may be a result of cyclic mucus production which favours attachment of suspended particles to the cells and causes their sedimentation. Cells are then likely to survive in low light on bottom sediments, because under controlled conditions at low light (30 μE m^{-2} s^{-1}) the cellular chlorophyll *a* content (6.18 pg $cell^{-1}$) of *Asterionellopsis glacialis* exceeds that of cells (1.52 pg $cell^{-1}$) cultured under high light (300 μE m^{-2} s^{-1}; Garcia and Rörig, unpubl.).

The presence of potentially toxic dinoflagellate blooms in the surf zone, such as *Gyrodinium* cf. *aureolum*, *Dinophysis acuminata*, and *Noctiluca scintillans* have been observed on several occasions and may be largely a result of rapid offshore cell growth during calm weather and their concentration in the surfzone during cold front passages (Odebrecht et al. 1995b).

These conditions may have led to the massive mortality of the intertidal molluscs *Mesodesma macroides* and *Donax hanleyanus* and the crustacean *Emerita brasiliensis* along several hundred kilometers of southern Brazilian beaches (Machado 1979; Rosa and Buselato 1981; Tommasi 1983; Garcia et al. 1994; Odebrecht et al. 1995b).

6.8 Protozooplankton

C. Odebrecht

Protozooplankton is composed of apochlorotic flagellates, ciliates, and ameboid organisms. These heterotrophic protista are important ecological intermediates of the microbial loop because they act on nutrient recycling, as predators of other micro-organisms, and serve as prey for larger consumers (Sieburth et al. 1978). Early citations of protozooplankton organisms in southern Brazilian shelf waters included tintinnid ciliates (*Tintinnopsis campanula*, *Tintinnopsis beroidea*) and heterotrophic dinoflagellates (*Noctiluca scintillans*, *Peridinium* = *Protoperidinium divergens*; Cunha and Fonseca 1918). Estimates of abundance in formaldehyde-fixed bottle samples from various depths demonstrated that zooflagellates and naked apochlorotic dinoflagellates (Gymnodiniales) were the most abundant forms (10^3 cells l^{-1}; Hentschel 1932).

Protozooplankton organisms appear to be an important component of the pelagic shelf system (Appendix). Both the vertical and horizontal distribution patterns of their cell concentrations tend to coincide with phytoplankton and chlorophyll *a* concentrations, and are higher in either surface layers or under the influence of nutrient-rich SAW, STW, and CW. Because of CW, cell concentrations of *Noctiluca scintillans*, *Gyrodinium* spp., *Polykrikos schwartzii*, unidentified dinoflagellates, and naked ciliates tend to be high ($>10^4$ cells l^{-1}) near the Patos Lagoon mouth (Castello et al. 1990). Because of SAW, large (40-80 μm) naked ciliates (*Strombidium*), tintinnids (*Tintinnopsis*, *Eutintinnus*), and large heterotrophic dinoflagellates (*Gyrodinium*, *Protoperidinium*) exceed cell concentrations of 10^4 cells l^{-1} in shelf waters (Ciotti 1990). In contrast, in oceanic regions, protozooplanktonic organisms are largely represented by small (<20 μm) naked ciliates (10^3-10^4 cells l^{-1}) in oligotrophic warm waters of the Brazil Current (Dohms 1983; Ciotti 1990).

6.9 Zooplankton

M. Montú, I. M. Gloeden, A. K. Duarte, and C. Resgalla Jr.

Taxonomic Composition and Distribution

Coastal and marine zooplankton is represented by foraminiferans (37), medusae (53), siphonophores (32), polychaetes (23), molluscs (29), cladocerans (7), ostracods (67), copepods (152), mysids (5), euphausiids (26), chaetognaths (20), and tunicates (39; Appendix). The species composition and the spatial and temporal distribution of the zooplankton organisms are directly related to the dominance of different water masses over the southern Brazilian shelf and slope. Although planktonic foraminiferans are good indicators of different water masses, they are generally less abundant, except for the more frequent *Globigerinoides ruber*, *Globorotalia inflata*, *Globorotalia menardii*, and *Globoquadrina dutertrei* (Boltovskoy 1959, 1970, 1981; Closs and Barberena 1960, 1962; Roettger 1970; Bé and Tolderlund 1971).

The abundance and diversity of medusae increase during the spring and summmer, with *Liriope tetraphylla* and *Proboscidactyla ornata* being dominant (Meneghetti 1973). While some species (*Aglaura hemistoma*, *Rhopalonema velatum*, *Aegina citrea*, *Liriope tetraphylla*, *Proboscidactyla ornata*) have a widespread distribution (Ramirez and Zamponi 1981), others are more restricted to either coastal (*Obelia* spp., *Eucheilota ventricularis*) (Navas-Pereira 1973) or shelf regions (*Zanclea costata*, *Cytaeis tetrastyla*, *Bougainvillia ramosa*, *Bougainvillia platygaster*, *Sarsia eximia*, *Turritopsis nutricula*, *Proboscidactyla ornata*, *Cunina peregrina*). Both *Aglaura hemistoma* and *Cunina octonaria* are typical of the Brazil Current (Vannucci 1957, 1963; Navas-Pereira 1973).

Like medusae, siphonophores (*Chelophyes appendiculata*, *Diphyes bojani*, *Bassia bassensis*, *Abylopsis eschscholtzi*, *Muggiaea kochi*, *Enneagonum hyalinum*) are most abundant in high temperature and salinity waters during the spring and summer. *Muggiaea kochi* and *Enneagonum hyalinum* occur preferentially in nearshore waters, whilst *Chelophyes appendiculata*, *Eudoxoides spiralis*, *Diphyes dispar*, *Diphyes bojani*, and *Bassia bassensis* appear in the fall. Under the influence of different water masses in winter, *Lensia subtilis*, *Diphyopsis mitra*, *Sulculeolaria quadrivalvis*, and *Abylopsis tetragona* occur. Only *Hippopodius hippopus* can be found throughout the year in oceanic regions (Gloeden, unpubl.). In addition, shelled planktonic pteropod molluscs (*Limacina trochiformis*, *Creseis virgula*), some characteristic of the Brazil Current (*Limacina lesueuri*, *Limacina bulimoides*) and others of wider distribution (*Cavolinia inflexa*, *Clio pyramidata*, *Hyalocylis striata*), are most abundant during the summer. Species of the genera *Creseis* and *Cavolinia* as well as *Limacina inflata*, *Limacina bulimoides*, *Limacina lesueuri*, and the predominant *Limacina* f. *retroversa* are restricted to southern shelf waters in the winter (Resgalla and Montú 1994).

The cladocerans are essentially neritic, generally with higher concentrations close to the coast (Navas-Pereira 1973). The species richness rises in the winter when *Penilia avirostris*, *Pseudoevadne tergestina*, *Evadne spinifera*, and *Podon intermedius* become common. Species like *Pleopys polyphemoides* and *Evadne nordmanni* are indicators of cold and *Podon schmackeri* of warm coastal waters, whilst *Penilia avirostris* is characteristic of the Brazil Current (Resgalla and Montú 1993).

As elsewhere, copepods (up to 2000 org. m^{-3}) constitute the most abundant group of zooplankton in southern Brazilian neritic and oceanic waters. The dominance of SAW during the winter favours the presence of cold water species (*Calanus simillimus*, *Ctenocalanus vanus*, *Calanoides carinatus*, *Eucalanus elongatus*, *Centropages brachiatus*, *Pontella marplatensis*, *Oncaea conifera*, *Oithona similis*) over the shelf and the occurrence of *Rhyncalanus nasutus*, *Conea rapax*, and *Acartia danae* in oceanic regions. During the summer, typical TW species (*Paracalanus quasimodo*, *Paracalanus parvus*, *Calocalanus pavoninus*, *Clausocalanus furcatus*, *Temora stylifera*, *Oithona nana*) occur. During this period *Paracalanus campaneri*, *Mecynocera clausi*, *Calocalanus pavo*, *Lucicutia flavicornis*, *Acartia negligens*, *Oncaea venusta*, *Corycaeus furcifer*, *Corycaeus latus*, *Sapphirina auronitens*, *Sapphirina ovatolanceolata*, and *Sapphirina scarlata* are found more frequently in oceanic regions, and other species like *Acartia tonsa*, *Paracalanus parvus*, *Temora stylifera*, *Eucalanus sewelli*, *Euterpina acutifrons*, and *Clytemnestra rostrata* are common to the shelf (Björnberg 1963; Duarte and Montú, unpubl.).

Species of mysids tend to be abundant in coastal waters, and *Metamysidopsis elongata* and *Neomysis americana* may constitute up to 90% of the zooplankton (Bersano 1994). In contrast, euphausiids are generally more common in slope and oceanic regions, and especially the adult forms of some dominant species (*Euphausia similis*, *Euphausia recurva*, *Euphausia lucens*, *Thysanoessa gregaria*) appear to have a preference for these waters (Gorri 1995). Several Euphausiacea (*Euphausia lucens*, *Nematoscelis megalops*, *Thysanoessa gregaria*) occur in SAW, whilst others (*Stylocheiron abbreviatum*, *Stylocheiron affine*, *Thysanopoda aequalis*, *Thysanopoda monacantha*, *Thysanopoda obtusifrons*, *Thysanopoda orientalis*, *Euphausia brevis*, *Euphausia hemigibba*, *Euphausia recurva*, *Euphausia spinifera*, *Nematobrachion flexipes*) are present in TW (Seguin 1965; Barth 1966; Mauchline and Fischer 1969; Ramirez 1971; Brinton 1975; Mauchline 1980; Gorri 1995).

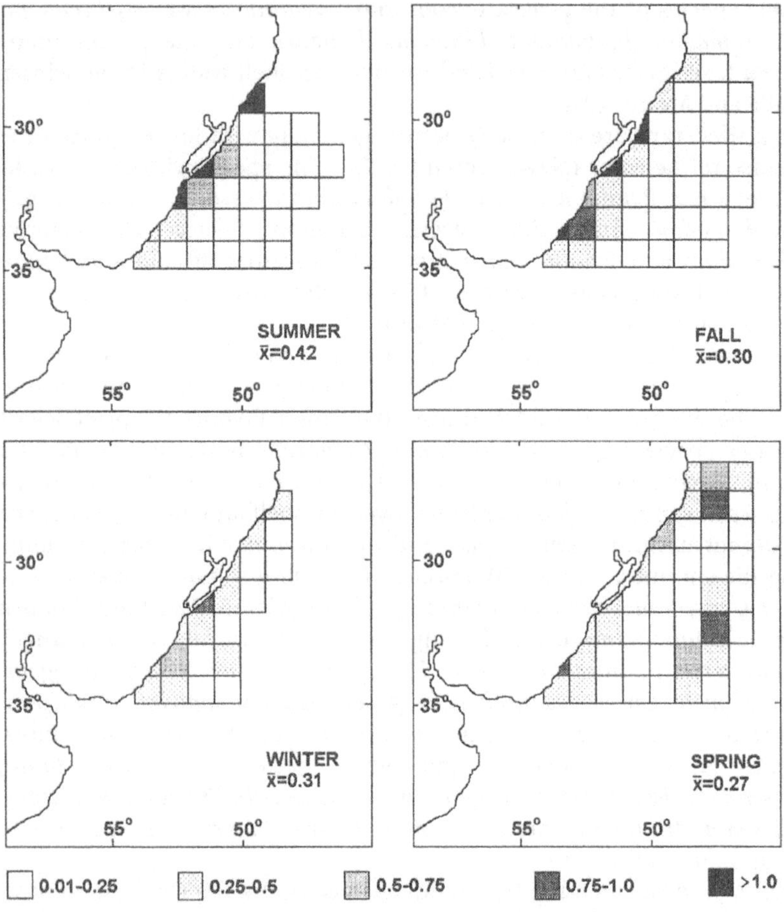

Fig. 6.9.1. Distribution of zooplankton biomass in volume (ml m^{-3}) over the continental shelf

Several species of the Chaetognatha are indicators of TW (*Sagitta serratodentata*), STW (*Krohnitta subtilis, Sagitta decipiens*), CW (*Sagitta tenuis*), and SAW (*Sagitta tasmanica;* Resgalla 1993). Other common cold water species include *Sagitta planctonis, Sagitta maxima,* and *Eukrohnia hamata* (Boltovskoy and Jankilevich 1981; Duarte and Montú, unpubl.). Throughout the year, *Sagitta tenuis* is the most abundant species close to the coast and over the shelf, whilst *Sagitta enflata, Sagitta serratodentata, Sagitta minima, Sagitta bipunctata, Sagitta hispida,* and *Pterosagitta draco* occur in shelf and slope water. Similarly, different species of tunicates (Appendicularia) are more abundant in nearshore water (*Oikopleura dioica*), shelf and slope water (*Oikopleura longicauda, Oikopleura fusiformis, Oikopleura rufescens, Fritillaria borealis, Fritillaria pellucida*), or occur predominantly in oceanic water (*Oikopleura albicans, Oikopleura cophocerca, Stegosoma magnum, Fritillaria formica*). Species of the Thaliaceae (*Ihlea punctata, Ritteriella retracta, Salpa aspera, Salpa fusiformis, Cyclosalpa polae, Brooksia rostrata, Thalia democratica*) are of tropical or subtropical origin and abundant during the spring and summer, whilst the subantartic *Ihlea magalhanica* and *Salpa thompsoni* are common to epipelagic layers (Amaral, unpubl.).

Migration

The thermocline appears to control the vertical migration patterns of several zooplankton species, such as the migration of medusae to layers above (*Bougainvillia platygaster, Euphysora furcata*) and below (*Cunina peregrina*) the thermocline (Correia 1983) as well as the migrations of Siphonophora (*Lensia campanella, Lensia cossack, Sulcoleolaria quadrivalvis, Sulcoleolaria turgida, Sulcoleolaria chuni*) from deeper layers to the surface. Although Pteropoda are known for their pronounced vertical migration (Haagensen 1976), short distance migration of *Limacina* species between epipelagic and mesopelagic layers are common in southern Brazilian waters. During the winter, nictimeral migrations of Chaetognatha over the slope result in high concentrations in surface (*Sagitta tenuis, Sagitta tasmanica*) and epipelagic layers (100 m; *Eukrohnia hamata, Krohnitta subtilis, Sagitta decipiens*) at night, whilst *Sagitta enflata, Sagitta hexaptera, Sagitta lyra*, and *Pterosagitta draco* concentrate in surface layers during the day. In contrast, direct migration of the three latter species during the summer results in high concentrations at night (Resgalla 1993).

Biomass

Zooplankton biomass evaluations provide important information to help clarify community structure and energy flow among trophic levels. Just as for species composition, different water masses control the spatial and temporal distribution patterns of zooplankton biomass. Generally, biomass values tend to decrease northward and with increasing distance from the coast. High carbon values over the shelf are related to the influence of SAW and CW, with an important contribution from continental runoff from the La Plata River and the Patos Lagoon during the coldest months. However, the highest carbon values (up to 98.47 mg C m^{-3}) occur during the spring and summer, principally in coastal surface waters. Especially in the surf zone, mysids may achieve concentration peaks of 8142 mg C m^{-3} (Bersano 1994).

Annual variations of estimated zooplankton volumes (Bongo net samples) are highest (>1 ml m^{-3}) during the summer and fall in areas close to the coast, owing to the abundance of *Thaliacea* species which are large organisms with a low energetic content (Meneghetti 1973; Navas-Pereira 1973; Hubold 1980a, b). In contrast, during the winter and spring, the highest values (0.25-1 ml m^{-3}) occur in offshore areas under the influence of SAW and Subtropical Convergence (Fig. 6.9.1). Owing to the diversity of the zooplankton community, the summer and fall biomass peaks are frequently associated with gelatinous plankton, whilst the winter and spring biomass is largely a function of copepods and occasionally of euphausids.

6.10 Benthic Beach Invertebrates

N. M. Gianuca

The extensive, gently sloping, fine grained sandy beaches of Brazil's extreme south favour the occurrence of an abundant and diverse macrobenthic fauna (Appendix). Their vertical distribution follows a typical zonation pattern, including a backshore or supratidal zone, an intertidal zone, and inner and outer surfzones (Fig. 6.10.1). Early studies of beach invertebrates were concerned with the assessment of commercially valuable stocks (Gianuca 1971) or concentrated on taxonomic and qualitative surveys (Orensanz and Gianuca 1974; Escofet et al. 1979). More recent quali-quantitative studies characterized the fauna of the intertidal zone (Gianuca 1983, 1985, 1987; Santos 1990) and the inner (Paes 1989; Soares 1992) and outer (Borzone and Gianuca 1990) surfzone, revealing that they are among the richest and most productive sandy beach environments anywhere.

Supratidal Organisms

The supratidal zone, or backshore, is only flooded by seawater during extreme storm tide events. Among the organisms that inhabit this zone, *Orchestoidea brasiliensis* is restricted to the lower fringe, whilst *Ocypode quadrata* occupies the entire upper part of the beach (Gianuca 1985, 1988). However, most of the resident species are of terrestrial origin. Insects constitute the dominant fauna and are represented by well-adapted coleopterans, like the beetles *Phaleria testacea, Bledius bonariensis, Bledius microcephalus, Bledius fernandezi, Cicindela nivea conspersa*, and *Cicindela patagonica*. Several predators (Carabidae) are especially attracted by the high densities (2350 ind. m^{-2}) of *Bledius* spp. (Gianuca 1987). A particular association of insects concentrates in moist sands around freshwater creeks, which drain marshes behind the foredunes. The most conspicuous

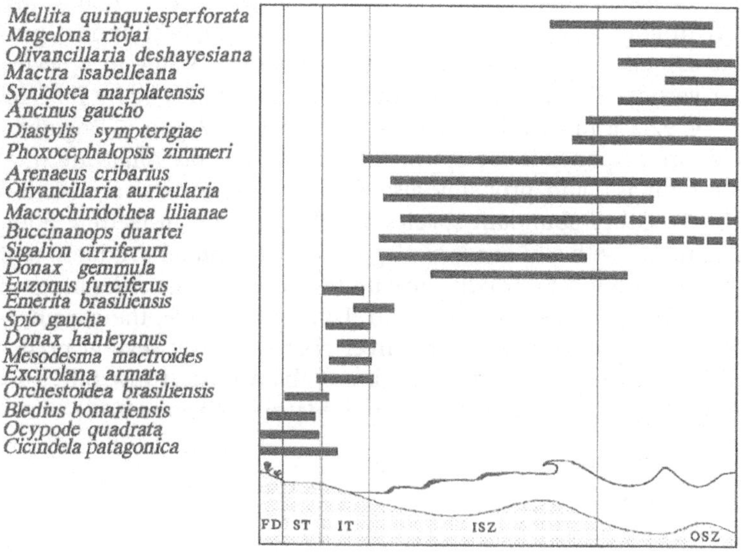

Fig. 6.10.1. Zonation of principal invertebrates between foredunes and surf zone: foredune (*FD*); supratidal (*ST*); intertidal (*IT*); inner surf zone (*ISZ*); outer surf zone (*OSZ*)

species are the beetles *Efflagitatus freudei* and *Paracymus rufocinctus* and the orthopteran *Neotridactylus carbonelli.* All feed on organic detritus, microscopic psammic algae, and associated protozoans in the surface layer of the sand. The principal predators of most supratidal insects are the resident birds *Charadrius collaris* and *Anthus correndera.*

Intertidal Organisms

A wide intertidal zone is characteristic of intermediate and dissipative beaches. Most resident species exhibit vertical distribution patterns, but since their vertical migration depends on tidal flooding, their distributional range follows seawater level and beach profile changes (Gianuca 1987). Along dissipative beaches, some of the subtidal species, like *Neocallichirus mirim* and associated crustaceans (*Pinnixa patagoniensis, Hemicyclops subadhaerens, Yoninae* sp.), polychaetes (*Grubeulepis bracteata, Leocrates* sp.), the turbelarian *Stylochoplana* sp., and an actiniarian which inhabit its tunnels, extend into the lower intertidal zone (Gianuca 1985).

The biomass and production of benthic invertebrates are highest in the intertidal zone (Gianuca 1983, 1985). Accumulated annual production (ash-free dry weight) of *Emerita brasiliensis* and the bivalves *Mesodesma mactroides* and *Donax hanleyanus* equals 238.5, 185.3, and 13.6 g m^{-2}, respectively. In contrast, in a 1 m wide transect, the highest annual production was associated with *Mesodesma mactroides* (3251.76 g m^{-1}), owing to the wider vertical distribution range of this species, followed by *Emerita brasiliensis* (499.32 g m^{-1}) and *Donax hanleyanus* (55.57 g m^{-1}). Together, these species are responsible for approximately 95% of the total intertidal invertebrate biomass, and probably also for the total annual production (Gianuca 1985). Other abundant species are the suspension-feeding polychaete *Spio gaucha*, with a biomass of 35 g m^{-2} and an annual production of 114 g m^{-1}, and the isopod *Excirolana armata* (8 g m^{-2}) with densities of up to 7075 ind. m^{-2} (Gianuca 1985; Santos 1990). During high tide, the considerable density and biomass of intertidal invertebrates attract several predators, like fishes, gastropods, and swimming crabs, whilst at low tide resident and migrating birds become the main predators (Gianuca 1983).

Surfzone Organisms

The shallow and turbulent sandy bottoms of the inner surfzone, between the beach and about 2 m depth, are under the influence of regular wave action and strong currents. Frequent winter storms further increase the stressful hydrodynamic and substrate conditions. As a result, only 10 of the 60 benthic inner surfzone species occur during the entire year. The bivalve *Donax gemmula* is the most abundant species (Gianuca 1985; Paes 1989). Other common organisms are the crustaceans *Pinnixa patagoniensis*, which inhabits the galleries of *Neocallichirus mirim*, *Macrochiridothea lilianae*, *Macrochiridothea giambiagiae*, *Arenaeus cribarius*, and *Phoxocephalopsis zimmeri*, gastropods (*Buccinanops duartei*, *Olivancillaria auricularia*, *Olivancillaria uretai*), and polychaetes (*Sigalion cirriferum*, *Nephtys simoni;* Gianuca 1983, 1985, 1988). Benthic macrofaunal densities of up to 431 000 ind. m^{-1} near Cassino Beach (32^{0} 14' S) are among the highest ever recorded for inner surfzones (Soares 1992).

The outer surfzone (2-10 m depth), although exposed to high wave turbulence during storms, is inhabited by a total of 126 species (Borzone and Gianuca 1990). Crustaceans make up 39%, with the isopods *Ancinus gaucho* and *Synidotea marplatensis* and the cumacean *Diastylis sympterigiae* being the most abundant species. Among the molluscs, which contribute 32%, small species (<15 mm) like *Donax gemmula* and *Parvanachis isabellei* and larger species (15-70 mm) such as *Olivancillaria deshayesiana*, *Mactra isabelleana*, and *Dorsanum moniliferum* are predominant. Thirty species of polychaetes represent 26% of the benthic community. The most abundant ones are *Stheneloplis oculata*, *Magelona riojai*, *Parandalia americana*, *Sthenelais limicola*, *Kinbergonuphis difficilis*, *Nephtys simoni*, *Odontosyllis hetero-falcheata*, and *Owenia fusiformis*. *Mellita quinquiesperforata* and

Amphioplus lucyae are the dominant echinoderms, which represent 3% of the total number of benthic species.

The vertical distribution of the benthic macrofauna in the surf zone is largely a function of depth. Cluster analysis identified three associations (Borzone and Gianuca 1990). The first corresponds to shallower areas (2-5 m) and is characterized by *Olivancillaria auricularia, Buccinanops duartei, Corbula caribaea, Donax gemmula, Macrochiridothea lilianae, Sigalion cirriferum, Diopatra viridis*, and *Mellita quinquesperforata*. A second association, with higher diversity but few exclusive species, like *Olivancillaria carcellesi, Olivancillaria urceus, Acteon pelecais, Epitonium angulatum, Apoprionospio pygmaea*, and *Amphicteis* sp., occurs at depths between 5 and 8 m. In deeper waters (8-10 m) of the outer surfzone, a third association is dominated by *Kinbergonuphis difficilis, Paraprionospio pinnata, Mactra isabelleana, Tellina petitiana, Tellina sandix*, and *Terebra gemmulata*. However, several species (*Diastylis symppterigiae, Synidotea marplatensis, Ancinus gaucho, Abra lioica, Parvanachis isabellei, Parandalia americana*) have a wider distribution and can be found in high densities throughout the outer surfzone.

6.11 Continental Shelf Benthos

R. R. Capítoli

Inner Shelf Fauna

Shallower inner shelf regions (10-50 m depth), which are under the influence of highly turbid freshwater runoff from the Patos Lagoon, suffer periodic perturbations because of elevated silt and clay deposition (Calliari and Fachin 1993). The resulting environmental instability induces a state of disequilibrium (Borzone 1988) since recolonization by species without pelagic life-cycle phases is frequently unsuccessful. Common species in this area include molluscs (*Mactra isabelleana, Natica limbata, Parvanachis isabellei, Olivancillaria urceus, Abra lioica*), polychaetes (*Kinbergonuphis difficilis, Paraprionospio pinnata, Neanthes bruaca, Parandalia americana*), cumaceans (*Diastilis sympterigiae*), and isopods (*Synidotea marplatensis, Ancinus gaucho*). Based on depth and sediment type the 93 molluscs form five associations between 10 and 50 m depth (Appendix; Absalão 1986). At a depth of 10-20 m, the omnivorous *Parvanachis isabellei* and opportunists like *Mactra isabelleana* and *Solen tehuelchus* dominate an association with low diversity in depositional areas where physical processes tend to rework the muddy (30%) sediments. A deeper placed (20-40 m) association in predominantly fine and very fine sandy substrate is characterized by *Tellina gibber, Cadulus tetrachistus*, and *Olivella puelcha*. Co-occurring species

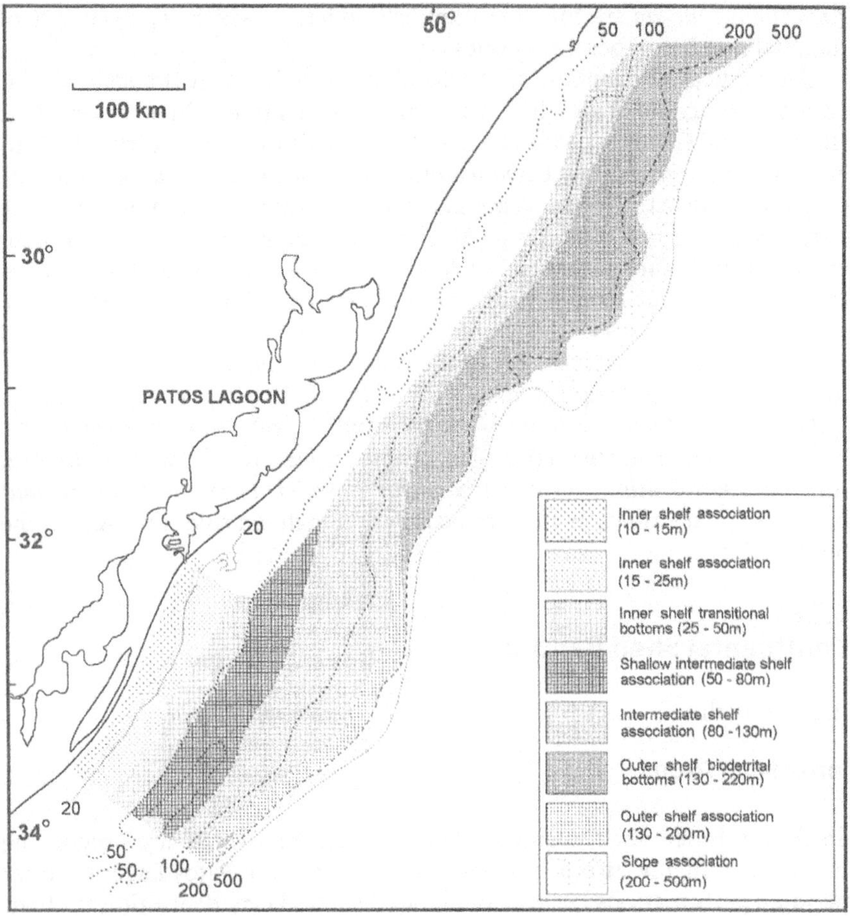

Fig. 6.11.1. Macrobenthic invertebrate associations of the continental shelf and slope

include *Pitar rostatus*, *Mactra petiti*, *Adrana electa*, the predators *Natica limbata* and *Buccinanops lamarcki*, the polychaete *Owenia* sp., and the echinoderms *Astropecten armatus*, *Olivancillaria articulatus*, and *Encope emarginata*. *Nucula puelcha*, *Corbula patagonica*, and *Dentalium infractum*, together with ophiuroids (*Amphiura joubini*) represent an association in sandy mud at approximately 40 m depth. Two less extensive areas of variable sandy substrate types, proximate to beachrock outcrops, are occupied by *Halistylus columna* and *Crassinella lunulata* and by *Olivancillaria urceus* and *Glycymeris longior*, respectively (Absalão 1986).

More distant inner shelf areas south of 32° S are not affected by runoff from the Patos Lagoon. The shallower (10-15 m) sandy areas (Fig. 6.11.1) are characterized by the presence of cumaceans, isopods, the penaeid *Artemesia longinaris*, and the molluscs *Dorsanum moniliferum*, *Olivancillaria urceus*, *Adelomelon brasiliana*, and *Mactra isabelleana*, the last becoming more abundant with increasing silt concentrations. Deeper (15-25 m) areas with fine sand (Fig. 6.11.1) are occupied by *Encope emarginata*, *Buccinanops lamarcki*, *Loxopagurus loxochelis*, *Astropecten armatus*, *Adrana electa*, isopods of the genus *Serolis*, and polychaetes of the family Owenidae. Transitional sandy mud bottoms (25-50 m; Fig. 6.11.1) between the inner and intermediate shelf regions are characterized by *Pitar rostratus*, *Corbula patagonica*, *Buccinanops gradatum*, *Mactra petiti*, *Dardanus arrosor*, *Persephona punctata*, *Hepatus pudibundus*, and ophiurids. *Phyllochaetopterus socialis*, *Astrangia rathbuni*, serpulid polychaetes, and *Crepidula protea* are conspicuous organisms in areas with frequent beachrock outcrops, consolidated substrates, and accumulation of shell fragments. *Chlamys tehuelcha*, *Acroscalpellum bubaloceros*, *Crepidula aculeata*, *Atrina seminuda*, and *Pteria hirundo* may co-occur in less extensive areas.

Intermediate and Outer Shelf and Slope Fauna

Surveys of the benthic macrofauna of the southern Brazilian intermediate and outer shelf and slope (50-500 m) have largely focused on the general distribution of species (Tommasi 1969; Tommasi et al. 1973; Rios 1994).

South of 32° S an association, represented by *Pitar rostratus*, polychaetes (*Asychis* sp., *Owenia* sp.), amphipods, and ophiurids, is relatively abundant in muddy sand bottoms of the shallower (50-90 m) intermediate shelf (Capítoli, unpubl.). These areas also display the highest relative abundance of demersal and benthic fish among all shelf areas (Haimovici et al. 1994a, 1995). The association of the extensive sandy mud bottoms of the intermediate shelf (80-130 m; Fig. 6.11.1) displays the highest faunal abundance, with densities and biomass of 500-1000 ind. m^{-2} and 10-100 g m^{-2}, respectively. The macrobenthic assemblage is largely represented by *Diopatra tridentata*, *Chasmocarcinus typicus*, *Terebellides* sp., *Onuphis tenuis*, *Astropecten cingulatus*, *Squilla brasiliensis*, and *Portunus spinicarpus*. In muddy bottoms (100-130 m) the abundance of characteristic assemblage species decreases, and the polychaetes *Eupanthalis rudipalpa*, *Panthalis oerstedi*, *Spiochaetopterus* sp. and the sipunculid *Nephasomma* sp. become more abundant. Low faunal abundance coincides with regions of biodetrital bottoms (Tommasi et al. 1973), which occupy large areas between 130 and 220 m depth north of 32° S but become narrower around shelf break areas (170-200 m) further south (Fig. 6.11.1). Biodetrital bottoms are composed of a mixture of sand/silt with either exoskeleton debris of ahermatipic corals (*Cladocora debilis, Trochocyathus* sp.) in the north or shell fragments and debris of the coral *Rhizopsammia manuelensis* in the south. The macrobenthic invertebrate association of outer shelf (130-200 m) bottoms (Fig. 6.11.1) is poor (100-500 org. m^{-2}, <10 g m^{-2}) and is largely composed of gorgonians (*Primnoella bisseialis, Primnoella compressa, Elisella barbadensis, Elisella elongata, Muriceides hirtus, Bebryce cinerea, Acanthogorgia scharammi, Muricea midas*), echinoderms (*Anthenoides brasiliensis, Ophiomisidium pulchellum, Brissopsis atlantica, Stylocidaris affinis, Crinometra* sp.), isopods (*Cirolana exigua*), barnacles (*Rostratoverruca nexa*), and serpulid polychaetes. Slope areas (200-500 m) display faunal heterogeneity and low density (<100 ind. m^{-2}) and biomass (<10 g m^{-2}). However, different depth strata (200-300 m and 300-500 m) with varying substrate composition appear to select for different species dominance. Consolidated substrates of upper slope regions are typically occupied by *Eunice frauenfeldi, Zoanthidea* sp., and species of Porifera, Actinaria, and ophiurids. Co-occurring species include *Myriopsis quinquespinosa, Stylocidaris affinis, Verruca caribea, Brissopsis atlantica*, and *Spiochaetopterus* sp. Lower slope regions are characterized by *Acanella eburnea. Flabellum brasiliensis* (coral), *Ophiurolepis* sp. (ophiurid), and *Priapulus* sp. are relatively constant components. *Molpadia* sp. (holothurian), *Batyplax typhla* (brachyurid), *Bathynomus giganteus* (isopod), *Antipathes* sp. (black coral), and *Cephalodiscus* (pterobranchia) are occasionally present.

6.12 Ichthyoplankton

C. Sinque and J. H. Muelbert

Taxonomic Composition

Initial ichthyoplankton surveys over the southern Brazilian shelf and slope (Aboussouan 1969) aimed to elucidate aspects related to fishery biology. Eggs and larvae of about 88 fish species occur in the plankton community and are distributed among 48 families (Appendix). Three species of clupeids (*Sardinella brasiliensis, Harengula jaguana, Brevoortia pectinata;* Hubold and Ehrlich 1981) and of Engraulididae (*Lycengraulis grossidens =L. olidus, Engraulis anchoita, Anchoa marinii;* Weiss et al. 1976; Weiss and Souza 1977a,b; Phonlor 1984a,b) occur in this region. Stomiiformes are recognized as the most important component of oceanic mid-water ichthyofauna (Bonecker and Hubold 1990). They are represented by the families Sternoptychidae (*Valenciennellus tripunctulatus*), Photichthyidae (*Vinciguerria nimbaria, Vinciguerria poweriae, Vinciguerria attenuata, Pollichthys mauli*), Stomiidae (*Stomias* sp.), Idiacanthidae (*Idiacanthus* sp.) and Gonostomatidae. The Gonostomatidae are represented by five species of

Cyclothone, *Diplophos taenia*, *Gonostoma* spp., *Magarethia obtusirostra*, and *Maurolicus muelleri* which is the most abundant species among the Stomiiformes (Weiss et al. 1988; Bonecker and Hubold 1990).

Larvae of Myctophidae occur throughout the year and are occasionally the dominant component of the ichthyoplankton. The most important species are *Diogenichthys atlanticus*, *Hygophum reinhardtii*, *Myctophum* spp., and *Diaphus* spp. which may account for up to 60% of myctophid larvae. Among the Gadiformes, Bregmacerotidae is the most abundant family, with *Bregmaceros* sp. making up 44% of the larvae. Although commercially important, planktonic stages of *Urophycis* spp. (Gadidae) and *Merluccius hubbsi* (Merluccidae) are not abundant, and larvae of *Coelorhynchus* sp. (Macrouridae) are rare.

Eight species of Sciaenidae occur in the ichthyoplankton (Weiss 1981; Muelbert and Weiss 1991; Ibagy and Sinque 1995), and the abundant *Micropogonias furnieri*, *Paralonchurus brasiliensis*, and *Macrodon ancylodon* are important local fishery resources. The suborder Scombroidei is represented by larvae of Gempylidae (*Diplospinus multistriatus*, *Thyrsitops lepidopoides*, *Neolatus tripes*) and Scombridae (*Scomber japonicus*, *Sarda sarda*, *Katsuwonus pelamis*, *Euthynnus alletteratus*, *Auxis* sp.) and by the eggs and larvae of Trichiuridae (Mafalda 1989). Among the Trichiuridae, the eggs and larvae of *Trichiurus lepturus* are the most abundant ones, while larvae of *Lepidopus caudatus* are less common. Pleuronectiformes, represented by eggs and larvae of Pleuronectidae and larvae of Bothidae, Cynoglossidae, and Soleidae, also comprise an important fisheries resource. Larvae of Anguilliformes, Exocoetidae, Atherinidae, Zeidae, Caproidae, Triglidae, Serranidae, Carangidae, Mugilidae, Gobiidae, and Balistidae are less abundant in the ichthyoplankton (Weiss et al. 1976, 1987).

Spatial and Temporal Distribution

Although correlations between feeding, growth, predation, transport, and retention of eggs and larvae and primary and secondary production patterns are still lacking, it is clear that the spawning periods of different fish species and the dynamics of different water masses influence both species composition and large scale spatial-temporal distribution patterns of eggs and larvae.

With the exception of the winter months, clupeid eggs are present throughout the year. Their abundance is highest along the 30 m isobath and decreases towards the open ocean. *Brevoortia pectinata* spawns in coastal areas, and larvae of different developmental stages enter the nursery areas of the Patos Lagoon estuary (Weiss et al. 1976). *Engraulis anchoita* occurs in subtropical waters (STW) and in mixed waters of the La Plata River between 29° and 35° S (Weiss and Souza 1977b), but the greatest amount of spawning occurs at the southern limits of this area under optimal temperature (9-20°C) and salinity (30-36) conditions (Phonlor 1984b). A dominance of *Engraulis anchoita* during the winter and spring can be attributed to the northward displacement of the Malvinas Current (MC), whilst in the summer and fall, owing to the influence of the warmer Brazil Current (BC), other species dominate. *Lycengraulis grossidens* spawns in shallower coastal and less saline waters between 12.9 and 18.4°C (Hubold and Ehrlich 1981) south of 28° 58' S; the planktonic stages then grow further in estuarine areas (Weiss 1981; Muelbert and Weiss 1991). *Anchoa marinii* is a typical marine species which is virtually continuously present and widely distributed geographically, though the eggs and larvae are concentrated over the southern region of the shelf (Weiss et al. 1976; Hubold and Ehrlich 1981).

The eggs and larvae of Stomiiformes are distributed in tropical and temperate pelagic oceans (Bonecker and Hubold 1990). The eggs of *Maurolicus muelleri* occur throughout the year in deeper waters (<250 m) but are most abundant in the fall, whilst larvae are common between 100 and 150 m depth (Weiss et al. 1988) and reach their highest densities during the winter and spring at mean temperatures between 15 and 25°C and salinities of about 36. The larvae of *Cyclothone* spp. and *Vinciguerria* spp. are confined to oceanic waters of the BC below 200 m depth. Their highest concentrations are reached in the winter and spring between 18 and 23°C and a salinity of 36. Because of their low abundance, the distribution pattern of other Gonostomatidae is less well defined. With the exception of *Pollicthys mauli* in shelf waters of Cabo St. Marta (29° S), the species are always distributed in oceanic waters (Bonecker and Hubold 1990).

The larvae of Myctophidae also have an oceanic distribution and occur throughout the year along the shelf break. However, they are most abundant over the southernmost Brazilian shelf region along the 100 m isobath in the winter and summer (*Diaphus* spp.) when they may account for 18-25% of the ichthyoplankton. The larvae of *Bregmaceros* sp. are most abundant in the fall. *Bregmaceros cantori* and *Bregmaceros atlanticus* dominate over the outer shelf north of 29° S, and the highest densities of *Bregmaceros cantori* larvae are associated with the maximum chlorophyll layer (Matsuura et al. 1993). The distribution of *Merluccius hubbsi*, which is highly influenced by temperature (12-13°C), salinity (26-28), and depth (60-90 m), is determined by the northward movement of the MC over the shelf during the winter and spring (Pereira 1983).

Sciaenids are demersal fish that inhabit coastal waters. In general, spawning occurs close to the shore, and several species use the Patos Lagoon estuary as a nursery area (Weiss 1981; Muelbert and Weiss 1991; Ibagy and Sinque 1995). Although eggs and larvae are present throughout the year, the highest densities are registered during the spring and summer. Reproduction of *Antigonias capros* and *Zenopsis conchifer* occurs over the shelf at about 200 m depth under the influence of the warm (17.5-23.4°C) and saline (32.2-36.3) waters of the BC during the fall and spring, respectively. Eggs and larvae of the eurythermal and euryhaline species *Trichiurus lepturus* are present throughout the year, with the highest densities appearing during the summer and fall in TW, STW, and CW (Mafalda 1989). During the summer and fall they occur closer to the shore and also in the estuaris (Weiss 1981; Pereira 1986; Muelbert and Weiss 1991). *Lepidopus caudatus* larvae concentrate during the spring and fall in TW and STW north of this region.

The highest larval densities of the Gempylidae (*Diplospinus multistriatus, Neolatus tripes, Thyrsitops lepidopoides*) occur during the winter in oceanic waters and during the summer in coastal waters (Mafalda 1989). Larvae of Scombridae (*Scomber japonicus, Sarda sarda, Katsuwonus pelamis, Euthynnus alletteratus, Auxis* sp.) tend to occur during the summer in coastal regions under the influence of TW. Similarly, Pleuronectiformes are most abundant during the summer, when their eggs may account for 7% of planktonic fish eggs and their larvae for 17% of the ichthyoplankton. Larvae of Cynoglossidae appear during the fall. Eggs and larvae of less abundant species of Exocoetidae and Bregmacerotidae are noted in the summer, those of Triglidae in the summer and fall, whilst Atherinidae larvae are common in the winter. Coastal waters of the southern part of this region favour the presence of larvae of Carangidae (*Trachurus lathami =T. picturatus, Parona signata, Chloroscombrus crysurus*) and Mugilidae during the summer and fall, respectively (Weiss et al. 1976).

6.13 Pelagic Teleosts

J. P. Castello

Based on their distribution and ecological and economic importance, pelagic fish (Appendix) of the southern Brazilian shelf and slope regions can be classified as (1) inhabitants of nearshore and shelf waters down to approximately 30 m depth, (2) inhabitants of the shelf between the coast and the break at about 200 m depth, and (3) oceanic inhabitants of the slope and deeper waters.

Shallow Coastal Water Inhabitants

Typical inhabitants of shallow waters are *Trachinotus marginatus, Anchoa marinii, Lycengraulis grossidens (=L. olidus), Brevoortia pectinata*, and juveniles of *Engraulis anchoita* (Cunha 1987), whilst *Centengraulis edentulus* and *Anchoa tricolor* are found rarely (Buckup 1984). Among the different *Trachinotus* species (*T. marginatus, T. goodei, T. carolinus,*

T. falcatus), characteristic of turbulent surfzones with variable salinity, *Trachinotus marginatus*, with a total length (TL) of 15-60 mm, is the most abundant one (Cunha 1987). During their first year, the fast growing juveniles feed on molluscs (*Mesodesma mactroides, Donax hanleyanus*) and the crustacean *Emerita brasiliensis* and then migrate to deeper waters. The two engraulidids *Anchoa marinii* (<150 mm TL) and *Lycengraulis grossidens* (60-100 mm TL <180 mm; Castello and Krug 1978) and the clupeid *Brevoortia pectinata* (<350 mm TL; Figueiredo and Menezes 1978) occur in coastal waters of less than 25 m depth, though the distributional range of *Lycengraulis grossidens* is narrower than that of the other two species. Both, *Lycengraulis grossidens* and *Brevoortia pectinata* reproduce principally from the spring to fall and the early developmental and juvenile stages of their life-cycle are associated with estuarine waters (Weiss and Krug 1977; Castello and Krug 1978; Weiss 1981). During the spring, *Anchoa marinii* commonly occurs together with *Pomatomus saltatrix*, *Balistes* spp., *Peprilus paru*, *Sarda sarda*, and *Stromateus* spp. (Mello et al. 1992).

Shelf Inhabitants

Pomatomus saltatrix, Sarda sarda, Scomber japonicus, Trachurus lathami, Katsuwonus pelamis, three species of Mugil, and adults of *Engraulis anchoita* are abundant pelagic species found over the shelf. Owing to its wide distribution, great abundance, and role as a secondary consumer, *Engraulis anchoita* is an important pelagic species in the Southwestern Atlantic (23-45^0 S; Castello and Habiaga 1982; Habiaga and Castello 1986; Acuña and Castello 1986). The distribution of adult shoals follows the coastal branch of the Malvinas Current and is limited to the east by the thermohaline front with the Brazil Current. Freshwater runoff from the Patos Lagoon and La Plata River during the winter (Castello et al. 1990; Lima 1992) provides stable water column conditions in the coastal zone for the development of larvae which are retained in this region by Ekman transport mechanisms (Lima and Castello 1995). Coastal zones north of 31^0 S are preferentially occupied by larvae and juveniles, but further south larvae, juneniles, and adults are associated with low-temperature waters of sub-antartic origin (Castello and Habiaga 1982; Lima and Castello 1995). Near sub-surface upwelling areas (29^0 S), the daily growth rates (0.41-0.73 mm day^{-1}) of larvae (8-32 mm CP; Castello and Vasconcellos 1995) are similar to those in Argentine waters (41^0 S) (Sanchez et al. 1991). Since males and females abandon shallower coastal waters after reaching first maturation at 70-85 mm TL (Phonlor 1984b; Acuña and Castello 1986), the size of individuals tends to increase towards deeper waters. The species grows rapidly, and more than 50% of the asymptotic length (175 mm) is achieved during the first year. Adults congregate over the inner and intermediate shelves during reproduction in the winter and spring, and at least four spawning groups have been identified for southeastern and southern Brazilian shelf regions (Weiss and Souza 1977b; Nakatani 1982). Batch spawning occurs at night (22.00-4.00 h) in near-bottom water (Acuña and Castello 1986) every 12-17 days (Christiansen and Cousseau 1985). In the late spring and early summer, adults migrate south to Uruguayan and Argentine waters, returning in the winter. Adults of *Engraulis anchoita* feed during the day, and even though their diet preference may overlap with *Trachurus lathami* and *Scomber japonicus*, their diet diversity is considerably larger. About 90% of the food is composed of calanoid copepods, euphausids, and hyperid amphipods, though occasionally phytoplankton is also ingested (Schwingel 1991; Schwingel and Castello 1994). The production of the stock, with up to 90% during the first year of life, varies between seasons. A winter biomass of 60 g m^{-2} results in 44.2 g m^{-2} production, whilst the spring biomass (14 g m^{-2}) produces 9.3 g m^{-2}, the natural mortality of *Engraulis anchoita* being the principal limiting factor. Catches are largely composed of 1- and 2- year-old individuals and equal ratios of males and females, though females tend to dominate in mid-water trawls during peak production (Acuña and Castello 1986). This considerable stock which, owing to differential recruitment and fluctuating oceanographic conditions (Lima and Castello 1995), may vary interannually between 35 000 and 1 900 000 tons, is still not being commercially exploited.

Of the three *Mugil* species, only *M. curema* and *M. gaimardianus* are essentially warm water representatives and are seen during the summer and fall, when high water temperature and salinity conditions prevail. In contrast, juveniles of the more common *Mugil platanus* persist throughout most of the year, though they are more abundant in the winter and spring in nursery areas of the Patos Lagoon estuary and adjacent coastal waters (Vieira 1985). In the fall, which is the major period of catch, adults of *Mugil platanus* leave the estuary, owing to gradually decreasing water temperatures and frequent seawater penetration, and initiate their reproductive migration. Spawning occurs in warmer (19^0-21^0C) offshore waters at about 27^0 S between the end of the fall and winter (Vieira 1985; Viera and Scalabrin 1991). Eggs and larvae are generally transported by currents towards the surfzone, followed by long-shore migration to the Patos Lagoon estuary (Vieira 1991b).

The bluefish *Pomatomus saltatrix* has a wide biogeographic distribution and is present throughout the year (Krug 1984). Decreasing temperatures in Argentine waters (38^0 S) in late fall appear to stimulate migration to the southern and southeastern Brazilian shelf, where the highest abundance is reached in the winter. Approximately 50% of growth occurs during the first 2 years, although males and females may reach a maximum age of 7 and 8 years, respectively. Batch spawning occurs during the late spring and summer after males and females (35-40 cm TL) reach first sexual maturity at an age of approximately 2 years. The period of reproductive activity coincides with a second peak of abundance at approximately 32° S (Krug and Haimovici 1991). While juveniles of *Pomatomus saltatrix* (<25 cm TL) feed on small pelagic fish and crustaceans, adults (>25 cm TL) prefer anchovy, Atlantic sabretooth anchovy, horse mackerel, and less frequently squids, cutlass fish, and weakfish (Haimovici and Krug 1992). The scombriform *Sarda sarda* (mean 53 cm TL, maximum 78 cm TL), which largely feeds on anchovy, is a migratory species. Migrations from Argentina (38° S) into southern Brazilian shelf waters are limited by superficial isotherms of 15-16°C (Aasen and Castello 1968a,b). After reproduction in southern Brazilian shelf waters from November to January, *Sarda sarda* returns to waters near Mar del Plata (38° S), where the stocks form a resource for the Argentine purse seine fishery.

Scomber japonicus is present during the entire year and migrates latitudinally in the Southwestern Atlantic. Reproduction occurs in the winter over the southeastern Brazilian shelf. In spite of the assumption that only one population exists between 22° and 42° S (Seckendorff and Zavala-Camin 1985), evidence based on morphometric and meristic characteristics (Perrotta et al. 1990) and on reproduction in the spring at 38° S (Perrotta 1987) suggests the presence of two populations. Also, shoals of different size classes reach the Mar del Plata region (Gagliardi and Cousseau 1970) and the interannual abundance varies strongly (Castello and Cousseau 1976). In southern Brazilian waters, the abundance of *Scomber japonicus* may alternate with the abundance of the widely distributed horse mackerel (*Trachurus lathami;* Saccardo 1980, 1987), thus modifying the composition of the pelagic association (Mello et al. 1992). *Trachurus lathami* from Brazilian and *Trachurus picturatus australis* from Argentine waters (38° S) have been considered as two different taxa (Cousseau 1967; Ciechomski and Cassia 1980), however, since they are morphologically identical the presence of two taxa is not justified (Saccardo 1980). The size of *Trachurus lathami* (85-210 mm TL) in southeastern and southern Brazilian shelf waters is similar. Sex maturation occurs at 115-119 mm TL, and even though the reproductive period is prolonged peak reproduction is reached in November (Saccardo 1980, 1987; Castello, unpubl.). The species can live to a maximum age of 8 years, but an average of 2-3 and 3-4 years is found for the southern and southeastern Brazilian shelf regions, respectively. During the day, shoals of *Trachurus lathami* form at 50-90 m depth in southeastern and at 20-75 m depth in southern shelf waters (Saccardo 1980, 1987; Castello, unpubl.). At night, the species disperses in surface water and frequently mixes with *Engraulis anchoita* (Mello et al. 1992).

The abundant small to medium-sized skipjack (*Katsuwonus pelamis)* is widely distributed in tropical and subtropical oceans and represents an important fishery resource (Matsuura 1982a; Jablonski and Matsuura 1985; Vilela 1990; Vilela and Castello 1991, 1993). Although *Katsuwonus pelamis* undergoes extensive latitudinal migrations, the species is found throughout the year over the southeastern Brazilian shelf (Maluf et al. 1984). Schools migrate to more southern regions in the late spring, where their distributional limits coincide with the western margin of subtropical waters (18°C, 3.5 ml l^{-1} dissolved oxygen) (Habiaga and Castello 1986). In these areas, *Katsuwonus pelamis* feeds from November to May largely on *Maurolicus muelleri, Engraulis anchoita,* and *Euphausia similis.* Other pelagic species of the southern Brazilian shelf include *Thyrsitops lepidopoides, Balistes* spp., *Lagocephalus laevigatus, Peprilus paru, Stromateus maculatus, Stromateus brasiliensis, Parona signata, Saurida brasiliensis, Decapterus* sp. (Castello, unpubl.), and *Auxis thazard* (Miyake and Hayasi 1978). The species *Cynoscion guatucupa, Trichiurus lepturus,* and *Merluccius hubbsi* only become pelagic as juveniles or during nocturnal feeding (Mello et al. 1992). Mid-water trawls produce a characteristic assemblage of pelagic species, *Engraulis anchoita, Trachurus lathami, Loligo sanpaulensis, Trichiurus lepturus, Cynoscion guatucupa,* and *Merluccius hubbsi* being the most frequent ones in the winter and spring. A second assemblage during spring is composed of *Peprilus paru, Balistes* sp., *Pomatomus saltatrix, Sarda sarda, Katsuwonus pelamis* and *Anchoa marinii* (Fig. 6.13.1; Mello et al. 1992).

Slope and Deep Water Inhabitants

Tunafish and similar species, pelagic sharks, and other smaller oceanic fish occur abundantly over the slope or, under favourable environmental conditions, also over the outer shelf. The tunafish *Thunnus albacares, Thunnus alalunga, Thunnus thynnus,* and *Thunnus obesus* are commercially valuable species, of which *Thunnus albacares* is the most important one. The concentration of shoals is controlled by the influence of the Subtropical Convergence. *Thunnus albacares* as well as *Thunnus alalunga* and *Thunnus obesus* occur in southern Brazilian waters (27-35° S) from August to February and from April to September, respectively (Zavala-Camin 1974, 1978; Zavala-Camin and Antero da Silva 1991). Since planktonic eggs and larvae are absent from southeastern regions (Matsuura 1986; Mafalda 1989), reproduction of *Thunnus albacares* occurs in the tropics (Zavala-Camin and Antero da Silva 1991). In southern Brazilian waters, the age structure of *Thunnus albacares* varies between 1 and 6 years (Vaske 1992), probably as a result of the migration of two or three cohorts twice a year (Zavala-Camin 1978). Juveniles form large schools in surface waters and often mix with other species, like *Katsuwonus pelamis,* whilst adults occur in subsurface and upper mesopelagic waters (Zavala-Camin 1981). Both *Thunnus albacares* and *Thunnus alalunga* prey during the winter and spring on fish, principally *Brama* sp., cephalopods (Ommastrephidae and *Ornitotheuthis antillarum*), and hyperid crustaceans (Zavala-Camin 1981; Vaske 1992). Other tunafish species (*Thunnus atlanticus, Thunnus maccoyii)* are rare.

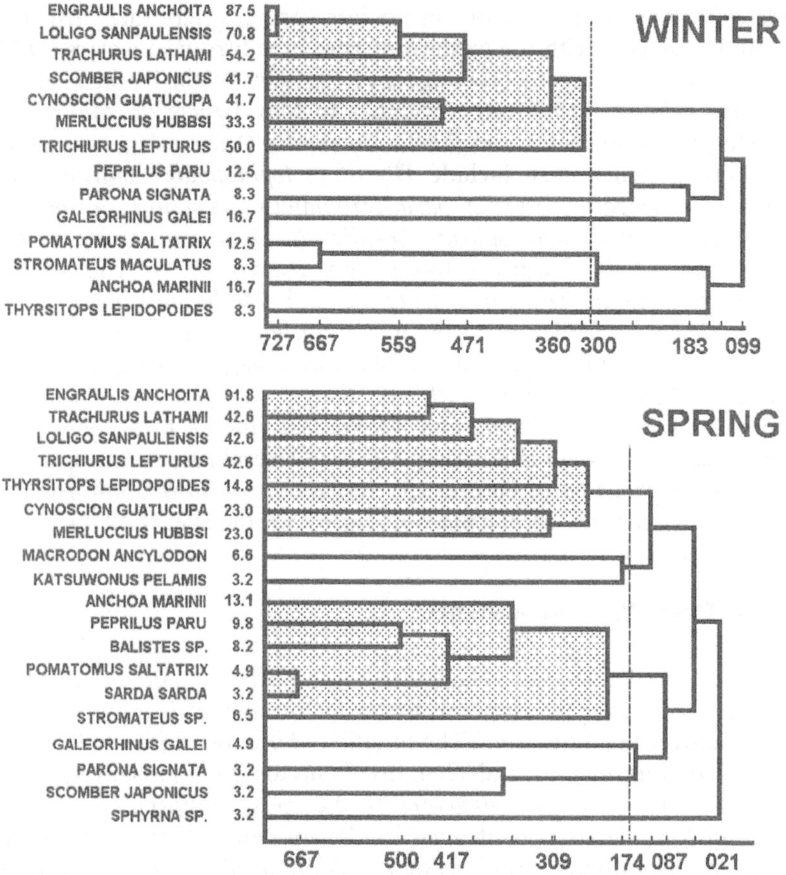

Fig. 6.13.1. Winter and spring dendrograms of pelagic species assemblages (*shaded areas*) with frequency of occurrence

Several commercially valuable tunafish-like species, represented by bill-fish, swordfish, and marlin (*Euthynnus alletteratus*, *Coryphaena hippurus*, *Xiphias gladius*, *Isthiophorus albicans*, *Tetrapturus albidus*, *Makaira nigricans*) are caught in southern Brazilian waters (Zavala-Camin 1974, 1978; Mello et al. 1993). In particular, the swordfish (*Xiphias gladius*), which potentially competes with *Thunnus obesus*, is an important fishery resource. The preferential diet of this species is *Illex argentinus* (Mello 1992). Finally, smaller fish species, like *Maurolicus muelleri*, genera of the Mycthophidae (*Diaphus*, *Symbolophorus*, *Myctophum*, *Diogenichthys*), and Lampanyctinae, which undergo large vertical migrations, occasionally reach a high level of abundance (Castello and Habiaga 1982, Souto, unpubl.).

6.14 Demersal and Benthic Teleosts

M. Haimovici

Over the past few decades, the search for new fishery resources stimulated several surveys of demersal and benthic teleost fish over the southern Brazilian shelf and slope (Nakamura 1963; Vazzoler and Iwai 1971; Menezes 1971; Roux 1973; Yesaki 1973; Vazzoler 1975; Yesaki et al. 1976; Vooren et al. 1988). Today, over 200 species from more than 100 families have been identified (Menezes 1971; Benvegnu-Lé 1973; Figueiredo 1977; Figueiredo and Menezes 1978, 1980; Menezes and Figueiredo 1980, 1985; Haimovici et al. 1994a; Appendix).

Seasonal Patterns of Species Richness and Abundance

The demersal ichthyofauna of the southern Brazilian shelf, transitional between tropical and temperate zones, is part of the Argentine zoogeographic province, which extends between 22^0 and 42^0 S (Figueiredo 1981). The fauna can be divided into species with tropical and temperate water affinities of both the northern and the southern hemisphere. Most species are eurybathic and eurythermal, and few endemics are found. Although the species composition varies little with the seasons, annual variations of species richness and relative abundance follow annual bottom temperature variations in coastal waters (>10^0C) and at the shelf break (4-5^0C), as a result of the latitudinal oscillation of the Subtropical Convergence between the warm Brazil Current (BC) and the cold Malvinas Current (MC; Miranda et al. 1973; Haimovici et al. 1996). In winter, the highest biomass at 40-60 m (*Micropogonias furnieri*, *Umbrina canosai*, *Cynoscion guatucupa* = *Cynoscion striatus*) coincides with the northward flowing cold waters of the coastal branch of the MC, though the relative abundance decreases from coastal to deeper waters. In contrast, at the end of the summer, abundance increases from the inner shelf towards the shelf break, probably owing to the existence of vortices and to partial upwelling of subtropical water (Podestá 1990; Lima and Castello 1995). The total number of species varies little between seasons. The greatest species richness over the shelf break is probably a result of both different types of substrate and different water layers, whilst a pronounced decrease in total species number over the slope below 350 m appears to be associated with the poor benthic invertebrate fauna (Capítoli, Sect. 6.11).

Genera and Species	Family		Absolute FO	Warm Season %	Cold Season %
Menticirrhus littoralis	Sciaenidae	*	25	61	39
Paralonchurus brasiliensis	Sciaenidae		103	49	51
Macrodon ancylodon	Sciaenidae	*	83	53	47
Peprilus paru	Stromateidae	*	103	89	11
Micropogonias furnieri	Sciaenidae	*	172	52	48
Urophycis brasiliensis	Gadidae	*	201	58	42
Netuma sp.	Ariidae	*	83	64	36
Conger orbignyanus	Congridae	*	108	40	60
Cynoscion guatucupa	Sciaenidae	*	222	50	50
Trichiurus lepturus	Trichuridae	*	231	64	36
Balistes capriscus	Balistidae	*	41	68	32
Umbrina canosai	Sciaenidae	*	190	47	53
Prionotus punctatus	Triglidae		164	57	43
Paralichthys patagonicus	Bothidae	*	136	55	45
Porichthys porosissimus	Batrachoididae		151	48	52
Cynoscion jamaicensis	Sciaenidae	*	88	76	24
Ctenosciaena gracilicirrhus	Sciaenidae		87	72	28
Trachurus lathami	Carangidae	*	120	42	58
Saurida caribbaea	Synodontidae		44	74	26
Thyrsitops lepidopoides	Gempylidae		58	38	62
Pagrus pagrus	Sparidae	*	40	50	50
Prionotus nudigula	Triglidae		60	55	45
Merluccius hubbsi	Merlucciidae	*	135	31	69
Mullus argentinae	Mullidae	*	74	50	50
Paralichthys isosceles	Bothidae		56	23	77
Lophius gastrophysus	Lophiidae	*	39	45	55
Zenopsis conchifera	Zeidae		59	47	53
Antigonia capros	Caproidae		21	45	55
Synagrops spinosus	Acropomatidae		64	52	48
Urophycis mystacea	Gadidae		60	36	64
Lopholatilus villarii	Branchiostegidae	*	5		100
Genypterus brasiliensis	Ophidiidae	*	30	44	56
Bembrops heterurus	Percophidae		24	30	70
Benthodesmus elongatus	Trichiuridae		18	33	67
Ariomma bondi	Ariommatidae		31	48	52
Synagrops bellus	Acropomatidae		16	19	81
Helicolenus lahillei	Scorpaenidae		25	42	58
Argentina striata	Argentinidae		19	42	58
Polyprion americanus	Polyprionidae	*	10	77	23
Diaphus dumerilii	Myctophidae		15	40	60
Polymixia lowei	Polymixiidae		29	41	59
Malacocephalus occidentalis	Macrouridae		17	47	53
Coelorinchus marinii	Macrouridae		20	50	50

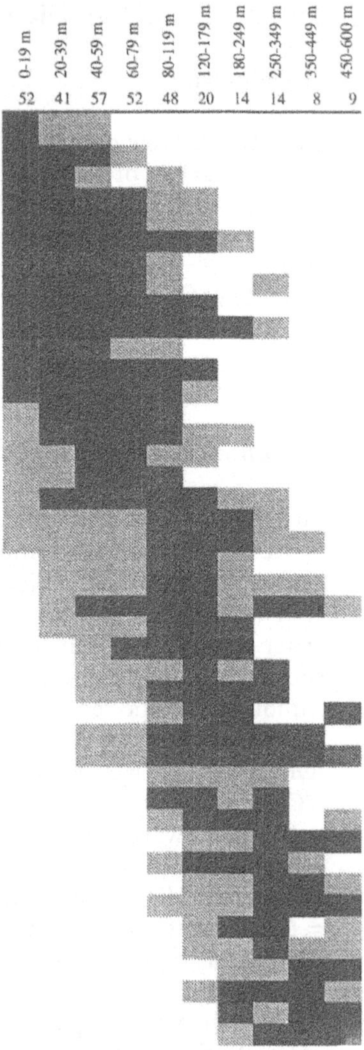

Fig. 6.14.1. Most common teleost species in bottom trawl surveys (1981-1987) from the southern Brazilian shelf and upper slope with absolute frequency of occurrence (*FO*) in 315 hauls. Occurrence of species (*black* >50%, *dark gray* 25-50%, *light gray* <25%) during northward (cold season) and southward (warm season) shifts of the Subtropical Convergence (* commercially important species)

Composition and Depth Distribution

The wide and gently sloping southern Brazilian shelf is predominantly covered by soft-bottom substrates (Zembruski et al. 1972; Furtado 1973). As a result, the gradual substitution of species is largely associated with depth (Benvegnu-Lé 1978; Haimovici et al. 1994a; Fig. 6.14.1). Over the shelf down to 120 m depth, the families with the larger number of species are Sciaenidae, Carangidae, Clupeidae, Bothidae, and Serranidae. Based on percent catch per unit effort (CPUE), Sciaenidae (80.9%) and Trichiuridae (14.5%) have the highest relative abundance. Four species alone (*Micropogonias furnieri, Cynoscion guatucupa, Umbrina canosai, Trichiurus lepturus*) are responsible for 71.9% of the total biomass and together with another 11 abundant species (*Macrodon ancylodon, Prionotus punctatus, Merluccius hubbsi, Peprilus paru, Ctenosciaena gracilichirrus, Paralonchurus brasiliensis, Cynoscion jamaicensis, Trachurus lathami, Urophycis brasiliensis, Porichthys porosissimus, Balistes capriscus*) represent as much as 95.5% of the biomass (Haimovici et al. 1996). The generally low abundance of small Anguilliformes and Pleuronectiformes may be a result of inefficient sampling (Arena et al. 1992), whilst the low abundance of *Pagrus pagrus, Pogonias cromis,* and *Netuma barba* reflects the depletion of their stocks between 1970 and 1990 (Haimovici et al. 1989a; IBAMA 1993). Over the outer shelf between 120 and 180 m, *Trachurus lathami, Trichiurus lepturus,* and juveniles of *Cynoscion guatucupa, Umbrina canosai,* and *Scomber japonicus* are the dominant species and together total 83% of the teleost biomass. Over the shelf break (180-249 m depth), *Trichiurus lepturus* and *Thyrsitops lepidopoides* are abundant, whilst in areas with irregular coralline substrates *Antigonia capros* and *Priacanthus arenatus* occur, which together represent 63% of the biomass. The demersal fish fauna of upper slope regions (250-587 m) with extensive irregular bottoms is largely composed of Serranidae, Macrouridae, Trichiuridae, and Myctophidae. The monospecific families Ariommatidae and Polyprionidae are most abundant, with *Ariomma bondi, Helicolenus dactylopterus,* and *Polyprion americanus* making up 77% of the total biomass (Haimovici et al. 1994a).

Ecology of Principal Species

Macrodon ancylodon occurs in shallow coastal waters between Venezuela (10° N) and Argentina (40° S). The species is most abundant over the southern Brazilian shelf, and the local stock mixes little with northern populations (Yamaguti 1979). Spawning occurs in late spring and summer in coastal areas adjacent to the Patos Lagoon mouth (32° S), followed by migration to more northern latitudes (28° S) in winter (Santos and

Yamaguti 1965; Yamaguti 1979). Intense exploitation of *Macrodon ancylodon* in southern Brazil has altered the density and consequently the growth and age structure of the stock. Prior to 1970, first sexual maturity was reached between 3 and 4 years and at 25 cm total length of maturity (Tl_m; Yamaguti 1967), with a maximum age of adults around 12 years (45 cm; Yamaguti and Santos 1966). At the end of the decade, older age classes (>5 years) were absent, but the length of age classes had increased (Martins Juras 1980; Haimovici 1988a), and first sexual maturity was reached at 1-2 years (TL_m females 27.4 cm, males 21.5 cm; Juras and Yamaguti 1989). Subadults and adults of *Macrodon ancylodon* feed preferentially on shrimp (*Artemesia longinaris*), cephalopod (*Loligo sanpaulensis*), and fish (*Paralonchurus brasiliensis*, *Macrodon ancylodon*), but fish become more important in the diet as size increases (Juras and Yamaguti 1985).

Micropogonias furnieri is widely distributed along the Atlantic coast of South America down to 40^0 S (Isaac 1988; Vazzoler 1991). Although genetically distinct populations are absent from southeastern and southern Brazilian waters (Maggione et al. 1994), the exchange between stocks from southern (29-34^0 S), southeastern (22-29^0 S), and northeastern (<22^0 S) latitudes is limited (Vazzoler 1971; Valentini et al. 1991a; Haimovici and Umpierre, unpubl.). However, stocks which reproduce south of this region (34^0 S) migrate into southern Brazilian waters during the winter, following the coastal branch of the MC. Maximum spawning in southern Brazilian waters occurs in the spring and summer in coastal water under the influence of freshwater runoff from the the Patos Lagoon (Vazzoler 1971), and the estuary of the lagoon constitutes an important nursery ground (Weiss 1981; Barbieri 1986; Castello 1986; Pereira 1994). Although some juveniles reach first sexual maturity at 18-20 cm TL_m in the estuary, most mature at 30-40 cm after leaving the estuary (Vazzoler 1971; Haimovici, unpubl.). Subadults (<30 cm TL) of *Micropogonias furnieri* occur throughout the year near the coast, whilst adults disperse over the shelf, where they feed on polychaetes, crustaceans, molluscs, ophiuroids, and fish (Tanji 1974; Vazzoler 1975). During the fall and winter, adults concentrate in waters of up to 100 m depth and migrate from Uruguay to southern Brazil. In contrast, during the summer they are seen in shallower waters (<50 m; Haimovici 1987; Haimovici et al. 1996) and migrate southward (Vazzoler 1991; Reis 1992). In southern Brazilian waters *Micropogonias furnieri* attains over 70 cm TL. Age determinations, based on scales (Vazzoler 1971; Castello 1986) and transverse cut of otholits (Schwingel and Castello 1990; Haimovici and Umpierre 1996) show a correspondence between the number of scale rings and opaque otholit bands for individuals younger than 5 years (Haimovici 1977). Although the age of individuals frequently exceeded 30 years prior (1980) to intensive exploitation (Schwingel and Castello 1990), today individuals more than 20 years old are rarely found.

Netuma barba is a slow-growing species with a calculated life span of 23.1 years and a mortality of 0.13, though adults may occasionally attain 36 years of age and a total length of 98 cm. Sexual maturity of females (41.5 cm) and males (43 cm) is reached at about 7 years. At the end of the winter *Netuma barba* migrates into the Patos Lagoon estuary. Gonad maturation and reproduction, with low female fecundity (32-272 intraovaric oocytes of 12-19 mm), take place in early spring in the estuary followed by spawning in the coastal waters. The males incubate the eggs for up to 2 months in the buccal cavity (Reis 1986a,b). Between spawning seasons, adults disperse over the entire shelf (Haimovici et al. 1996). The first juveniles (6 cm TL) appear in late summer near the Patos Lagoon inlet and also in the estuary, where they feed on pereiopods of *Callinectes* spp. and polychaetes (Araújo 1984). The diet of adults is composed of molluscs and polychaetes in the estuary, epibenthic crustaceans in adjacent coastal waters, and principally sciaenid fish in both environments.

Although *Pogonias cromis* ranges from 45° N to 40° S (Fischer 1978), information on the species from the Southwestern Atlantic is scarce. Over the southern Brazilian shelf the species occurs at less than 40 m depth near the inlet of the Patos Lagoon, and juveniles of 1-2 years are commonly found in the estuary. Between the end of the spring and summer, the gonads of females reach an advanced stage of maturity, and partial spawning takes place. Growth appears to be rapid, since a total length of more than 1 m is reached after 10 years, though the age of adults (TL 140 cm, 40 kg) may exceed 40 years (Haimovici et al. 1989a; Haimovici, unpubl.). Juveniles feed on *Heleobia* (*Litoridina*), *Erodona*, and *Balanus improvisus* in the estuary, whilst the diet of subadults and adults in coastal waters is composed of molluscs (*Mactra patagonica*, *Mactra isabelleana*, *Glycymeris longior*, *Corbula* sp.), gastropods (*Buccinanops gradatum*), and crustaceans (*Hepatus pudibundus*, *Loxopagurus loxochelis*, *Libina spinosa*).

Cynoscion guatucupa (=*C. striatus*) is a demersal-pelagic sciaenid between the latitudes 22° and 40° S. The species is most abundant from 29° to 40° S where it appears to belong to one stock (Vargas-Boldrin 1980; Vieira 1990). Adults of *Cynoscion guatucupa* occur at depths of less than 50 m, especially between the spring and fall. In southern Brazilian shelf waters, multiple spawning peaks coincide with the spring and the end of the summer or the beginning of the fall. Principally during the summer and fall, small juveniles (<10 cm) are restricted to shallow coastal waters, whilst larger juveniles and subadults (10-30 cm) are abundant throughout the year at depths between 25 and 100 m. First sexual maturity of females (32-35 cm) and males (29-31 cm) is reached at about 4 years, and the maximum age approaches 15 and 13 years, respectively (Vieira 1990; Vieira and Haimovici 1993). The diet preference of *Cynoscion guatucupa* changes with age. Juveniles preferentially feed on planktonic crustaceans (hiperiid amphipods, euphausids, copepods, mysidaceans), whilst adults ingest a diet composed of fish (*Anchoa marinii, Engraulis anchoita, Paralonchurus brasiliensis, Macrodon ancylodon, Cynoscion guatucupa*) and shrimp (*Artemesia longinaris, Pleoticus muelleri;* Vazzoler 1975, Vieira 1990).

The southern Brazilian stock of the demersal sciaenid *Umbrina canosai*, which represents a major part of bottom trawl catches during the winter and spring, exhibits a clearly defined migratory cycle. Partial spawning initiates with larger females at the end of the winter as far north as 29^0 S and continues southward until the end of the spring. The large females are also the first to migrate to southern feeding grounds in Uruguay and Argentina, followed by males and smaller females during the summer and fall (Gonzalez Alberdi and Nani 1967; Haimovici and Cousin 1989). During their first year, juveniles remain over the southern Brazilian intermediate shelf (Zaneti Prado 1979; Haimovici 1982). Males and females reach first sexual maturity at 18.4 and 21.9 cm, respectively. At age II, 60% of the males and 27.4% of the females are sexually mature. Adult females tend to be larger than males and occasionally surpass an age of 20 years (Haimovici and Reis 1984), although intense exploitation of stocks caused a significant age-dependent average weight increase (Haimovici 1988b). *Umbrina canosai* displays its highest feeding activity during the summer (Haimovici et al. 1989b) and feeds principally on benthic organisms. Smaller individuals prefer a diet of amphipods and mysids, whilst larger individuals feed on polychaetes, ophiuroids, bivalves, gastropods, decapods, and juvenile fish (Vazzoler 1975; Haimovici et al. 1989b). Adaptive characteristics like high fecundity, prolonged spawning over a large area, great longevity, age-dependent partitioning of feeding areas, and migratory patterns appear to explain the abundance of *Umbrina canosai* in southern Brazil (Haimovici 1982; Haimovici and Cousin 1989).

The cosmopolitan *Trichiurus lepturus* is an abundant species over the southern Brazilian shelf (Haimovici et al. 1996), and the eggs and larvae of this species are common (Mafalda 1989). Under the influence of tropical waters in the summer, spawning females occupy areas near the coast, but they are also found over the shelf break throughout the year. First sexual maturity is reached at approximately 70 cm TL_m. The diet preference of *Trichiurus lepturus* changes from copepods to mysidaceans (*Promysys atlantica*), sergestids (*Lucifer faxoni*), and euphausids (*Eufausia similis*) during the early juvenile stages. Feeding diversity augments with increasing size, including juvenile sciaenids, *Engraulis anchoita*, benthic macro-crustaceans (*Artemesia longinaris, Squilla brasiliensis*) and cephalopods (*Loligo sanpaulensis, Illex argentinus;* Martins 1992).

Although only *Prionotus punctatus* is of some commercial importance, both *Prionotus punctatus* and *Prionotus nudigula* inhabit the southern Brazilian shelf and slope throughout the year, becoming more abundant during the spring. Both species reproduce and spawn from the spring to the fall. *Prionotus punctatus* reaches first sexual maturity at 26.2 cm, and adults

(40 cm TL) are restricted to depths of less than 135 m between 11 and 22.5°C. *Prionotus nudigula* matures sexually at 16 cm, and adults occur at depths of up to 378 m and bottom temperatures of 9.8-21.2°C. Although the diet of both species is largely composed of crustaceans, food preferences differ between species and change with size, depth, and season. *Prionotus punctatus* displays higher feeding activity during the afternoon and in the summer and ingests bottom dwellers of reduced motility. With increasing size, the diet preference changes from cumaceans, amphipods, *Leptochela serratorbita*, and mysids, over *Portunus spinicarpus*, other brachyurans, and *Pleoticus muelleri*, to larger brachyurans, stomatopods, and benthic fish (*Porychthys porosissimus*). *Prionotus nudigula*, on the other hand, also preys on organisms in the water column and, during the summer and fall, ingests *Munida* sp. and the fish *Bregmaceros atlanticus* at depths exceeding 100 m, but in the spring it feeds on *Euphausia similis* and amphipods in shallower depths (Teixeira and Haimovici 1989).

Peprilus paru is a less abundant species which seems to follow the displacement of the Subtropical Convergence. Although *Peprilus paru* is present throughout the year, it is less important in bottom trawl catches in the winter and spring, and juveniles only occur during the summer and fall. Multiple spawning occurs from the end of the spring to early fall, and females and males reach first sexual maturity at 120 and 130 mm fork length, respectively. Juveniles seem to feed principally on gelatinous organisms, like jellyfish and salps, but also on small crustaceans, polychaetes, and ascidians (Cerqueira and Haimovici 1990).

The distribution of *Merluccius hubbsi* ranges from the upwelling area of Cabo Frio (22° S) to the extreme south of the American continent (Fagundes Netto and Oliveira 1991). The southern Brazilian shelf, where the species occurs at bottom temperatures of 6.8-21° C between 22 and 490 m depth, appears to serve as a growth region for locally reproducing individuals and those which reproduce further south. Multiple spawning occurs during the winter and early spring, and first sexual maturity of both sexes is likely to be reached at the end of the third year (30-40 cm). Juveniles are present throughout the year but tend to be associated with subtropical water of the outer shelf and slope. In contrast, adults occur mainly in the winter. Both juveniles and adults are most frequently found at depths between 40 and 120 m. *Engraulis anchoita* is an important component of the diet, though juveniles of *Merluccius hubbsi* also ingest mysids and amphipods. Larger individuals prey on *Trachurus lathami*, *Merluccius hubbsi*, and cephalopods (*Illex argentinus*, *Loligo sanpaulensis;* Haimovici et al. 1993).

Pagrus pagrus is a hermaphrodite species which is widely distributed in the Atlantic and Mediterranean (Manooch and Hassler 1978). Based on growth rings, two groups of this species have been identified for southern Brazilian shelf waters. During the winter, individuals with well-defined rings concentrate over the intermediate shelf south at 33° 40' S, whilst a group without obvious rings is less abundant and occurs over the entire southern shelf during the summer. Juveniles (<2 years) are rare over the southern Brazilian shelf. The growth of males and females is similar, but bisexual individuals tend to be smaller. Advanced stages of sexual maturity occur from late spring to the end of summer, and adults may exceed an age of 15 years. The diet of *Pagrus pagrus* is principally composed of fish (*Engraulis anchoita*, *Trachurus lathami*, *Umbrina canosai*, *Synagrops spinosus*, *Cynoscion guatucupa*) and squid (*Loligo sanpaulensis*), but in coastal areas this changes to predominantly benthic invertebrates (Capítoli and Haimovici 1993).

Despite slow growth, the serranid *Polyprion americanus* reaches a total length of 140 cm and up to 80 kg weight. In contrast to other serranids, the sexes of *Polyprion americanus* are separate, and reproduction occurs in the winter. Gonad maturation is common in females and males of more than 80-90 cm, and vitellogenesis of oocytes is restricted to May, July, and September. The diet of *Polyprion americanus* is largely composed of demersal fish (*Helicolenus dactylopterus*, *Merluccius hubbsi*, *Urophycis mystaceus*), but the squid *Illex argentinus* and to a lesser extend *Lycoteuthis diadema*, *Ancistrocheiros lesueuri*, and the deep water crab *Chaceon notialis* are also ingested (Peres and Haimovici, unpubl.).

Flatfishes of the genus *Paralichthys* are represented by *P. orbignyanus*, *P. patagonicus*, and *P. isosceles*. The first is largely restricted to the Patos Lagoon estuary, the second is most abundant over the shelf, and the third is the smallest species which occurs mostly over the outer shelf (Haimovici et al. 1996). All three species spawn during the spring and summer and are active predators on shrimp, fish, and cephalods (Kawakami 1975).

6.15 Early Life History of Silverside Fishes

G. Phonlor and J. C. B. Cousin

Several species of the family Atherinidae are abundant in southern Brazilian coastal nearshore waters (*Odontesthes incisa*, *Odontesthes argentinensis*) and the Patos Lagoon estuary (*Atherinella brasiliensis*, *Odontesthes* sp.), as well as in coastal freshwater bodies (*Odontesthes bonariensis*, *Odontesthes humensis*, *Odontesthes retropinis;* Kleerekoper 1945; Figueiredo and Menezes 1978; Chao et al. 1982; Bemvenuti 1987, 1993; Prodohl and Levy 1989; Phonlor and Vinagre 1990). Owing to a considerable morphological similarity between species, the taxonomy of the family is confused (Figueiredo and Menezes 1978). Furthermore, since marine representatives also invade and colonize brackish and freshwater environments, different phenotypes of the same species are common (Boyd and

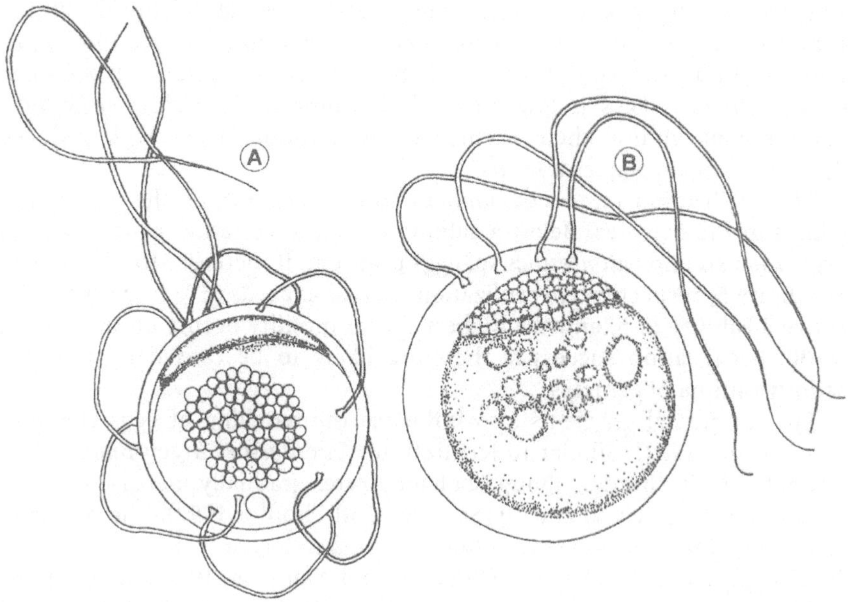

Fig. 6.15.1. Eggs of marine (A) and estuarine (B) morphotypes of *Odontesthes argentinensis*

Simmons 1974; Prodohl and Levy 1989). Brackish and marine morphotypes of *Odontesthes argentinensis* (Bemvenuti 1987, 1990, 1993) differ in egg size. They either exhibit a smooth chorion with 3-6 filaments which surrounds the micropyle (estuarine morphotype) or 17-19 looped filaments, attached by both ends to the egg, and 3 filaments attached at one end to the chorion (marine morphotype; Fig. 6.15.1). Although karyotype and electrophoretic studies suggest the presence of one species, differences in chorion structure and filament number in other species (*Atherinops affinis;* White et al. 1983) have been interpreted as either species-specific (Ivankov and Kurdyayena 1973; Lonning 1981) or as a result of abiotic gradients (*Fundulus heteroclitus;* Brummet and Dumont 1981).

Spawning, Fertilization and Hatching in the Genus *Odontesthes*

The spawning periods of different *Odontesthes* species vary between habitats. In coastal freshwater lagoons, *O. humensis* and *O. bonariensis* spawn between the end of fall and the end of winter, though spawning periods tend to overlap (Phonlor and Vinagre 1990). In brackish Patos Lagoon waters *Odontesthes* sp. spawns from late fall to early summer, whilst spawning of *O. argentinensis* in coastal waters lasts from the end of winter to early summer (Bemvenuti 1987; Phonlor and Vinagre 1989, 1990; Phonlor and Sampaio 1992). The oxygen consumption of spawned *O. argentinensis* eggs is low, probably because of their slow embryonic development and/or their occurrence in bottom layers with reduced dissolved oxygen concentrations.

The fertilization success of *Odontesthes* eggs depends on the salinity tolerance of the gametes. Below a salinity of 2.5, fertilization of *Odontesthes argentinensis* and *Odontesthes* sp. eggs is absent; it increases to 12 and 39% at salinity 5, respectively. Fertilization success exceeds 85% in both species above a salinity of 5, and the highest sperm motility occurs at salinities of 20-30. Fertilization success is therefore likely to be caused by different salinity optima of gametes.

Environmental gradients also influence the hatching of eggs (Phonlor and Vinagre 1989). Similar to fertilization, *Odontesthes argentinensis* does not hatch in freshwater, though hatching rates gradually increase to 100% in salinities between 10 and 30 (Sampaio and Phonlor 1992). The period of larval survival (6 days) is much lower in freshwater, but virtually all larvae survive in salinities above 5 (Phonlor and Sampaio 1992). In contrast, hatching rates of freshwater representatives (*O. humensis*, *O. bonariensis*) increase from 40 to 100% between salinities of 0 and 10, but decrease to 38% at higher salinities (20). In addition, temperature shocks (3^{0}C) increase hatching of *Odontesthes bonariensis* significantly. The hatching time of *Odontesthes argentinensis* eggs is reduced by low temperature (15^{0}C) and light intensities. Furthermore, *Odontesthes argentinensis* eggs tend to be exposed to prolonged desiccation after their deposition on the beaches during exceptionally high tides (Phonlor and Vinagre 1989). Under experimental conditions, desiccation shortens the hatching period, and all embryos survived 96-120 h of desiccation. However, embryos in early stages of development (16 cells) are more susceptible than those just prior to hatching. Since the mortality of hatched larvae is insignificant and their locomotion and feeding behaviour remain unchanged, survival chances after desiccation under natural conditions appear to be high. Physiological adaptations of *Odontesthes* species of marine origin, principally to different salinity regimes in brackish and freshwaters (Phonlor and Sampaio 1992), thus confirm the genetic plasticity of Atherinidae (Bamber and Henderson 1988).

Larval Development and Growth of *Odontesthes argentinensis*

The euryhaline characteristics of *Odontesthes argentinensis* not only extend the distributional limits of the species, but also distinguish it as an excellent organism for bioassay studies and, together with high larval survival, make it a potential candidate for large-scale culture. Under controlled culture conditions (20°C), larval feeding commences on the first day. Weight increases and aerobic metabolism of hatched larvae (6.94 mm, 0.6 mg dry weight) are a function of time. The weight specific growth rate decreases from 0.14 to 0.07 mm day^{-1} after 4 weeks (Phonlor and Sampaio 1996). A decrease in food intake per unit larval weight after 14 days is possibly compensated for by an increased assimilation efficiency, accompanied by higher oxygen consumption and growth rates.

Organogenesis and processes of cellular differentiation accelerate after hatching. As the sole energy source, the vitellus (Fig. 6.15.2a) assures the formation and function of transient larval organs. Major changes of the digestive system include the expansion of the intestinal tract and division into anterior, middle, and posterior parts (Fig. 6.15.2b,c), the formation of a sphincter that separates the oesophagus from the anterior intestine, and the complete stratification of the oesophagus and pharynx epithelium, with an increase in number of teeth, taste buds, and goblet cells (Fig. 6.15.2d). Owing to the formation of new hepatocytes and capillarization, the liver volume increases rapidly and substitutes for the vitellus (Fig. 6.15.2c). In the pancreas, groups of exocrine cells assume an acini form, and the basophil cells become replete with zymogen granules. The gas gland and the *rete mirabile* differentiate into the swim bladder, which inflates and remains connected to the digestive tract by a short pneumatic duct. Since newly hatched larvae increase in body weight and notocord length during exposure to thyroxine, it is likely that after the development of the thyroid gland, increased thyroxine secretion influences tissue differentiation and provokes adaptations to higher salinities (Santos and Santos 1989).

During a period of starvation, larvae consume most of the vitellus in the first 4.5 days, with some remaining until the 8th day (Fig. 6.15.2e). At that age, low mortality (10%) emphasizes the resistance of larvae to starvation, though cellular atrophy leads to volume reduction and structural changes of tissues and organs. The digestive system in particular undergoes significant modifications. Intestinal folds begin to disappear, the entire intestinal wall becomes thinner (Fig. 6.15.2f) and cellular basophilia increases, owing to cytoplasmic density. After 10 days of starvation, the space between the organs in the abdominal cavity grows, the gall bladder expands, the pancreatic acini begin to disintegrate, and localized necrosis of cells commences. In spite of these tissue alterations, surviving larvae (3%) resume growth upon being supplied with food, which retards the "point of no return" (Blaxter and Hempel 1963) and establishes a somewhat flexible "critical period" (Fabre-Domérgue and Biétrix 1897) for this species. Starvation beyond 10 days causes general cellular atrophy, peeling of the intestinal epithelium, a virtual absence of zymogen granules in the pancreas, and increasing necrosis of the liver. Similar lesions have been reported for other species of teleosts after starvation (Ehrlich et al. 1976; Theilacker 1978; Watanabe 1985; Cousin et al. 1986). *Odontesthes argentinensis* larvae do not survive 12.5 days of starvation.

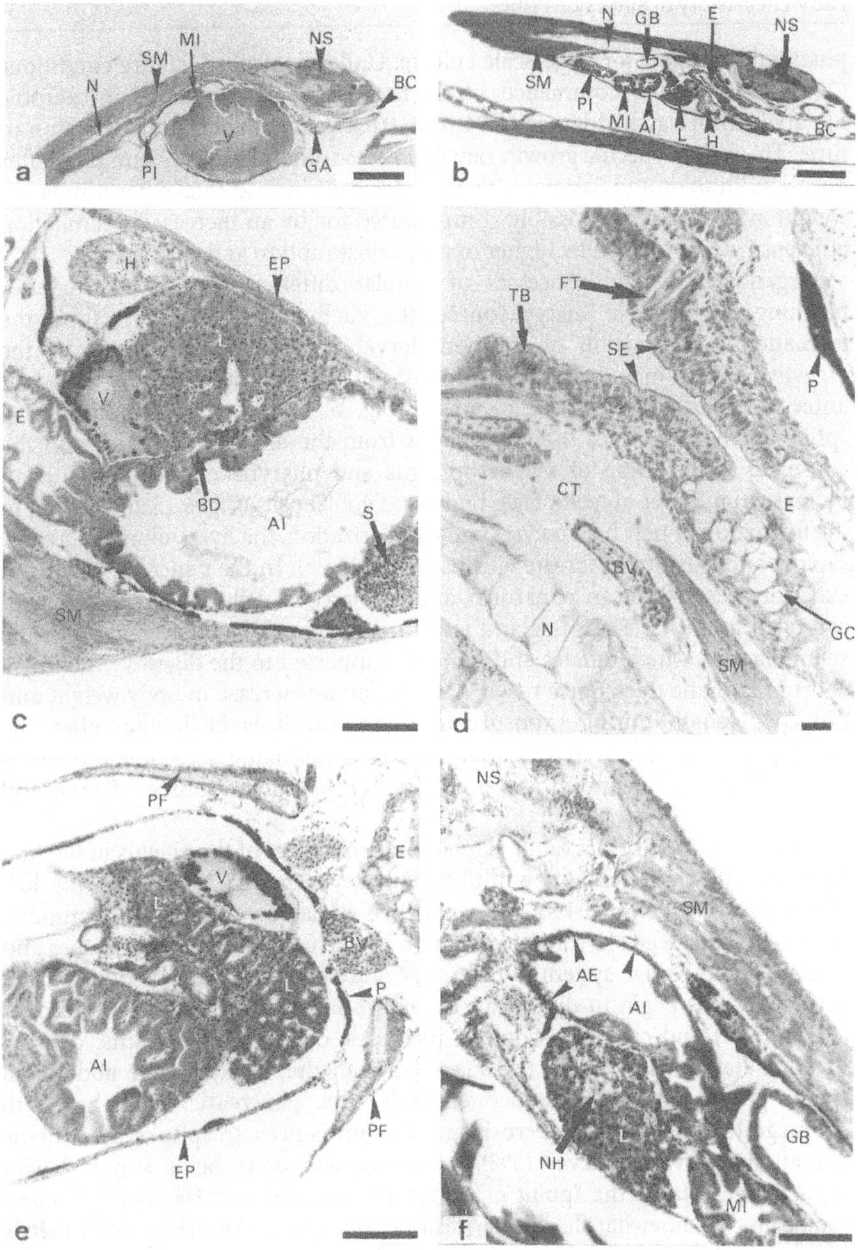

Fig. 6.15.2. Histological sections of starved *Odontesthes argentinensis* larvae at hatching (**a**) and after 6 (**b**), 4 (**c,d**), 8 (**e**), and 12 days (**f**). (*AE* Atrophied epithelium; *AI* anterior intestine; *BC* buccal cavity; *BD* bile duct; *BV* blood vessel; *CT* connective tissue; *E* esophagus; *EP* epidermis; *FT* faringeal teeth; *GB* gas bladder; *GA* gill arches; *GC* goblet cells; *H* heart; *L* liver; *MI* middle intestine; *N* notochord; *NH* necrotized hepatocytes; *NS* nervous system; *PF* pectoral fins; *P* pigment; *PI* posterior intestine; *S* spleen; *SE* stratified epithelium; *SM* striated skeletal muscle; *TB* taste buds; *V* vitellus). *Bars* equal 400 μm (**a,b**) and 100 μm (**c,d,e,f**)

In culture, the survival and growth of *Odontesthes argentinensis* larvae are largely a function of their density and of the available food's nutritional quality. A diet composed of *Artemia* nauplii (20 org. ml^{-1}) satisfies nutritional requirements and permits optimal growth (0.11 mm day^{-1}) of larvae (15 org. l^{-1}) up to the 2nd month of life (28.6 mm TL) with less than 5% mortality. Proportional increases of larval and *Artemia* densities, or *Artemia* supplied *ad libitum*, alone or in combination with rotifera, do not improve the growth rates further. Although artificial diets potentially reduce costs and may simplify fish culture procedures (Tandler 1985) and larvae of *Odontesthes argentinensis* larvae ingest this diet immediately after hatching, their limited nutritional value only leads to suboptimal growth.

6.16 Demersal Elasmobranchs

C. M. Vooren

In terms of abundance and species diversity, elasmobranchs are important components of the ichthyofauna of the southern Brazilian shelf and upper slope. Bottom trawl surveys between 1968 and 1972 (Sadowsky 1973; Vazzoler et al. 1973; Figueiredo 1977) and from 1980 to 1986 (Peres and Vooren 1991; Vooren 1992) provided mostly information on demersal species, but occasional catches of pelagic sharks also yielded data on the genera *Carcharhinus*, *Sphyrna*, *Alopias*, *Squaliolus*, and *Etmopterus*. Sharks of the pelagic environment are frequently caught by the commercial tuna longline fishery on the continental slope, but their biology and ecology have not been studied.

Species Composition and Abundance

Fifty-seven species have been recorded over the shelf and upper slope, at depths from 10 to 500 m between 28° 40' and 34° 34' S (Cape Santa Marta Grande to Chuí; Appendix). The sharks are represented by 32 species (3 Squatiniformes, 2 Lamniformes, 8 Squaliformes, 16 Carcharhiniformes, 3 Hexanchiformes) and 25 species belong to the rays and skates (14 Rajiformes, 7 Myliobatiformes, 4 Torpediniformes), though several identifications are uncertain at the species level. Sadowsky (1973) and Figueiredo (1977) cite *Myliobatis goodei* Garman, 1885 in the area, but in later studies of material which fitted the description of this species, two morphotypes were found, with different morphometric dimensions of the head and the tooth plates, which were classified as *Myliobatis* NT (narrow teeth) and *Myliobatis* BT (broad teeth), respectively. Levy and Conceição (1989) showed by biochemical methods that the morphotypes represent two separate species, but their identification was not possible because available taxonomic descriptions are lacking in relevant detail. Of the genus *Squatina*, Figueiredo (1977) and Compagno (1984) cite only *S. argentina* in the area, but Solé-Cava et al. (1983) and Solé-Cava and Levy (1987) found three morphotypes which were characterized as separate species through frequency of isoenzymes. Vooren and Silva (1991) identified these as *S. guggenheim*, *S. argentina*, and the new species *S. occulta*. Of the genus *Squalus*, Calderón (1994) identified *S. megalops* as common in the area, while *S. mitsukurii* and *S. acanthias* occurred sporadically. The occurrence of *S. cubensis* in the area, cited by Sadowsky (1973), Figueiredo (1977), and Compagno (1984), was not confirmed in later studies.

In the years 1981 to 1983, the abundance of demersal elasmobranchs as a group on the continental shelf at depths between 10 and 100 m varied seasonally from 43 000 tons in the summer to 96 000 tons in the late winter and spring. Twenty-five elasmobranch species account for 94% of the total biomass. Angel sharks, rays, and skates constituted about 90% of the demersal elasmobranch biomass. *Squatina guggenheim*, *Squatina occulta*, *Sympterigia acuta*, *Sympterigia bonapartei*, *Myliobatis* NT, *Myliobatis* BT, and *Raja castelnaui* together made up about 80% of the biomass of bottom-dwelling elasmobranchs, while *Galeorhinus galeus* and *Mustelus schmitti* accounted for 75% of the biomass of free-swimming demersal elasmobranchs in the winter. The decrease of the catch of elasmobranchs per unit effort in the commercial otter-trawl fishery is evidence that since 1985 the abundance of many species has decreased drastically. Indeed, the population of the guitar fish *Rhinobatos horkelii* is in danger of extinction (IBAMA 1993; Vooren, unpubl.).

Depth Distribution and Migrations

According to their temporal distribution, species are classified as permanent residents, winter or summer migrants, and sporadically occurring species. *Psammobatis rutrum*, *Psammobatis glansdissimilis*, and *Psammobatis bergi* are not included in this classification because they have only been identified in preserved material and not routinely in the field.

Among the 18 permanent residents, 12 are dominant species which are seen throughout the year without predictable changes in abundance, and complete their entire life-cycle in the waters off southern Brazil: *Squatina guggenheim*, *Mustelus fasciatus*, *Rhinobatos horkelii*, *Zapteryx brevirostris*, *Raja castelnaui*, *Raja agassizi*, *Raja platana*, *Raja cyclophora*, *Sympterigia acuta*, *Myliobatis* NT, *Myliobatis* BT, and *Gymnura altavela*. Four species occur constantly but in small numbers: *Carcharhinus plumbeus*, *Carcharhinus obscurus*, *Carcharhinus signatus*, and *Rhizoprionodon lalandei*. *Squatina argentina* and *Squatina occulta* are common all year round, but gravid females and neonates are rarely found, so gestation and birth cannot take place in that area. The constant presence of *Myliobatis* NT and *Myliobatis* BT with a peak abundance in the spring and fall is a result of complex migrations and the passage of different populations through the area. *Sympterigia acuta* and the four dominant species of the genus *Raja* are sedentary, each species occupying a distinct depth range (Fig. 6.16.1). The gravid females of the species *Squatina guggenheim*, *Rhinobatos horkelii* and *Mustelus fasciatus* live at depths greater than 20 m for most of the year but migrate during the spring and summer to shallow coastal waters to give birth, and the juveniles remain there during their first year of life. Intense commercial fishing in these waters interferes with reproduction and recruitment and therefore represents a threat to the survival of these species (Queiroz 1986, 1995; Lessa et al. 1986; Vooren et al. 1988; Vooren and Silva 1991; IBAMA 1993).

The 12 winter migrants are the skate *Sympterigia bonaparte*, the ray *Discopyge tschudii*, and the sharks *Galeorhinus galeus*, *Mustelus schmitti*, *Mustelus canis*, *Sphyrna zygaena*, *Eugomphodus taurus*, *Squalus megalops*, *Squalus mitsukurii*, *Squalus acanthias*, *Notorhynchus cepedianus*, and *Heptranchias perlo*. These species reproduce in Uruguayan and Argentine waters during the summer, and migrate in the winter to the continental shelf of southern Brazil. All sharks of this group represent important regional fishery resources during the winter (Barcellos 1957; Souto 1986; Batista 1988; Vooren et al. 1988, 1990; Ferreira and Vooren 1991; Peres and Vooren 1991; IBAMA 1993; Calderón 1994; Queiroz 1995).

The six summer migrants are the shark *Sphyrna lewini* and the rays *Myliobatis freminvillei*, *Dasyatis say*, *Dasyatis centroura*, *Rhinoptera bonasus*, and *Narcine brasiliensis*. These warm water species migrate with tropical water masses during the summer to the shelf of southern Brazil where *Sphyrna lewini*, *Dasyatis say*, and *Myliobatis freminvillei* give birth in nearshore waters. With the exception of *Narcine brasiliensis* and *Rhinoptera bonasus*, all summer migrants represent commercial fishery resources. Since large juveniles of *Carcharhinus plumbeus* are fished with beach seines during the summer in southern Brazil (Vooren, unpubl.) and Uruguay (De Buen 1952), it is likely that this species is a summer migrant to inshore waters of less than 10 m depth.

Eighteen species occur sporadically: *Galeocerdo cuvier*, *Carcharhinus brevipinna*, *Carcharhinus acronotus*, *Carcharhinus brachyurus*, *Scyliorhinus besnardi*, *Schroederichthys bivius*, *Alopias vulpinus*, *Squaliolus laticaudus*, *Etmopterus hillianus*, *Etmopterus pusillus*, *Centroscymnus cryptacanthus*, *Echinorhinus brucus*, *Hexanchus griseus*, *Raja leptocauda*, *Psammobatis lentiginosa*, *Gurgesiella dorsalifera*, *Torpedo puelcha*, and *Benthobatis* sp. (Barcellos 1963; Barcellos and Pinedo 1980; Vooren, unpubl.). Most species occur in small numbers at depths from 100 to 500 m. However, the abundance of pelagic sharks such as members of the genera *Etmopterus* and *Squaliolus* may be greater than suggested by from bottom-trawling.

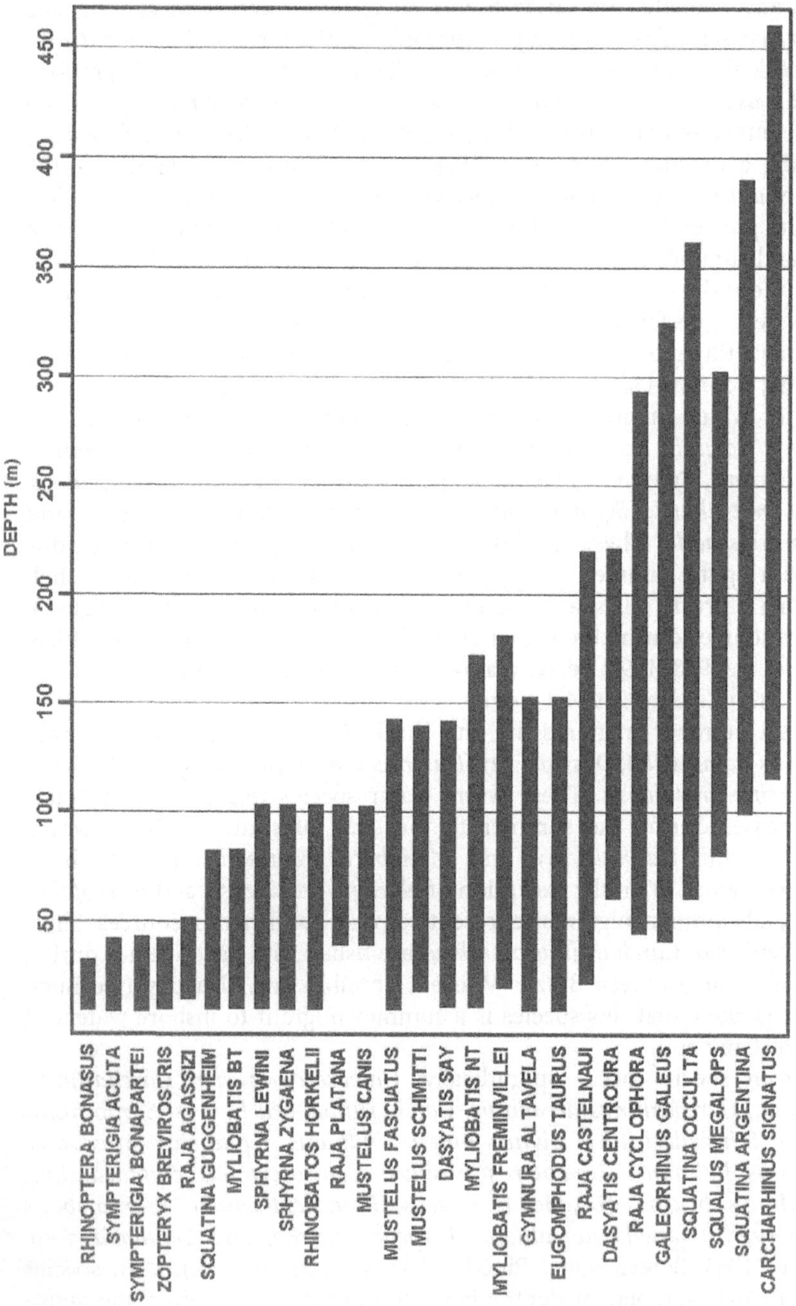

Fig. 6.16.1. Depth ranges of dominant demersal elasmobranch species over the continental shelf and upper slope between 28° 40' and 34° 34' S

Reproductive Strategies and Evolution

The elasmobranch fauna of southern Brazil is composed of 13 oviparous species (Rajidae and *Scyliorhinus besnardi*) and 44 viviparous forms (all others). All types of trophic relation between mother and embryo as classified by Wourms et al. (1988) are observed among the viviparous species, and these fishes display different reproductive strategies in terms of sexual development, reproductive cycle, body size at birth, and fecundity. Among the 10 species that have been studied in detail, lecithotrophic gestation occurs in *Rhinobatos horkelii*, *Squalus megalops*, *Squalus mitsukurii*, *Squatina guggenheim*, *Squatina occulta*, *Squatina argentina*, and *Galeorhinus galeus*, while *Mustelus schmitti*, *Mustelus canis*, and *Mustelus fasciatus* have matrotrophic gestation (Vooren 1992). In most species the reproductive cycle of the female is annual, with vitellogenesis and gestation occurring simultaneously. However, the female reproductive cycle of *Squatina guggenheim*, *Squatina occulta*, and *Galeorhinus galeus* lasts 3 years: 1 year for gestation, and 2 years for gametogenesis and yolk formation (Peres and Vooren 1991; Vooren 1992; Sunyé 1993; Silva 1996). In matrotrophic species, the annual production of litter mass as a proportion of maternal body mass ranges from 11% in *Mustelus fasciatus* to 26% in *Mustelus canis*. In lecithotrophic species the values range from 5 to 7% in *Rhinobatos horkelii*, *Squatina guggenheim*, *Squatina occulta*, and *Galeorhinus galeus*, but reach 19% in *Squalus megalops* and *Squalus mitsukurii*. If the production of yolk is included, annual reproductive investment as a proportion of maternal body mass reaches 32% in *Squalus megalops* and *Squalus mitsukurii* (Vooren 1992; Calderón 1994). Thus, there are considerable differences between species in the relative magnitude of the annual investment of metabolic energy in reproduction by the female. The partition of metabolic investment between reproduction, growth, and maintenance is genetically determined and results in the mutual adjustment between recruitment and mortality at the population level. Future research on population energetics and mortality will elucidate the mechanisms and constraints which determine the process.

Rhinobatidae, Squatinidae, and Hexanchidae have existed since the Jurassic, Sphyrnidae appeared during the Miocene, while other families such as Rajidae, Myliobatidae and Carcharhinidae date back to the Cretaceous (Cappetta 1987). Thus, the elasmobranch fauna of southern Brazil contains elements of widely different degrees of antiquity. This circumstance favours the comparative investigation of the biology and adaptation of elasmobranchs at different evolutionary stages, and permits study of the evolution of viviparity among the group. For instance, the cloacal gestation of the genus *Squatina* is unique among elasmobranchs and may represent an early stage of the evolution of viviparity, though uncertainty persists as to whether this mode of reproduction evolved from or gave rise to oviparity

(Sunyé 1993). The elasmobranch fauna of southern Brazil is a "laboratory of evolution", and there is an urgent need to preserve this fauna, which is being severely threatened by the fishery industry.

6.17 Cephalopods

M. Haimovici

Early reviews on the taxonomy and zoogeography of Brazilian cephalopods are based on museum collections of expeditions during the 19th and the beginning of the 20th century as collated by Palacio (1977). More recent studies have re-examined the octopus *Eledone massyae* (Voss 1964) and the abundant *Loligo sanpaulensis* (=*L. brasiliensis;* Brackoniecki 1984), as well as described a new genus and one member (*Vosseledone charrua;* Palacio 1978) and a new species of *Eledone* (*E. gaucha;* Haimovici 1988c). Based on the revision of cephalopod collections (Haimovici et al. 1989c; Perez and Haimovici 1991a; Haimovici and Perez 1991a,b), the study of cephalopod beaks in the stomachs of predators (Santos 1992), as well as macrozooplankton samples of paralarvae and juveniles (Haimovici and Perez 1991a; Haimovici et al. 1994b), the knowledge concerning southern Brazilian cephalopods has improved (Palacio 1977; Juanicó 1980; Roper et al. 1984; Haimovici 1985; Haimovici and Perez 1991b; Haimovici et al. 1994b).

Faunistic Composition and Ecology of Dominant Species

For much of the cephalopod fauna, the Brazilian coast between 20° and 34° S represents a biogeographic transition zone. This region appears to be the centre of temperate coastal species with a relatively restricted distribution, like the frequent *Loligo sanpaulensis*, *Eledone massyae*, *Eledone gaucha*, and *Octopus tehuelchus.* Temperate species with a wide distribution (tropical species *Loligo plei* and *Semirossia tenera*, and the cosmopolitan *Octopus vulgaris*) are rare (Palacio 1982). Compared with more northern regions, the southern Brazilian shelf and slope exhibit an apparently impoverished cephalopod fauna (Haimovici and Andriguetto 1986; Haimovici et al. 1989c; Haimovici and Perez 1991a). The low diversity of benthic shelf habitats and pronounced seasonal water temperature oscillations, owing to largely sandy bottoms and to the influence of the Brazil and Malvinas Currents, respectively, do not favour the occurrence of tropical neritic species like *Loligo plei*, *Pickfordiateuthis*, and *Lolliguncula brevis*, nor of cold temperate coastal species like *Loligo gahi.*

Over 40 cephalopod species have been recorded for the southern Brazilian shelf and slope between Santa Marta Grande Cape and Chuí (28° 30'-34° 40' S; Appendix; Haimovici and Andriguetto 1986; Haimovici and Perez 1991b; Perez and Haimovici 1993; Fig. 6.17.1). *Octopus tehuelchus* is the most abundant benthic species in shelf waters down to 50 m depth, whilst *Semirossia tenera* and *Octopus vulgaris* occur in deeper waters. Between 50 m depth and the shelf break (200 m), the Octopodidae *Eledone massyae* and *Eledone gaucha* are the most frequent and dominant benthic species (Haimovici and Andriguetto 1986; Haimovici and Perez 1991a). The life-ycle of *Eledone massyae* extends approximately 2 years. Males reach sexual maturity prior to females in the fall, followed by reproduction in the spring. Intraovaric fecundity with 27-126 ovocytes in *Eledone massyae* (Perez and Haimovici 1991b) and 5-58 in *Eledone gaucha* (Perez et al., unpubl.) is low. In contrast to other Octopodidae, the oviductal gland in females of both species lacks the function of a spermatheca since sperms are stocked in filaments of the intraovaric maturing oocytes (Perez et al. 1990). Adults of *Eledone massyae* and *Eledone gaucha* appear to explore different food sources. While the diet preference of *Eledone massyae* changes with age from amphipods to portunid crabs and polychaetes, *Eledone gaucha* continues to feed on amphipods and isopods (Perez and Haimovici 1995). In the deeper slope regions species of subantarctic origin (Voss 1988), like *Pareledone charrua*, *Pareledone turqueti*, and eledonids occur on sandy bottoms, in contrast to *Vosseledone charrua*, which is probably restricted to consolidated bottoms of relict corals.

Loligo sanpaulensis is the most abundant neritic species in shelf waters during the entire year (Haimovici and Andriguetto 1986; Andriguetto and Haimovici 1991). The highest densities and largest individuals occur south of 32° S in waters between 40 and 80 m depth at less than 18°C. Since the spring biomass may reach 3500 tons, *Loligo sanpaulensis* stocks are of potential commercial value. The southern Brazilian shelf waters appear to be a spawning area of the species, with smaller and larger individuals spawning in shallower (<40 m) and deeper waters, respectively. Smaller individuals are more common in the summer and fall, whilst mature females are found throughout the entire year but are less frequent in the fall (Andriguetto and Haimovici, in press). The availability of food, largely composed of fish and crustaceans, is likely to influence the distribution patterns of *Loligo sanpaulensis* (Juanicó 1979; Costa and Fernandes 1993). Other pelagic species of the shelf water, especially during the summer, are *Argonauta nodosa* and *Loligo plei*.

Illex argentinus is the most frequent and abundant species over the southern Brazilian outer shelf and upper slope (Haimovici and Perez 1991b). The species has a life-cycle of 1 year (Arkhipkin 1990) and reproduces essentially during the winter and spring when spawning adults are commonly found along the shelf break. Based on their size and the absence

Species and genera	Families	Habitat
Order SEPIOIDEA		
Spirula spirula	Spirulidae	Mesopelagic
Semirossia tenera	Sepiolidae	Demersal,muddy and sandy bottoms
Heteroteuthis atlantis	Sepiolidae	Mesopelagic
Order TEUTHOIDEA		
Loligo sanpaulensis	Lolliginidae	Demersal-pelagic
Loligo plei	Lolliginidae	Demersal-pelagic in the warm season
Illex argentinus	Ommastrephidae	Demersal-pelagic
Ornithoteuthis volatilis	Ommastrephidae	Epipelagic
Ornithoteuthis antillarum	Ommastrephidae	Epipelagic
Ommastrephes bartramii	Ommastrephidae	Epipelagic
Todarodes angolensis	Ommastrephidae	Demersal-pelagic
Thysanoteuthis rhombus	Thysanoteuthidae	Epipelagic
Onychoteuthis banksi	Onychoteuthidae	Epipelagic
Moroteuthis ingens	Onychoteuthidae	Demersal-pelagic
Moroteuthis robsoni	Onychoteuthidae	Demersal-pelagic
Onychia carribaea	Onychoteuthidae	Mesopelagic
Abraliopsis pfefferi	Enoploteuthidae	Mesopelagic-bathypelagic
Abralia veranyi	Enoploteuthidae	Mesopelagic-bathypelagic
Abralia redfieldi	Enoploteuthidae	Mesopelagic-bathypelagic
Ancistrocheirus alessandrini	Enoploteuthidae	Mesopelagic-bathypelagic
Pterygioteuthis giardi	Enoploteuthidae	Mesopelagic-bathypelagic
Histioteuthis dofleini	Histioteuthidae	Mesopelagic
Histioteuthis atlantica	Histioteuthidae	Mesopelagic

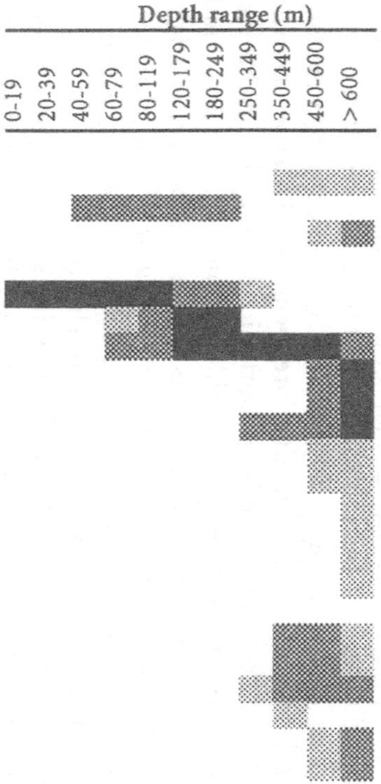

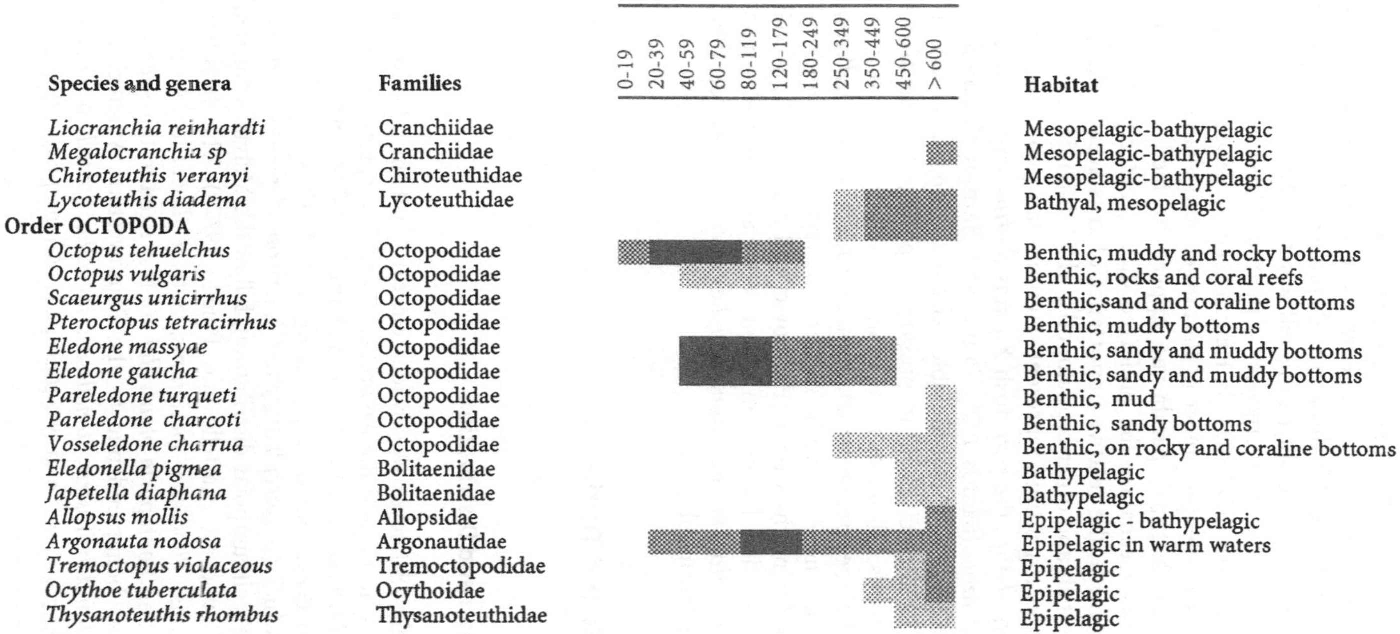

Fig. 6.17.1. Depth ranges and habitats of neritic and upper slope cephalopods along the southern Brazil coast (*black* abundant; *dark gray* frequent; *light grey* infrequent)

of didymozoid parasites in their digestive tract, spawning adults from southern Brazil, Uruguay, and northern Argentine may belong to the same group (Santos and Haimovici, in press). Paralarvae and juveniles of less than 50 mm mantle length exhibit three growth phases. They concentrate principally during the winter and spring in shelf waters (>100 m, >12°C) of tropical origin and at the convergence with the Malvinas Current (Haimovici and Perez 1990; Vidal 1994a,b; Haimovici et al. 1995; Vidal and Haimovici, unpubl.). The diet preference of *Illex argentinus* changes during development. Recently hatched paralarvae appear to feed on micro-organisms, whilst crustaceans are the preferred diet of older paralarvae. During continued growth, the relative importance of crustaceans changes in favour of a diet of fish and other cephalopods, which includes cannibalism (Santos 1992). *Illex argentinus* represents an important food source for the tunafish *Thunnus obesus*, the swordfish *Xiphias gladius*, and the wreckfish *Polyprion americanus* (Santos 1992; Santos and Haimovici, unpubl.). Although *Illex argentinus* is of considerable economic importance worldwide (Csirke 1987; Haimovici et al., in press), the potential value of the southern Brazilian stocks still needs to be evaluated (Haimovici and Perez 1991a). Other typical slope species include *Abralia verany*, *Abralia redfieldi*, *Lychoteuthis diadema*, and *Ancistrocheirus lesueuri* (syn. *A. alessandrini*) which are abundant during the winter (Haimovici and Perez 1990), as well as oceanic species with wide distribution in temperate and tropical waters, like *Ornithoteuthis antillarum* and *Ommastrephes bartrami* (Santos 1992; Santos and Haimovici, unpubl.).

6.18 Marine Mammals and Turtles

M. C. Pinedo

Species Composition and Distribution

Of the approximately 78 known worldwide cetaceans and the 13 endemic pinnipeds of South America and the Antarctic, 27 and 7, respectively, occur along the southern Brazilian coast between the latitudes 25° 14' and 33° 45' S (Castello and Pinedo 1977b; Pinedo 1990; Pinedo et al. 1992; Appendix). The pinnipeds, some of which (*Arctocephalus gazella*, *Hydruga leptonyx*, *Lobodon carcinophagus*) have a circumpolar distribution, follow the Malvinas Current to reach the Brazilian coast (Castello and Pinedo 1977a). The cetaceans *Hyperoodon planifrons*, *Phocoena spinipinnis*, *Sotalia fluviatilis*, *Australophocaena dioptrica*, and *Pontoporia blainvillei* are restricted to the Southern Hemisphere, the last being endemic to the Southwestern Atlantic. Five cetaceans (*Mesoplodon densirostris*, *Stenella coeruleoalba*, *Stenella attenuata*, *Steno bredanensis*, *Kogia simus*) and two pinnipeds (*Arctocephalus tropicalis*,

Arctocephalus gazella) have been cited for the first time in the southwestern Atlantic (Castello and Pinedo 1977b; Pinedo and Castello 1980; Pinedo 1987; Pinedo et al. 1992), whilst five others (*Megaptera novaeangliae*, *Phocoena spinipinnis*, *Pseudorca crassidens*, *Kogia breviceps*, *Australophocaena dioptrica*) have been registered for the first time in the southern Brazilian region (25° 14'-33° 45' S; Pinedo 1985, 1989; Rosas and Pinedo 1989). Among the 27 cetaceans, *Balaenoptera musculus* is an endangered species, and *Megaptera novaeangliae* and *Eubalaena australis* are classified as vulnerable because overexploitation and degradation of habitats has drastically reduced the original stocks. The remaining 24 cetacean species are categorized as "insufficiently known" (IUCN 1991), which impedes their characterization as endangered or vulnerable. Populations of nine of these species run a serious risk of being accidentally caught in fishing nets (Perrin 1988). Apart from the rich marine mammal fauna, four (*Caretta caretta*, *Chelonia mydas*, *Eretmochelys imbricata*, *Dermochelys coriacea*) of the eight known marine turtle species have extended their distribution into southern Brazilian waters (de Menezes 1972; Frazier 1984; Márquez 1990).

Ecology of Principal Cetaceans

Most of the cetaceans over the southern Brazilian shelf are coastal species, though 10 (*Delphinus delphis*, *Balaenoptera musculus*, *Physeter macrocephalus*, *Hyperoodon planifrons*, *Mesoplodon densirostris*, *Ziphius cavirostris*, *Grampus griseus*, *Globicephala melas*, *Stenella clymene*, *Stenella coeruleoalba*) are oceanic and another four (*Balaenoptera acutorostrata*, *Megaptera novaeangliae*, *Stenella frontalis*, *Stenella longirostris*) commonly occur in both environments. The more frequently encountered species are *Pontoporia blainvillei*, *Tursiops truncatus*, *Eubalaena australis*, and *Delphinus delphis* (Pinedo 1985, 1986; Castello and Pinedo 1986).

Pontoporia blainvillei occurs in waters of up to 30 m depth near the coast (<50 km; Praderi et al. 1989) between 18° and 39° S (Pérez-Macri and Crespo 1989; Pinedo 1994a). Male and female adults reach a total length of 158 cm and 177 cm, respectively (Brownell 1989), though geographically distinct forms of smaller and larger sized adult individuals are seen north and south of this region at 22-27° S and 32-38° S, respectively (Pinedo 1991, 1995). First sexual maturation occurs between 2 and 3 years (Kasuya and Brownell 1979), followed by reproduction in the late spring. Gestation generally exceeds a period of 9 months, and newly born individuals measure about 75 cm (Pinedo et al. 1989). Both sexes display early development of fins, respiratory system, and sensory organs. The life span of *Pontoporia blainvillei* is relatively short since physical maturity occurs between 4 and 8 years (Pinedo 1991), and the longevity of males and females is about 16 (Brownell 1989) and 21 years (Pinedo 1994b), respectively.

Pontoporia blainvillei is often caught accidentally in beach seines along the southeastern and southern Brazilian coast, which represents the principal cause of mortality (Pérez-Macri and Crespo 1989; Pinedo et al. 1989; Praderi et al. 1989; Pinedo 1994a; Corcuera et al. 1994). It is the most common species interacting with fisheries between approximately 26° and 34° S (Pinedo 1994a). Since most of the 1085 dead specimens registered over a 10-year period along the southern coast (29° 20'-33° 45' S) were still sexually immature (<3 years), their mortality is likely to have had an impact on the age structure of the population (Pinedo 1994a,b). *Pontoporia blainvillei* preys on a large variety of benthic organisms and fish. The fish diet is largely composed of Sciaenidae, particularly juveniles (<50 mm) of *Cynoscion guatucupa* (Brownell 1975; Pinedo et al. 1989). The cephalopod *Loligo sanpaulensis* is ingested especially by females, and young individuals also feed on the crustaceans *Pleoticus muelleri*, *Artemesia longinaris*, and *Penaeus paulensis* (Pinedo 1982).

Tursiops truncatus (Castello and Pinedo 1977a; Pinedo 1982, 1986; Pinedo et al. 1992) is common in coastal shelf waters and adults and young individuals are seen throughout the year in the Patos Lagoon estuary (Castello and Pinedo 1977a). Natural death appears to be the principal cause of mortality, followed by accidental capture in fishing nets (Pinedo 1986). Twenty of the 76 strandings (29° 20' and 33° 45' S) between 1976 and 1993 were attributed to fishery interactions, five being caught two miles offshore in gill-nets (22 cm mesh) set for sharks (Pinedo 1986, 1994a). Along this coast *Tursiops truncatus* does not ingest cephalopods or shrimp but essentially preys on fish (>150 mm TL). *Micropogonias furnieri* dominates among the 10 teleost species which compose the diet (79%; Pinedo 1982).

Other species, like *Globicephala melas*, *Steno bredanensis*, *Sotalia fluviatilis*, *Stenella frontalis*, and *Pseudorca crassidens*, are caught by long lines or nets. The coastal form of *Sotalia fluviatilis* is common between the equator and 29° S and is the principal species interacting with fisheries at about 25° S. *Pontoporia blainvillei* is the most common species along the coast between 26 and 34° S (Pinedo 1994a). Pelagic and oceanic species, like *Physeter macrocephalus*, *Kogia simus*, *Globicephala melas*, *Phocoena spinipinnis*, and *Pseudorca crassidens*, feed on a diverse diet of cephalopods, shrimps (Clarke et al. 1980; Pinedo 1987), and Sciaenidae and Serranidae (Pinedo 1989; Pinedo and Rosas 1989), whilst *Orcinus orca* sometimes ingests individuals of the swordfish *Xiphias gladius* which are hooked on long lines (Pinedo 1994a).

Every year groups of up to ten individuals of *Eubalaena australis* are seen during the winter and spring, with the highest incidence in October, over the southern Brazilian shelf (27° 23'-32° 10' S) between the coast and about 30 km offshore (Pinedo 1985; Pinedo, unpubl.). The seasonal occurrence and behaviour associated with copulation (Pinedo 1985) provide evidence that the southern Brazilian shelf may have greater importance as a region of reproduction between populations from northern (28^0 29' S) and southern (42^0 30' S) nursing areas in the southwestern Atlantic than previously considered (Best et al. 1993). It is likely that the *Eubalaena australis* population is recuperating from past overexploitation. In contrast, *Delphinus delphis* occurs in groups of up to 500 individuals throughout the year over the intermediate shelf and slope (29^0 22'-34^0 31' S) between 70 and 2500 m depth and is occasionally caught by fishing nets (Castello and Pinedo 1986; Pinedo 1994a).

Ecology of Principal Pinnipeds

Otaria flavescens and *Arctocephalus australis* are the most frequently encountered species, especially between the fall and spring, although recordings of *Arctocephalus tropicalis* have increased over the last 15 years (Pinedo 1986, 1990). The remaining pinniped species are only occasional visitors (Castello and Pinedo 1977c). After reproduction along the coast of Uruguay and Argentina during the summer, starved individuals of *Otaria flavescens* migrate to both southern (Vaz-Ferreira 1982a) and northern feeding grounds. The migratory range of females appears to be more restricted since they are most abundant at the southern limits of Brazil's coast (34^0 S) in the winter (Pinedo 1986). In contrast, subadult and adult males (3-5 years) migrate further north. The tip of the northern breakwater of the Patos Lagoon inlet (32^0 S) serves as a resting place for individuals of *Otaria flavescens* throughout the year, though there are more in the spring (Rosas 1989). Sexual maturation of females and males is reached between 3 and 4 years and after 5 years, respectively (Vaz-Ferreira 1982a). Both sexes attain 95% of the maximum length after 8 years (females 194.4 cm, males 254.4 cm; Rosas et al. 1993), although females may live to be 14 years and males 16 years. Especially in the winter and spring, the principal cause of mortality is due to aggression by fishermen during fishing activities, since approximately 30% of the 292 dead specimens analyzed over a 10-year period displayed mutilations or perforations caused by firearms (Pinedo 1986, 1990; Rosas 1989). *Otaria flavescens* preys on five different elasmobranchs and on at least 15 different teleost fish of which *Paralonchurus brasiliensis*, *Macrodon ancylodon*, *Cynoscion guatucupa*, *Micropogonias furnieri*, and *Trichiurus lepturus* represent the major diet items.
Like *Otaria flavescens*, the pinniped *Arctocephalus australis* leaves the reproductive colonies in Uruguay (35^0 10' S; Vaz-Ferreira 1982b) to migrate to more northern regions. Most of the 106 dead individuals encountered over 10 years between latitudes 290 20' -330 45' S were young males of 80-100 cm (Pinedo 1986, 1990). Dead specimens do not generally display mutilations since, unlike *Otaria flavescens*, *Arctocephalus australis* does not follow fishing vessels and tends to avoid nearshore waters. Natural death thus seems to be the major cause of mortality (Pinedo 1986). *Arctocephalus australis* preys upon a large

variety of organisms, composed of gastropods, shrimp (*Pleoticus muelleri*, *Artemesia longinaris*), cephalopod (*Loligo sanpaulensis*, *Illex argentinus*), and teleost fish, predominantly Engraulididae and Sciaenidae.

The islands South Georgia, Tristan da Cunha, and Gough are reproductive areas for *Arctocephalus tropicalis*. Probably owing to a growing population, the southern Brazilian coast is visited with increasing frequency by mostly sexually inactive adult males, especially during the winter and spring (Pinedo 1990), as has been observed for the coast of Chile (Torres et al. 1984). In southern Brazilian waters, *Arctocephalus tropicalis* feeds on *Anchoa marinii*, *Pomatomus saltatrix*, *Cynoscion guatucupa*, *Micropogonias furnieri*, *Trichiurus lepturus*, *Mugil* spp., *Odontesthes* spp., Ariidae, and on *Loligo sanpaulensis*. Although records of fishing interactions are lacking, at least 35% of the 65 dead specimens found from 1976 to 1994 seem to have died of unnatural causes (Pinedo 1990; Pinedo, unpubl.).

Marine Turtles

Spawning areas of marine turtles in Brazil are restricted to latitudes north of 19^0 38' S (Marcovaldi 1991). Although *Dermochelys coriacea*, *Chelonia mydas*, *Eretmochelys imbricata* and, more frequently *Caretta caretta* are seen from the spring to fall between 29^0 20' and 33^0 45' S (Pinedo, unpubl.), in the absence of detailed information, studies on population characteristics, feeding area and behaviour, displacement patterns, and mortality causes of marine turtles from this region are urgently needed (Ogren et al. 1989).

6.19 Sea and Shore Birds

C. M. Vooren

The coast of the warm-temperate southwestern Atlantic between 28^0 40' and 34^0 34' S (Santa Marta Grande Cape to Chuí) and its adjacent waters sustain 83 species of sea and shore birds (Appendix). Of these, 72 are visitors who breed in distant places and migrate to the area to feed. Two major environments can be distinguished: the beach and nearshore waters provide habitats for 50 species of shorebirds and coastal seabirds, including 44 Charadriiformes (gulls, terns, skimmer, waders, and sheathbill), while the offshore waters of the continental shelf and slope sustain 33 species of oceanic seabirds, including 25 Procellariiformes (albatrosses and petrels). Shorebirds find in the swash zone an almost constantly accessible supply of abundant food, mainly the bivalve molluscs *Mesodesma mactroides* and *Donax hanleyanus* and the crustaceans *Emerita brasiliensis* and *Excirolana armata*, while gulls and hawks feed on stranded debris such as marine organisms, wind-blown insects, and discards of the coastal fishery. Terns, skimmers, herons, and cormorants feed on shrimps and fish which abound all year round in the coastal waters, and the terns constitute a source of food for the kleptoparasitic skuas. Zooplankton, small pelagic fishes, and squid reach peak abundance in offshore waters during the winter and constitute at that time, together with discards of the fishery, the food supply for albatrosses, petrels, and penguins (Gianuca 1983; Belton 1984, 1985; Sick 1984; Harrington et al. 1986; Resende 1988; Vooren and Fernandes 1989; Vooren and Chiaradia 1990).

Migrations

Of the 11 "resident species", only the oystercatcher *Haematopus palliatus* and the plover *Charadrius collaris* breed on the beach in the spring and are permanent components of the coastal ecosystem. The terns *Sterna trudeaui* and *Sterna superciliaris*, the gull *Larus maculipennis*, and the stilt *Himantopus himantopus* breed in nearby inland habitats, are scarce on the shore in the spring, and move to the coast in the fall and winter. The herons *Egretta thula* and *Ardea cocoi* and the hawks *Milvago chimango* and *Polyborus plancus* forage all year round on the beach in small numbers and are probably non-breeders of populations which nest further inland (Sick 1984; Vooren and Chiaradia 1990).

Forty-two species are "southern migrants" which breed in the summer in temperate to antarctic latitudes of the Southern Hemisphere, and most of these reach peak numbers in southern Brazil during the winter. The category includes 27 oceanic seabirds (albatrosses, petrels, and penguins) and 15 coastal seabirds and shorebirds (mainly terns, gulls, and waders; Fig. 6.19.1). Large numbers of dead juveniles of the penguin *Spheniscus magellanicus* are a typical feature of the beach during the winter, and groups of live birds are at that time common on the continental shelf. Adults are scarce, but the shelf waters of southern Brazil appear to be the nursery for first-year juveniles of *Spheniscus magellanicus* during the winter, when food such as the anchovy *Engraulis anchoita* reaches peak abundance. Banded juveniles from the penguin rookeries at Valdez Peninsula (Patagonia, Argentina) have been found, and most of the birds probably come from this breeding ground. Among the albatrosses, *Diomedea melanophris* feeds in flocks on discards of the fishery industry on the continental shelf, while *Diomedea exulans* is restricted to the edge of the shelf, where squid such as *Illex argentinus* abound in winter and constitute a food supply for the adults of *Diomedea exulans* which are at that time rearing chicks on South Georgia. Among petrels, the most common winter visitors from latitudes between the Malvinas Islands and the Antarctic are *Macronectes* sp., *Fulmarus glacialoides*, *Puffinus griseus*, *Pterodroma incerta*,

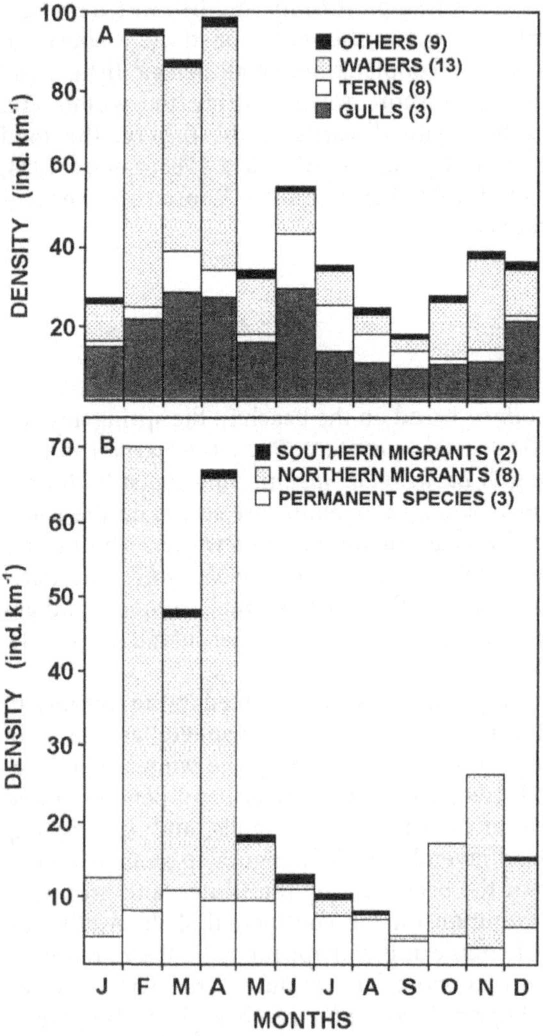

Fig. 6.19.1. Monthly mean density of shore and coastal seabirds (**A**) and of waders (**B**) on the Atlantic ocean beach at 32° S

Pachyptila belcheri, *Daption capense*, and *Procellaria aequinoctialis aequinoctialis*, the last two species feeding on discards of the fishery industry. *Puffinus gravis*, *Diomedea chlororhynchus*, and *Procellaria aequinoctialis conspicillata* from Tristan da Cunha Island are common all year round, the former two species being numerous on the continental shelf where they follow fishing boats, while *Procellaria aequinoctialis conspicillata* occurs only at the edge of the shelf. *Puffinus gravis* is usually said to migrate in the winter to the North Atlantic, but it is evident that part of its population remains in the southern hemisphere at that time. The same is true for *Oceanites oceanicus* which breeds in the Antarctic and is seen off southern Brazil from May to September. *Larus dominicanus* breeds on the coast of Uruguay in the spring and is one of the most numerous birds on the coast of Brazil's extreme south during the remainder of the year. *Sterna maxima*, *Sterna hirundinacea*, and *Sterna eurygnatha* breed in Argentina, the former two species wintering on the South Brazilian coast, while *Sterna eurygnatha* passes through the area in the spring and fall, with a few birds remaining during the winter. Flocks of non-breeding *Rynchops niger* roost during the summer at coastal lagoons, but it is not known where these birds come from. *Charadrius falklandicus* and *Charadrius modestus* are common on the beach in the winter and are evidently migrants from Patagonia, although a small number of the former species breeds in the southern Brazilian Peixe Lagoon (see Fig. 3.2). In addition, the Andean flamingo *Phoenicopterus chilensis* is habitually seen in numbers of up to 500 birds at the Peixe Lagoon (Escalante 1970; Belton 1984; Vooren and Fernandes 1989; Croxall and Prince 1990; Marchant and Higgins 1990; Vooren and Chiaradia 1990; Prince et al. 1992).

Thirty species are "northern migrants" which breed in temperate to arctic latitudes of the Northern Hemisphere during the boreal summer and reach peak numbers in southern Brazil from the austral spring to fall (Fig. 6.19.1). The category includes 23 waders, 3 skuas, 2 terns, and 2 petrels. Ten species occur in large numbers and are discussed below. *Pluvialis dominica*, *Pluvialis squatarola*, *Calidris alba*, *Calidris fuscicollis*, and *Calidris canutus* from Canada are the most common waders. *Calidris canutus* reaches peak numbers of up to 150 birds per km of shoreline from March to May, completing the pre-nuptial moult and then fattening up for the northward flight to Delaware Bay. The other five species spend the summer in the area, but populations of *Calidris fuscicollis* and *Calidris alba*, which spend the austral summer in Argentina, use the beach as a staging area in the fall in the manner of *Calidris canutus*. *Sterna hirundo* from North America roosts on selected places of the coast in groups of up to 14 000 birds during the summer and fall. *Stercorarius longicaudus* and *Stercorarius parasiticus* associate with flocks of terns at sea. *Puffinus puffinus* from the British Isles migrates southward through the area, mostly from August to October, as evidenced by the number of dead individuals washed ashore during those months. *Puffinus diomedea* is common over the edge of the continental shelf during the summer (Belton 1984; Harrington et al. 1986; Resende 1988; Vooren and Fernandes 1989; Vooren and Chiaradia 1990).

The coast and the waters of the continental shelf and slope of southern Brazil are of crucial importance for populations of migratory sea and shore birds which breed in Arctic Canada, the British Isles, oceanic islands off Spain and Portugal, the Tristan da Cunha Islands, Patagonia, the Malvinas

(Falkland) Islands, South Georgia and South Orkney Islands, and the Antarctic Peninsula. Through its role as a feeding and staging area on the migration routes of these birds, the region occupies a strategic position in a web of ecological relationships which spans the Americas and the Atlantic Ocean as a whole. The conservation of the coastal and marine environments of southern Brazil is of worldwide significance.

Feeding and Resource Partitioning on the Beach

On the lower beach the eight abundant waders *Haematopus palliatus, Pluvialis squatarola, Pluvialis dominica, Charadrius collaris, Charadrius falklandicus, Calidris canutus, Calidris alba,* and *Calidris fuscicollis* feed on invertebrates which live buried in the sand of the swash zone and which only emerge to migrate with the water level as this changes due to wind and lunar tides. As indicated by the shape and size of the bill and specialized searching behaviour, *Haematopus palliatus* preys preferentially on large individuals of the clam *Mesodesma mactroides* which remain buried at a depth of about 5 cm and are therefore not accessible to other birds, except when the clams emerge to migrate with the tide, and can be taken by the gull Larus dominicanus. From February to April, small juveniles of the clams *Mesodesma mactroides* and *Donax hanleyanus* and the crustaceans *Emerita brasiliensis* and *Excirolana armata* reach peak numbers in the swash zone, with up to 7000 to 20 000 individuals per m^2 for each of these species, and the arrival of the migrating sandpipers *Calidris canutus, Calidris fuscicollis,* and *Calidris alba* in the area is timed so as to coincide with the availability of this food supply. *Calidris canutus* and *Calidris fuscicollis* probe the slightly submerged or just exposed wet sand of the swash zone with their straight and fine bills, mostly in the manner known as stitching (Burton 1974), localizing prey through touch and taste. *Calidris canutus* searches in compact flocks which move along as a single unit, while *Calidris fuscicollis* does so in loosely scattered groups. The latter species also searches visually on the dry sand above the swash zone. *Calidris alba* has a thicker bill and feeds in dense flocks on the submerged substrate, running constantly with the advance and retreat of the waves and catching exposed prey on sight. In contrast to the sandpipers, the five common plovers of the genera *Pluvialis* and *Charadrius* have powerful running legs, short thick bills, and very large eyes. These birds search for their prey exclusively by visual means and in a solitary manner, over an intraspecific distance of at least 2 to 10 m. The birds feed mostly on the exposed substrate at the upper margin of the swash zone, and to a lesser extent on lightly submerged sands. *Charadrius falklandicus, Charadrius collaris, Pluvialis dominica*, and *Pluvialis squatarola* feed in a similar way in the same places, but while the latter two occur in the summer, *Charadrius falklandicus* is a winter visitor, and *Charadrius collaris* lives during the summer in the dunes and feeds on the lower beach in the winter. Thus, the seven abundant species of sandpipers and plovers differ in methods of prey searching and in the distribution in space and time of feeding within the swash zone, partitioning in this way the space available for feeding and the prey according to different states of behaviour, i.e. buried or migrating with the tide. Future research may decide whether through these differences in feeding the birds partition different species of prey or different population strata of the same species of abundant prey (Gianuca 1983; Harrington et al. 1986; Vooren and Chiaradia 1990).

Sterna superciliaris fishes mostly at creek mouths, *Sterna trudeaui* does so in the surf zone, while *Rynchops niger* fishes in lagoons and in quiet estuarine and inshore waters. The other terns fish further offshore, *Sterna hirundo* associating with schools of skipjack tuna. Thus, resource partitioning occurs through spatial separation of the fishing areas, as also happens among piscivorous birds of tropical oceanic islands (Ashmole and Ashmole 1967). The gulls *Larus dominicanus* and *Larus maculipennis* are unspecialized generalist feeders, taking animal debris on beaches and in coastal waters, *Larus maculipennis* also catching insects on the wing and small fishes at shallow creek mouths, while pellets of *Larus dominicanus* frequently contain shells of bivalve molluscs. In order to have access to their food supply, all these birds need undisturbed and safe places on the beach for roosting within reach of their feeding areas. Due to the long extent of the beaches and low level of human disturbance, the birds share bivalves without competitive interaction. Gulls and terns roost together on the beaches in multispecific flocks which are structured as an assemblage of monospecific groups (Vooren and Chiaradia 1990; Chiaradia 1991; Vooren, unpubl.).

6.20 Genetic-Biochemical Differentiation among Marine Organisms

J. A. Levy

Genetic-biochemical data have revealed patterns of phylogenetic relationships within the families of Sciaenidae (Cassano 1987) and Scombridae, and determined the taxonomic status in the genera *Squatina* (Chondrichthyes; Solé-Cava et al. 1983; Solé-Cava and Levy 1987), *Myliobatis* (Chondrichthyes; Levy and Conceição 1989), *Menticirrhus* (Cassano and Levy 1990), *Cynoscion* (Levy and Cassano 1994), *Eledone* (Cephalopoda; Levy et al. 1988) and *Odontesthes* (Atherinidae; Prodohl and Levy 1989) from the warm-temperate southwestern Atlantic.

Genetic-biochemical data have also provided important information for fisheries management (FAO/UNEP 1981) because aspects of population dynamics, speciation, and distribution can be interpreted by the degree of genetic differentiation within a species, which results from restricted genetic flow between individuals of the same population. Since populations of marine species are frequently divided into morphologically and ecologically different groups with distinct spawning periods and sites, reproductive isolation enhances the accumulation of genetic differences between groups, owing to mutation, genetic drift, or selection. Understanding the genetic variability within and genetic divergence between populations is paramount for the rational exploitation and conservation of biological resources (Pla et al. 1990), because decreasing variability in the natural populations increases the probability of their extinction (Meffe 1986).

Individuals of several economically important populations in the southwestern Atlantic, like *Micropogonias furnieri* (Maggioni 1992, Matthiensen et al. 1993; Maggioni et al. 1994), *Mesodesma mactroides* (Marins et al. 1995), *Artemesia longinaris* (Weber et al. 1993), *Penaeus paulensis*, and *Odontesthes argentinensis*, show normal genetic variability and high homogeneity among sampled allelic frequencies, and thus should be considered as belonging to genetically homogeneous populations. These results suggest that the warm-temperate southwestern Atlantic region might act as one large marine ecosystem, similar to those of the eastern coast of North America and the coast of South Africa (Sharp 1988).

7 Relationships and Function of Coastal and Marine Environments

7.1 The Foredune System

U. Seeliger

The coastal dunes of southern Brazil are remarkably similar because Holocene sea-level changes have influenced their genesis (Pfadenhauer 1980; Schwarzbold and Schafer 1984), resulting in the sequential formation of depressions and dune ridges parallel to the shore. The most recent transgression-regression events moulded the 300-1000 m wide foredunes, with seasonally flooded freshwater marshes immediately interior of them. The flat marine terrace supplies a considerable amount of fine quartz sand to the beach (Pfadenhauer 1980), and the sandy substrate of windblown foredunes contains a high proportion of shell fragments but little organic matter (Cordazzo and Seeliger 1987). Since the tidal ranges are low (<50 cm), the dominant southeasterly and northeasterly onshore winds in the winter and summer, respectively, control seawater flooding of the beaches and shore profiles (Pfadenhauer 1980; Costa et al. 1984, 1991; Bernardi et al. 1987). Different orientations of the coast north and south of 33^0 S imply variations in onshore wind approach angles and impact. On the northern coast, NE onshore winds induce considerable beach-dune sand transport. In contrast, along the southern coast beach-dune sandflow is reduced because NE winds blow parallel to the shore, whilst SE winds blow directly onshore and cause inundation of beaches and landward erosion of backshore and foredune sands. Additionally, winter cold front passages lead to severe beach overwash and differentially affect sand dynamics along the northern and southern coasts. As a result, the northern beaches are largely intermediate and occasionally dissipative, whilst the southern beaches with coarser-textured sand tend to become reflective (Calliari and Klein 1993).

Dune Habitats

In spite of latitudinal disturbance gradients, which are a source of spatial and temporal environmental heterogeneity (Pfadenhauer 1980; Bernardi and Seeliger 1989; Seeliger 1992b), the entire coastal foredune complex can

U. Seeliger et al.
Subtropical Convergence Environments

be regarded as one functional unit (*sensu* Doing 1985). Gradients of beach/foredune sediment transport and edaphic stability of foredunes not only induce different site-specific environmental conditions (Hesp 1991), but also influence the composition of vegetation and the abundance of species (Barbour 1992; Fahrig et al. 1993). Pioneer backshore and foredune plants are selected in accordance with their physiological tolerance and morphological and reproductive traits (Barbour et al. 1976, 1987; Hesp 1991). The environmental conditions and plant associations define six typical habitats on southern Brazilian backshores and foredunes. Plant associations, dominanted by *Blutaparon portulacoides*, *Panicum racemosum*, *Spartina ciliata*, *Hydrocotyle bonariensis*/*Paspalum vaginatum*, *Androtrichum trigynum*, and *Andropogon arenarius*, occur in response to mean low water (MLW) height and distance, and along salinity, substrate humidity and stability, and silt/clay contents gradients (Cordazzo and Seeliger 1993; Costa et al. 1995). The different environmental and vegetational characterstics of each habitat also provide the conditions for a diverse animal community in coastal foredunes.

Since tidemarks are wind-controlled, they are unstable and often overlap, thus leading frequently to life-cycle disruptions of the otherwise common annual driftline species *Cakile maritima*, *Calycera crassiflora*, and *Calystegia soldanella* (Barbour et al. 1976; Doing 1985; Moreno-Casasola and Espejel 1986; Fahrig et al. 1993). As a consequence, southern Brazilian beaches lack driftline habitats because species retreat to more stable and permanent tidemarks behind the backshore (Cordazzo and Seeliger 1993).

The backshores themselves are inhospitable environments and are seasonally unstable, owing to eolian sandflow and erosive seawater overwash during winter storms, respectively. The salt-tolerance and rhizome-propagated perennial growth of *Blutaparon portulacoides* enable this species to withstand the local conditions of saline stress and sand instability (Bernardi and Seeliger 1989). The plants capture the windblown sand (Pfadenhauer 1980) and promote the development of an extensive embryo dune habitat (Pfadenhauer 1978; Bernardi and Seeliger 1989). Although extreme storms occasionally completely erase this habitat, reestablishment takes less than 2 years (Bernardi et al. 1987). Only where creeks break the dunes does the outflowing water destroy the embryo dune habitat; under increasing freshwater inundation stress, *Blutaparon portulacoides* is replaced by a *Hydrocotyle*/*Paspalum* association (Cordazzo and Seeliger 1993). Owing to predominantly terrestrial conditions, the fauna of the embryo dune habitat is primarily composed of insects, although amphipods of marine origin (*Orchestoidea brasiliensis*) may also occur (Gianuca, Sect. 6.5).

Most sand cannot be fixed in the embryo dune habitat but is deposited in zones beyond seawater advance, where virtually monospecific stands of *Panicum racemosum* progressively capture wind-blown sand (Cordazzo and Seeliger 1993). Accretion rates of up to 70 cm year^{-1} (Costa et al. 1984) promote the formation of a 4-6 m high frontal dune ridge habitat (Cordazzo and Seeliger 1993). The continuous deposition of nutrient-rich sand favours the sprouting of rhizome fragments and vigorous growth of *Panicum racemosum* (Costa et al. 1991). On leeward dunes, with lower sand deposition and fewer nutrients (Costa et al. 1984, 1991), *Panicum racemosum* enters a regressive stage, and species like *Senecio crassiflorus* and *Gamochaeta americana* invade (Pfadenhauer 1978, 1980; Cordazzo and Seeliger 1987, 1988a). The relatively dense plant cover and the dry substrate offer niches for a large number of insects, though frogs, lizards, and small mammals like the tuco-tuco (*Ctenomys flamarioni;* Ctenomyidae) are also important components (Gianuca, Sect. 6.5).

Lower humidity areas beyond the backshore are exposed to disturbance by occasional seawater overwash. Here, its large number of highly viable seeds favours the growth of *Spartina ciliata* (Hueck 1955; Legrand 1959; Pfadenhauer 1980; Costa and Seeliger 1988a), and the plant's active excretion of salts compensates for sediment salinization (Bastos et al. 1993b). Since the clumpy growth of *Spartina* has little effect on sand-binding, a periodically recycled, typically 1-2 m high, rounded hummock dune habitat develops (Seeliger 1992a).

Behind the hummock dune and frontal dune ridge habitats, either gradual sand burial of lower freshwater marshes or the formation of blowouts establishes a mosaic of dry slack habitats and seasonally inundated wet slack habitats. Owing to greater substrate stability, the diversity of plant and insect species increases in these habitats (Pfadenhauer 1978, 1980; Costa et al. 1984; Cordazzo and Seeliger 1987, 1988a, 1993). The lower wet slack habitat is generally inundated during the winter and is dominated by *Androtrichum trigynum* and other inundation-tolerant species (*Phyla canensis, Bacopa monnieri, Pluchea sagittalis, Paspalum vaginatum, Hydrocotyle bonariensis*) and marsh remnants (*Juncus acutus, Typha domingensis;* Cordazzo and Seeliger 1987, 1993). The dry slack habitat, with reduced inundation stress, is characteristically occupied by *Andropogon arenarius.* Slacks and adjacent marshes serve as nesting sites for several shore birds and are inhabited by the skunk *Conepathus chinga*, the armadillo *Dasypus hybridus*, and the fox *Dusicyon gymnocercus* (Gianuca, Sect. 6.5).

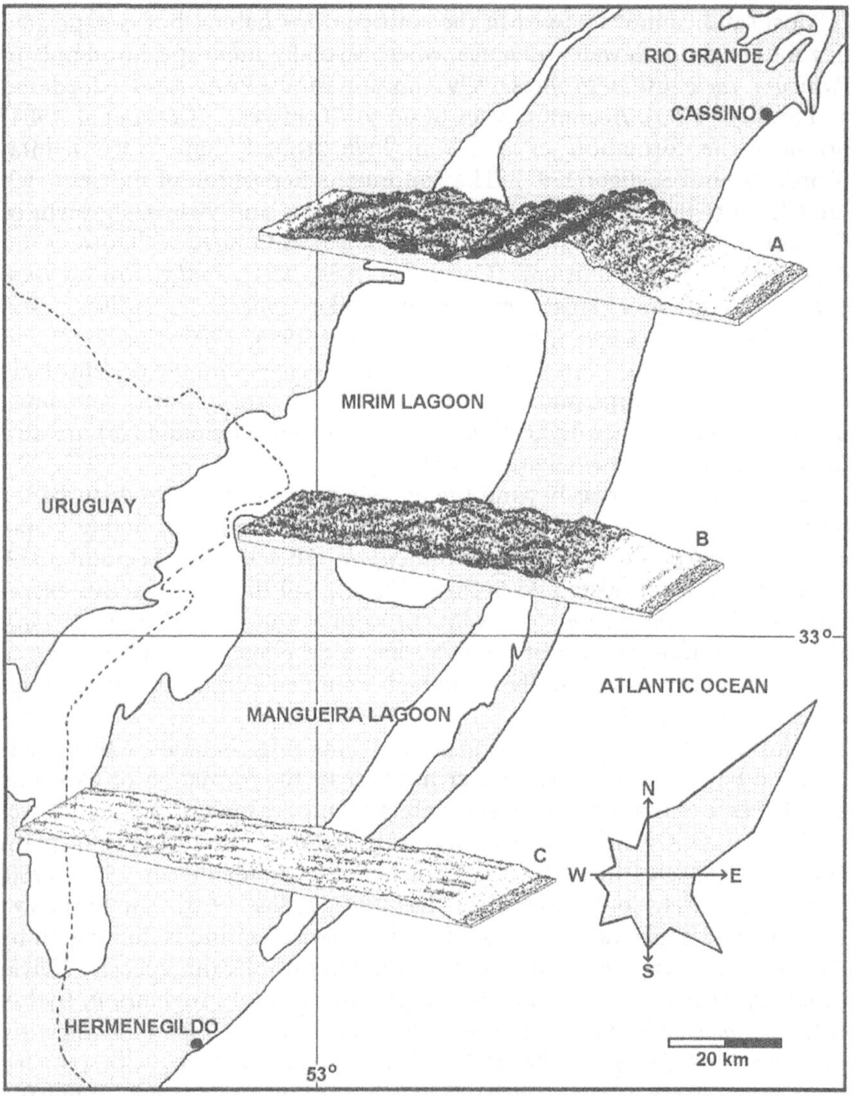

Fig. 7.1.1. Typical coastal foredune physiography changes from frontal dune ridges (**A**) over hummock dunes (**B**) to sand plains (**C**) with increasing shore disturbance

Dune Physiography

Ocean-shore disturbance gradients are a consequence of oceanic and climatic allogenic processes. Their relative significance depends on the contribution wave and wind energies has on both sand transport (Short and Hesp 1982) and the selection of beach/foredune plants. Disturbance not only selects for dominant plant species, but selected plants also modify the impact of disturbances. As has been observed elsewhere (Barbour et al. 1976, 1987; Godfrey 1977; Wiedemann 1984; Fahrig et al. 1993), shore disturbance gradients, either acting alone or in concert with the dominant plant cover, are likely to have influenced the genesis of different foredune physiographies along the southern Brazilian coast over time (Costa et al. 1991; Seeliger 1992b; Cordazzo and Seeliger 1993; Fig. 7.1.1).

A change in orientation of the southern Brazilian coastline causes a higher impact of winds and waves in the south (Calliari and Klein 1993), and thus disturbances owing to beach overwash and sand transport patterns increase in a north-south direction. Although disturbances tend to interact more severely with backshores than with foredunes, their modifying influence is more apparent in foredune areas. On backshores annual driftline species (*Cakile maritima*, *Calycera crassiflora*, *Calystegia soldanella*) and *Blutaparon portulacoides* tolerate intense disturbance. Owing to their reduced sand-binding capacity, the annual growth of driftline species, and the periodic obliteration of *Blutaparon portulacoides* populations by sea overwash (Bernardi et al. 1987; Bernardi and Seeliger 1989), their modifying influence on disturbance is likely to be negligible. However, different rates of sand flow from the beach and different sand accretion/erosion patterns may have selected for either the *Panicum* or *Spartina* associations in the foredune areas. Their different strategies of spatial occupation, growth, and sand-binding capacity (Costa et al. 1984, 1991; Costa and Seeliger 1988a,b; Cordazzo and Seeliger 1993) are likely to have ultimately affected the evolution of the foredune physiography.

At northern, less disturbed shores, above the influence of storm tide disturbances and saltwater flooding, the clonal spreading and vertical growth of *Panicum racemosum* promote efficient sand deposition, which counteracts sand erosion during periods of prevailing offshore winds (Pfadenhauer 1978, 1980; Costa et al. 1984, 1991; Cordazzo and Seeliger 1987, 1993). Over time, the continuous sand flow from the beach results in the formation of a prograding foredune ridge physiography which protects dry and wet slack habitats in the lee and prevents sand transgression over the inland marshes (Costa et al. 1991; Seeliger 1992a).

Further south, *Spartina ciliata* dominates lower shores characterized by increased disturbance from periodic wind erosion or occasional sea overwash. The clumpy clonal growth of *Spartina*, with limited sand-binding capacity, typically forms an extensive hummock dune physiography with interdispersed dry and wet slacks. Stands of *Acacia*, *Eucalyptus*, and *Pinus* bind persisting landward sand transport and occasionally form extensive retention ridges.

Along the severely disturbed coast in the extreme south of the region, the reduced sand flow from the beach and the continuous sand erosion in foredune areas by prevailing winds have essentially impeded the establishment of a plant cover (Costa et al. 1991; Costa and Seeliger 1988b, 1990), and allogenic factors appear to be entirely responsible for the shaping of the foredune landscape. As a consequence, the dominant foredune ridge and hummock dune physiographies of northern coastal regions have been replaced by a wind-eroded, humid, deflation plain physiography. Beyond the deflation plains, the continuous landward sandflow formed large transgressive dune fields which covered inland marshes and merged with older dunes (Gomes et al. 1987).

Since foredunes perform a major function in coastal protection and wildlife conservation, and human activities often interfere with this function, management approaches must consider ecological differences within and between physiographically distinct regions (Godfrey 1977) to preserve the value of any part of the coast for specific types of land use (van der Meulen et al. 1985) and to prevent heavy losses or complete destruction of pristine dune areas (Westhoff 1985).

7.2
The Beach and Surf Zone

V. M. T. Garcia and N. M. Gianuca

The beach and surf zone of intermediate and dissipative beaches, confined by the outer limits of circulation cells in the external surf zone and the upper limits of swash, may act as a single semi-closed and self-sustaining ecosystem, which is largely independent of food input from the adjacent sea (McLachlan 1980, 1983). Components of the necton and benthic, planktonic, and microbial communities tend to lend considerable diversity to such a system (Brown and McLachlan 1990). The extensive beaches of Brazil's southern coast are largely of the intermediate type (sensu Wright and Short 1984), though dissipative beaches dominate immediately south of 32° S (Gianuca 1983; Calliari and Klein 1993). Therefore, the majority of the coast is characterized by wide sandy beaches with a gentle slope and a well-developed surf zone with moderate to high wave energy, where sandbars, deposited parallel to the beach, are cut by rip currents. Sub-aerial and exposed beach profiles are seasonally modified as a result of wave energy-induced sand accretion during summer and erosion during the remainder of the year (Calliari and Klein 1993). During most of the year, waves break regularly at a similar distance from the shore and form the inner surf zone down to approximately 2 meters depth. Less frequent storm waves, particularly during cold-front passages, generate an external surf zone between 2 and 10 m depth (Borzone and Gianuca 1990).

Organic Matter Sources

Diatom patches are a typical feature of surf zones of intermediate and dissipative beaches and constitute a major source of organic matter (Schaefer and Lewin 1984; Brown and McLachlan 1990; Talbot et al. 1990). Along the southern Brazilian coast, the phytoplankton of the surf zone is characteristically dominated by frequent and dense accumulations of the diatom *Asterionellopsis glacialis*. Apart from mesoscale cycles (weekly or monthly), patch formation appears to follow diurnal cycles, with high cell concentrations in the water column during the day which decline at night. Although high silicate (30-60 μM) concentrations, highly variable nitrate (1-30 μM) levels, and low N:P ratios (7-8) in surf zone waters suggest a possible nitrogen limitation for phytoplankton growth, the intensity of mesoscale accumulations appears to be largely controlled by physical and meteorological factors and the dynamics of the system (Odebrecht et al. 1995a). Similar to patch formation during "storm-calm-storm" cycles elsewhere (Talbot and Bate 1989), the development of *Asterionellopsis glacialis* patches along the southern Brazilian coast is related to cyclic cold front passages with strong southerly winds, which force seawater onshore and generate high-energy waves and a wider surf zone (Gianuca 1983; Odebrecht et al. 1995a). Under these conditions, high concentrations of *Asterionellopsis glacialis* cells are resuspended from nearshore sediments and, owing to the high wave energy, accumulate in the surf zone where exposure to intense light stimulates further cell growth (Garcia, unpubl.). Rip currents and undertow return the diatom cells to nearshore sediments beyond the surf zone where they accumulate under calm conditions.

Mesoscale cycles of *Asterionellopsis glacialis* patch formations are accompanied by extreme fluctuations in chlorophyll *a* concentrations (1.2-1600 μg l^{-1}; Odebrecht et al. 1995a). Particularly in the summer, centric diatoms (*Skeletonema costatum* and large species of the genus *Coscinodiscus*) become important components of the phytoplankton community (Rörig, unpubl.) and contribute to average annual chlorophyll *a* concentrations of 2.7-350 μg l^{-1} in the water column. Finally, occasional nuisance blooms of potentially toxic dinoflagellates, like *Gyrodinium* cf. *aureolum*, *Dinophysis acuminata* and *Noctiluca scintillans* (Machado 1979; Rosa and Buselato 1981; Garcia et al. 1994; Odebrecht et al. 1995b) are associated with massive mortality of clams.

Although information on phytoplankton primary production levels is still lacking, peak chlorophyll *a* concentrations in the surf zone (Odebrecht et al. 1995a) and estimates based on experimental cultures of *Asterionellopsis glacialis* isolates with maximum rates of 3.74 mg C mg chlorophyll^{-1} h^{-1} (Garcia and Rörig, unpubl.) suggest that primary production rates during diatom accumulations may vary between 1300 and 6000 mg C m^{-3} h^{-1}. These data are comparable to values reported for surf accumulations of other diatom species elsewhere (Schaefer and Lewin 1984).

Average particulate organic carbon (POC) concentrations in the water column (460-20 050 μg l^{-1}) and chlorophyll *a* values provide an estimation of a mean C:Chl *a* ratio of 44 and a contribution of phytoplankton biomass between 20 and 70% to the total POC pool. Furthermore, elevated background levels (1817 ± 300 μg l^{-1}) of particulate organic detritus in surf zone waters may contribute between 20 and 95% to the particulate organic carbon pool (Garcia, unpubl.), and therefore comprise an important food source for detritivorous organisms and the microbial community.

The intense mucus secretion by *Asterionellopsis glacialis* cells (Odebrecht et al. 1995a) releases considerable amounts of dissolved organic matter into surf zone waters during patch formation. In addition, excretion products liberated by the abundant benthic intertidal fauna (Gianuca 1983, 1985) as well as decomposition processes are likely to increase dissolved organic carbon concentrations further. These compounds are either directly available to microbial organisms through the microbial loop (Brown and McLachlan 1990) or may aggregate via bubble formation to make particulate organic matter (Ogura 1972). As observed for other dissipative or intermediate beaches (Talbot and Bate 1988; Brown and McLachlan 1990), phytoplankton appears to be the principal source of organic matter in the surf zone along the southern Brazilian coast, although a constant supply of detrital matter may become important, especially in the absence of diatom patches.

Trophic Relations

Macrobenthic filter, suspension, and detritus feeders are the most important primary consumers of the beach and surf zone and are the major link between the abundant phytoplankton and organic detrital matter and higher level consumers (Gianuca 1985), though zooplanktonic organisms and some fish species are also important primary consumers.

Polychaetes and the filter-feeding molluscs *Amiantis purpuratus*, *Tivela ventricosa*, and *Mactra isabelleana* constitute the major macrobenthic biomass in the external surf zone, followed by detritus-feeding organisms, which ingest deposited diatoms and detrital matter. Among the 126 inner surf zone species (Borzone and Gianuca 1990), dense populations of filter and suspension feeders, like *Donax gemmula* (1809 ind. m^{-2}), *Magelona riojai* (7874 ind. m^{-2}), and *Phoxocephalopsis zimmeri* (2050 ind. m^{-2}), use phytoplankton and detrital particles as an important food source. Filter feeders, like *Mesodesma mactroides*, *Donax hanleyanus*, and *Emerita brasiliensis*, which total 95% of the secondary production in the lower intertidal zone, feed intensely on the surf diatom *Asterionellopsis glacialis* (Gianuca 1983, 1985, 1987). The abundant (91 581 ind. m^{-2}) suspension-feeding intertidal polychaete *Spio gaucha* can survive under culture conditions on a diet composed exclusively of *Asterionellopsis glacialis* (P. Santos, unpubl.). Even beyond the intertidal zone, large depositions of *Asterionellopsis glacialis* and detrital matter during storms in supratidal or overwash areas between foredunes serve as food for dense (2350 ind. m^{-2}) populations of beetles (*Bledius bonariensis, Bledius microcephalus, Bledius fernandezi*) and the ghost crab *Ocypode quadrata*, which occasionally complements its diet with diatoms (Gianuca 1985). The high density of supratidal beetles attracts several predators, both insects and birds, like the collared plover *Charadrius collaris* and the pipit *Anthus correndera*.

The zooplankton community of the surf zone in southern Brazil is dominated by mysids and copepods, although meroplankton, composed of gastropod, bivalve, cirriped, decapod, and echinoderm larvae, is also present (Bersano 1994). Particularly under the influence of longshore currents with increased hydrodynamic activity, mysids frequently exceed 60% relative abundance. *Metamysidopsis elongata atlantica* is the most abundant species and occurs in 95% of the samples. Copepods are less abundant but more diverse. The omnivorous *Acartia tonsa* and *Eucalanus pileatus* are the most common species (Bersano 1994). Owing to its high biomass of approximately 320 mg C m^{-3} (max. 8100 mg C m^{-3}), the zooplankton in the surf zone is likely to exert a strong grazing pressure on the phytoplankton community.

The fish assemblage of the surf zone, mainly represented by *Trachinotus marginatus, Mugil platanus, Mugil curema, Menticirrhus littoralis, Oncopterus darwini, Odontesthes bonariensis, Atherinella brasiliensis, Lycengraulis grossidens, Brevoortia pectinata*, and *Micropogonias furnieri*, differs from that of adjacent deeper (>5 m) coastal waters. Juveniles of many of these species utilize the surf zone as a nursery and feeding area (Cunha 1981). Juveniles of *Mugil platanus, Mugil curema*, and *Mugil gaimardianus* ingest cyanobacteria, dinoflagellates, and small crustaceans, though their principal diet is comprised of diatoms. The common surf zone diatoms *Asterionellopsis glacialis* and *Skeletonema costatum* may represent 90% of the stomach content of *Mugil platanus* and *Mugil gaimardianus*, respectively (Vieira 1985). *Odontesthes* sp. and *Atherinella brasiliensis* change their diet preference from diatoms to zooplankton during development (Bemvenuti 1990).

The extreme density (up to 430 000 ind. m^{-1}) of macrobenthic inner surf zone invertebrates (Soares 1992) supplies abundant food resources for a variety of secondary consumers. The diet of juvenile fishes of the species *Menticirrhus littoralis, Oncopterus darwini, Netuma barba*, and *Trachinotus marginatus* is almost entirely composed of benthic invertebrates (Gianuca 1985; Monteiro-Neto and Cunha 1990). Large individuals of *Pogonias cromis* feed on the bivalves *Amiantis purpuratus, Tivela ventricosa*, and *Mactra isabelleana* (Gianuca 1985). The surf zone gastropods *Buccinanops duartei, Olivancillaria auricularia, Olivancillaria uretai*, and the swimming crab *Arenaeus cribarius* prey on *Mesodesma mactroides, Donax hanleyanus*, and *Emerita brasiliensis* in the swash zone (Gianuca 1983, 1985). Macrobenthic invertebrates are also ingested by larger predators, like rays, skates, and small sharks.

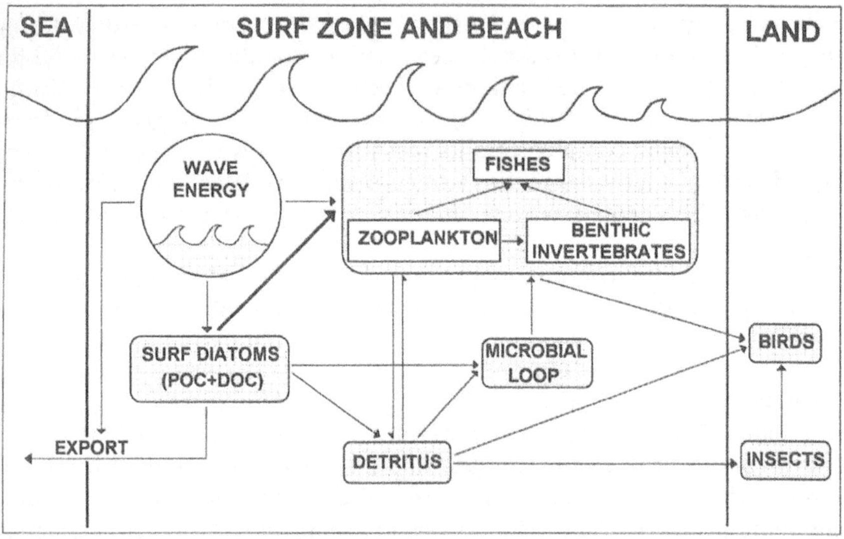

Fig. 7.2.1. Schematic diagram of trophic relationships in the beach and surf zone ecosystem of the warm-temperate southwestern Atlantic

The abundance of juvenile fish in shallow waters attracts sub-adults and adults of different ichthyophagus fish as well as gulls, terns, cormorants, and herons, which are top predators in the surf zone (Vooren and Chiaradia 1990). As the tide retreats, shore birds (plovers, oystercatchers, sandpipers) become the main predators of intertidal invertebrates. During the summer, 90% of the stomach content of *Calidris fuscicollis, Calidris canutus, Calidris alba, Pluvialis dominica, Larus dominicanus*, and *Larus maculipennis* is composed of the filter feeders *Mesodesma mactroides, Donax hanleyanus*, and *Emerita brasiliensis* (Gianuca 1983, 1985). The elevated biomass of benthic invertebrates constitutes an important food supply not only for resident bird species, but also for many other temporary visitors during their migrations north and southwards.

In conclusion, the environmental interactions and trophic relations between different community components suggest that the beach and surf zone in southern Brazil functions as a semi-closed and largely self-sustained ecosystem (Fig. 7.2.1), as has been proposed by McLachlan (1980) for exposed South African beaches.

7.3
The Continental Shelf and Slope

J.P. Castello, M. Haimovici, C. Odebrecht, and C.M. Vooren

The influence of the Subtropical Convergence (STC), with marked seasonal latitudinal displacement, characterizes the southern Brazilian continental shelf and slope regions (29^0-34^0 S) as a biogeographic transition zone (Sharp 1988) between the large neritic areas of Patagonia and tropical Brazil. The composition and abundance of species, the pelagic structure, the spatial distribution of communities and their trophic interactions, as well as biological production are largely controlled by the seasonal dominance of distinct water masses over shelf and slope. Coastal and shelf break upwelling of deep Subtropical Water (STW) (South Atlantic central water) is frequent in the spring/summer and winter/spring, respectively (Lima and Castello 1995). The influence of Subantarctic and Tropical Water is greatest in the winter and summer, respectively, though waters of subantarctic origin may also rise during the summer along the southernmost shelf break regions (Fig. 7.3.1). Freshwater runoff from the La Plata River and the Patos Lagoon become important in the winter and spring. Particularly between Rio Grande and Chuí, oceanographic conditions favour high biological production, which results in a considerable, still almost unexploited biomass of small pelagic fishes (Castello and Habiaga 1982; Lima and Castello 1995) and highly productive, but now overexploited, demersal fishery resources (Haimovici et al. 1989a; IBAMA 1993). During most of the year, the thermal front on the western side of the STC provides a suitable habitat for pelagic sharks, tunafish, and thelike. Together, pulsed sub-surface coastal upwelling, onshore Ekman transport, and large-scale influence of continental runoff seem to sustain favourable conditions for fish spawning grounds and larval survival.

Pelagic Production

Owing to the influence of nutrient-rich Subantartic Water (SAW) and freshwater runoff from the Patos Lagoon and La Plata River, chlorophyll *a* and primary production rates are highest in the late winter and spring. Although outflowing freshwater generally has low nitrate concentrations, owing mainly to in- or nearshore phytoplankton uptake (Ciotti et al. 1995; Abreu et al. 1995a), nearshore bottom turbulences add nutrients to the euphotic zone (Odebrecht and Djurfeldt 1996) indicating that recycled nutrients may efficiently return to the water column. The influence of SAW on biological production occurs especially over the southern central shelf, where horizontal and vertical fronts are related to areas of the highest integrated chlorophyll *a* concentrations (>100 mg m^{-2}; Ciotti et al. 1995). In the summer, the southward displacement of the oligotrophic Brazil Current leads to the lowest chlorophyll *a* and primary production rates. During this period, sporadically high chlorophyll *a* values and/or primary production rates depend on the upwelling intensity of nutrient-rich STW and the mixing regime, with high chlorophyll *a* concentrations occurring at sub-surface (Odebrecht and Djurfeldt 1996) or surface waters (Hubold 1980a). Interannual differences between phytoplankton chlorophyll *a* concentrations in shelf waters are strongly determined by continental runoff, which seems to be a function of the El Niño-Southern Oscillation cycle (ENSO).The highest chlorophyll *a* concentrations off southern Brazil coincide with a large freshwater outflow after periods of strong El Niño events in the Pacific Ocean (Ciotti et al. 1995). In addition, changes in the intensity of the Malvinas Current, probably as a consequence of climatic disturbances related to the ENSO phenomenum, may affect the variability of primary production from year to year.

A first evaluation of primary production data suggests moderate to high mean annual primary production rates (160 g C m^{-2} y^{-1}; Odebrecht and Garcia, Sect. 6.7) which probably sustain relative high secondary production (Cushing 1988). The high winter/spring biomass of zooplanktophagous (i.e. *Engraulis anchoita*), benthophagous, and predatory organisms, many of which, owing to the northward shift of the STC and the presence of SAW, migrate during this period from the south, supports this notion. Among the over 40 species of the pelagic necton community of the continental shelf, the fishes *Engraulis anchoita*, juvenile *Cynoscion guatucupa*, *Trichiurus lepturus*, and the squid *Loligo sampaulensis* are the most characteristic (Mello et al. 1992). Biomass estimates (in 1000 tons) for *Engraulis anchoita* range from 35 to 1928 in the summer and winter (Lima and Castello 1995), for *Trichiurus lepturus* from 3 to 30 in the winter and spring (Haimovici et al. 1996), and for *Loligo sanpaulensis* from 1.2 to 3.5 in the fall and spring (Andrigueto and Haimovici 1991). Over 50 species of cartilaginous fishes (mostly benthic feeding sharks, angel sharks, rays, and skates) and over 150 species of bony fishes (including 11 species of sciaenids) amount to over 80% of the total biomass, making up the demersal fish community of the shelf. Biomass estimates based on bottom-trawl surveys range from 43 000 to 96 000 tons for cartilaginous fishes (Vooren, Sect. 6.16) and from 137 000 to 340 000 tons for bony fishes (Haimovici et al. 1996) in the fall and winter/spring, respectively.

Fig. 7.3.1. Distribution of surface water masses over the shelf and vertical profiles at different latitudes duringthe winter (**A**) and summer (**B**). Coastal water (*CW*); South Atlantic Central Water (*SACW*); Tropical Water (*TW*); Subantartic Water (*SAW*); Antartic Intermediate Water (*AIW*); mixed Coastal Water with either SAW or TW (*MCW*). Mean ranges of chlorophyll *a* concentrations are given for some water masses

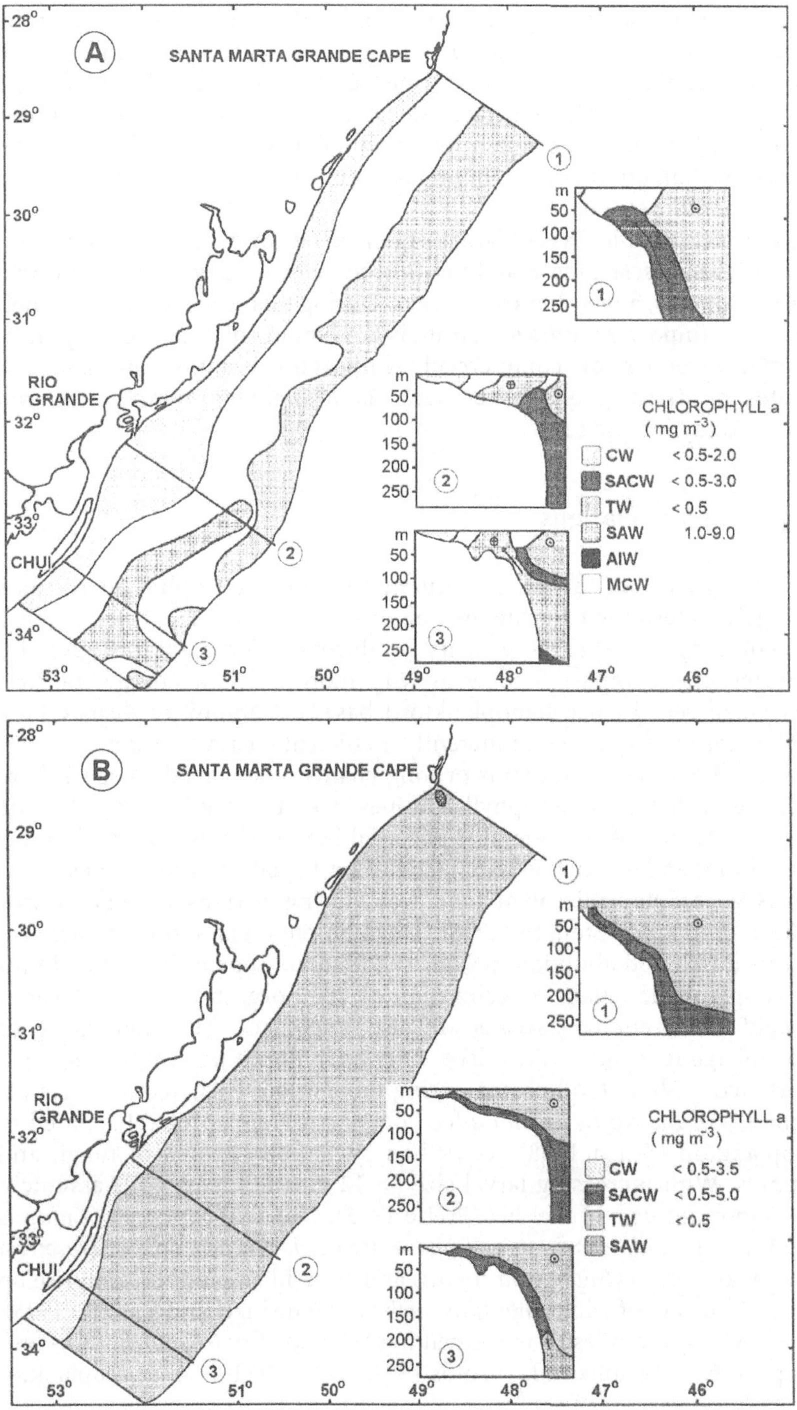
A
SANTA MARTA GRANDE CAPE
RIO GRANDE
CHUI
28°
29°
30°
31°
32°
33°
34°
53°
51°
50°
49°
48°
47°
46°
m
50
100
150
200
250
CHLOROPHYLL a
(mg m^{-3})
CW < 0.5-2.0
SACW < 0.5-3.0
TW < 0.5
SAW 1.0-9.0
AIW
MCW
B
SANTA MARTA GRANDE CAPE
RIO GRANDE
CHUI
CHLOROPHYLL a
(mg m^{-3})
CW < 0.5-3.5
SACW < 0.5-5.0
TW < 0.5
SAW

Fig. 7.3.1.

Biomass maxima of bony fish are associated with cold (<12°C) SAW of the coastal branch of the Malvinas Current over the central shelf and/or with the oceanographic front (>17°C) formed at the western side of the STC (Fig. 7.3.1). In general, demersal teleost diversity and species richness decrease with depth. The much lower abundance of bony fish over the upper slope than on the shelf (Haimovoci et al. 1994a) appears to be related to a poor benthic fauna.

In summer, the biomass of larger organisms is lower and is mostly composed of juvenile sciaenids, mainly *Trichiurus lepturus*, *Cynoscion guatucupa*, *Umbrina canosai*, angel sharks, and rays. During this season, the inner shelf areas form important nursery grounds for young of the year *Engraulis anchoita*, juveniles of commercially important bony fishes, and the neonates of cartilaginous fishes such as *Rhinobatos horkelii*, *Sphyrna lewini*, and myliobatid rays.

Pelagic Trophic Relations

The size structure of the pelagic community and the trophic interactions are largely determined by the size of primary producer species (Ryther 1969; Pomeroy 1974). Large primary producers (microplankton) give rise to shorter, less complex food webs with more efficient energy transfer, whilst small-celled (pico-nanoplankton) based communities display long and complex food webs with inherently inefficient energy transfer.

Large, single, centric diatoms or long chains of small-celled species are associated with high chlorophyll *a* values over the shelf during the late winter and spring. A considerable bacterial biomass, abundance of protozooplankton, and intense grazing activity by flagellates and ciliates point towards a complex microbial food web during periods of high phytoplankton biomass (Abreu, Sect. 6.6). The high phytoplankton biomass also supports a rich and abundant pelagic shelf fauna. Copepods are the dominant zooplankters (max. concentration of 2000 org. m^{-3}), but cladoceran and cold water euphausiaceans are also important. The dominance of planktophagous pelagic fish, like *Engraulis anchoita*, emphasizes the importance of short food chains during this period. Early larval stages (4-12 mm SL) of *Engraulis anchoita* feed on starch grains probably related to phytoplankton (Bursa 1968), as well as on nauplii, spores of fungi, and tintinnids. With increasing larval size (> 12 mm SL), copepods assume a more important part of the diet (Freire 1995). Calanoid copepods (*Calanus australis*, *Calanoides carinatus*, *Temora stylifera*, *Oncaea* sp., *Microsetella* sp.), euphausiacean (*Euphausia similis*) and hyperid amphipods comprise up to 90% of the diet of adult *Engraulis anchoita* (Schwingel and Castello 1995). Paralarvae and juveniles of the Argentinean squid *Illex argentinus* also feed on copepods and euphausids (Santos 1992; Vidal 1994b). Other zooplanktophagous species include the squid *Loligo sanpaulensis*, *Anchoa marinii*, *Trachurus lathami*, and juveniles of *Cynoscion guatucupa* and *Trichiurus lepturus* (Haimovici et al. 1996). Large individuals of *Trichiurus lepturus*, adult *Cynoscion guatucupa* (Vieira 1990; Haimovici et al. 1996), *Pagrus pagrus* (Capítoli and Haimovici 1993), *Galeorhinus galeus*, and the skates *Sympterigia acuta* and *Sympterigia bonapartei* prey on small fishes like *Engraulis anchoita*, *Paralonchurus brasiliensis*, and *Symphurus jenynsi* (Queiroz 1986). Large icthyophagous pelagic predators include the bluefish *Pomatomus saltatrix* (Haimovici and Krug 1992), angel sharks of the genus *Squatina*, and *Raja castelnaui*, which feeds both on pelagic and demersal fishes and squid. The feeding activity of other large predators, like the porpoise *Pontoporia blainvillei* and the sea lion *Otaria flavescens*, is largely restricted to coastal waters (Pinedo 1982). Over the outer shelf, pelagic zooplanktophagous species are represented by *Maurolicus muelleri*, myctophids, gonostomatids, juvenile *Merluccius hubbsi*, *Brama* spp., *Loligo sanpaulensis*, *Ornithoteuthis antillarum*, and *Illex argentinus*, which also preys on small pelagic and mesopelagic fishes (Santos 1992). Large ichthyophagous benthic fishes include *Galeorhinus galeus*, *Genypterus brasiliensis*, and *Polyprion americanus* which feed on young *Merluccius hubbsi* and *Illex argentinus* but also ingest macrocrustaceans. Large ichthyophagous pelagic predators, like the migratory *Xiphias gladius*, prey on *Illex argentinus* and other ommastrephids (Mello 1992), which also play an important role in the diet of *Thunnus obesus* and *Thunnus alalunga* (Santos and Haimovici, unpubl.), whilst *Thunnus albacares* preys on *Brama* sp., squids, and hyperid amphipods (Vaske 1992; Fig. 7.3.2).

In contrast, principally small primary producer organisms (picoplankton) compose the lower phytoplankton biomass during the summer (Odebrecht, unpubl.) when phytoplankton and bacterial carbon sources tend to be of similar magnitude. The pelagic community is more diverse than during the colder months and gives rise to complex feeding interactions, which is characteristic of oceanic oligotrophic areas where most zooplankton and all necton depend on secondary food sources (Cushing 1989; Legendre and Le Fèvre 1995). Filter feeders, like the Cladocera species *Penilia avirostris* and Thaliaceae, are abundant and consume bacteria and picoplankton. The large biovolume of Thaliaceae is common in subtropical seas because intensive grazing of large populations on small particles leads to fast growth. Planktonic molluscs (pteropods) feed on small particles, phytoplankters, protozoans, and zooplankters which adhere to their external mucus produced in the mantle cavity or in the foot (Parsons et al. 1984). The Euphasiaceae present diverse feeding mechanisms, like filter-feeding or "food basket" feeding. Practically all coelenterates are carnivores and prey on plankters, small fishes, and/or components of the necton. Zooplanktophagous fishes, like *Peprilus paru*, young *Trichiurus lepturus*, *Balistes capriscus*, *Cynoscion jamaicensis*, and the benthophagous

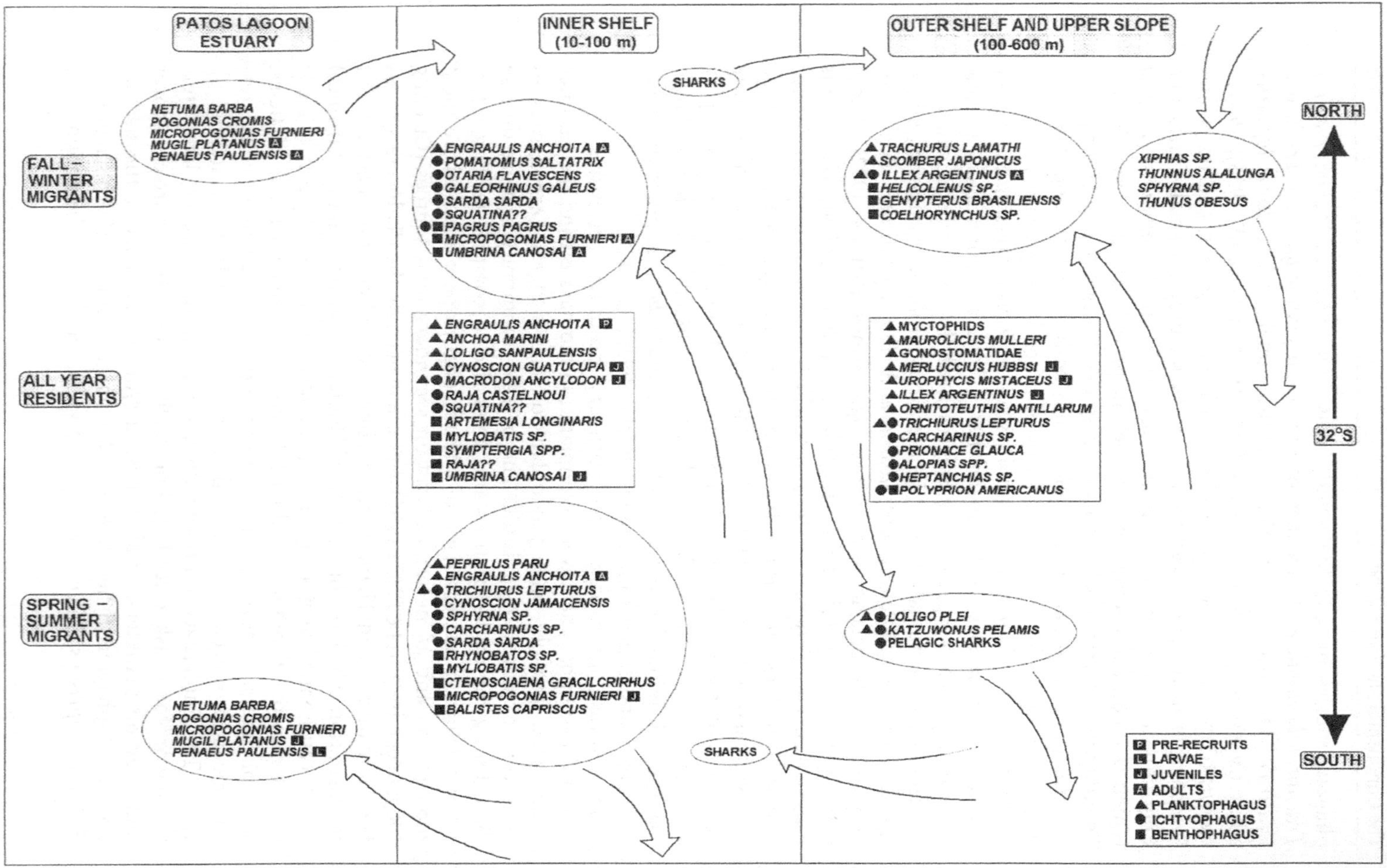
PATOS LAGOON ESTUARY
INNER SHELF (10-100 m)
OUTER SHELF AND UPPER SLOPE (100-600 m)
FALL – WINTER MIGRANTS
ALL YEAR RESIDENTS
SPRING – SUMMER MIGRANTS
NETUMA BARBA
POGONIAS CROMIS
MICROPOGONIAS FURNIERI
MUGIL PLATANUS A
PENAEUS PAULENSIS A
SHARKS
ENGRAULIS ANCHOITA A
POMATOMUS SALTATRIX
OTARIA FLAVESCENS
GALEORHINUS GALEUS
SARDA SARDA
SQUATINA??
PAGRUS PAGRUS
MICROPOGONIAS FURNIERI A
UMBRINA CANOSAI A
TRACHURUS LAMATHI
SCOMBER JAPONICUS
ILLEX ARGENTINUS A
HELICOLENUS SP.
GENYPTERUS BRASILIENSIS
COELHORYNCHUS SP.
XIPHIAS SP.
THUNNUS ALALUNGA
SPHYRNA SP.
THUNUS OBESUS
NORTH
32°S
SOUTH
ENGRAULIS ANCHOITA P
ANCHOA MARINI
LOLIGO SANPAULENSIS
CYNOSCION GUATUCUPA J
MACRODON ANCYLODON J
RAJA CASTELNOUI
SQUATINA??
ARTEMESIA LONGINARIS
MYLIOBATIS SP.
SYMPTERIGIA SPP.
RAJA??
UMBRINA CANOSAI J
MYCTOPHIDS
MAUROLICUS MULLERI
GONOSTOMATIDAE
MERLUCCIUS HUBBSI J
UROPHYCIS MISTACEUS J
ILLEX ARGENTINUS J
ORNITOTEUTHIS ANTILLARUM
TRICHIURUS LEPTURUS
CARCHARINUS SP.
PRIONACE GLAUCA
ALOPIAS SPP.
HEPTANCHIAS SP.
POLYPRION AMERICANUS
PEPRILUS PARU
ENGRAULIS ANCHOITA A
TRICHIURUS LEPTURUS
CYNOSCION JAMAICENSIS
SPHYRNA SP.
CARCHARINUS SP.
SARDA SARDA
RHYNOBATOS SP.
MYLIOBATIS SP.
CTENOSCIAENA GRACILCRIRHUS
MICROPOGONIAS FURNIERI J
BALISTES CAPRISCUS
LOLIGO PLEI
KATZUWONUS PELAMIS
PELAGIC SHARKS
NETUMA BARBA
POGONIAS CROMIS
MICROPOGONIAS FURNIERI
MUGIL PLATANUS J
PENAEUS PAULENSIS L
SHARKS
P PRE-RECRUITS
L LARVAE
J JUVENILES
A ADULTS
PLANKTOPHAGUS
ICHTYOPHAGUS
BENTHOPHAGUS

Fig. 7.3.2.

Ctenosciaenea gracilirrhus arrive from the north, whilst other ichthyophagous and/or benthic feeding species, including *Pomatomus saltatrix*, adult *Umbrina canosai*, *Cynoscion guatucupa*, *Micropogonias furnieri*, *Pagrus pagrus*, and *Mustelus schmitti* migrate southward. Owing to the southward migration of *Galeorhinus*, necton abundance decreases in slope regions, though dense schools of the skipjack tuna (*Katsuwonus pelamis*), which feed heavily on macro-zooplankton like *Euphausia similis*, the mesopelagic fish *Maurolicus muelleri*, and squids (Vilela 1990), arrive from the north (Fig. 7.3.2; Vilela and Castello 1991). Between November and May, large flocks of sea birds, like *Puffinus gravis*, *Sterna hirundo*, *Procellaria aequinoctialis conspicillata*, and *Diomedea chlororhynchos* (Chiaradia 1991), are associated with schools of *Katsuwonus pelamis* over the outer shelf. The rejected small bony fish from bottom-trawl fishing activities (Haimovici and Palacios 1981; Haimovici and Mendonça 1996) also provide an important food source for sea birds over the inner shelf (Vooren and Fernandes 1989).

Pelagic-Benthic Interactions

Pelagic-benthic interactions are either associated with the life-cycle of species with both planktonic and benthic stages or are related to their trophic behaviour. Trophic interactions may be bi-directional since either benthic organisms may benefit from sedimentation of particles from the water column or demersal and nectonic organisms may feed on the benthic fauna. Also, diel migrations favour an increase of pelagic-benthic interactions. For example, *Illex argentinus* feeds during the night at the surface but rests during the day near the bottom, where it is preyed upon by *Polyprion americanus* (Santos 1992) and probably by *Galeorhinus galeus*. Benthic food chains appear to be especially important over southern shelf regions, which coincides with intense bottom-trawl fishing (Haimovici et al. 1989a). Here, the coastal branch of the Malvinas Current supports high primary productivity, and areas with sandy and muddy bottoms favour an abundant invertebrate fauna of polychaetes. Among the benthophagous species, juvenile and adult *Umbrina canosai* (Haimovici et al. 1989b), *Micropogonias furnieri*, *Myliobatis* spp., *Sympterigia* spp., *Netuma* spp., *Mustelus schmitti*, *Raja* spp., and *Urophycis brasiliensis* are dominant over the shelf. Depending on the size of predators, the shrimps *Artemesia longinaris* and *Pleoticus muelleri*, the crabs *Portunis spinicarpus* and *Callinectes* sp., the bivalve mollusc *Solen tehuelches*, and a variety of polychaetes,

Fig. 7.3.2. Main feeding habits and seasonal movements of principal necton species of inner and outer shelf and upper slope regions isopods, amphipods, and cumaceans are the principal prey in shallow coastal waters (Queiroz 1986). Benthic feeding fish species of the outer shelf include *Urophycis mystaceus*, *Coelorhincus marinii*, and *Helicolenus lahillei*. Although many species feed exclusively on the epi- and infaunal organisms, some important species, including adults of *Cynoscion guatucupa* (Viera 1990) and *Pagrus pagrus* (Capítoli and Haimovici 1993), prey on both benthic and pelagic organisms.

8 Estuarine-Marine Interactions

P.C. Abreu and J.P. Castello

The Patos Lagoon estuary and the adjacent marine system are abiotically and biologically linked. In general, abiotic estuarine-coast interactions are exemplified by eutrophication of coastal waters, owing to nutrient and organic matter runoff which locally enhances primary and secondary production (Mann and Lazier 1991). Biological interactions are represented by the transport of neritic invertebrate and fish species which take advantage of the sheltered conditions and food availability in estuaries as nursery and feeding grounds (Knox 1986).

Physical Interactions

The topography of the southern Brazilian continental shelf controls the transport of oceanic water masses, but the dynamics of coastal waters as well as those of the Patos Lagoon follow short- and long-term meteorological cycles. The frequent passage of polar fronts augments the velocity of coastal currents and directs large amounts of water onshore (Fig. 8.1a). These conditions not only lead to high concentrations of phytoplankton cells (especially *Asterionellopsis glacialis*) in the surf zone which provide an important food source for the intertidal benthic fauna (Odebrecht et al. 1995a), they occasionally also cause accumulations of toxic dinoflagellates which result in the death of benthic organisms and fish (Odebrecht et al. 1995b). Since the tidal amplitudes are small, the exchange of water between the Patos Lagoon and coastal regions is also influenced by the passage of polar fronts (Costa et al. 1988a; Möller et al. 1991). SE winds force saltwater into the lagoon, which may penetrate up to 200 km upstream during drought periods (Niencheski et al. 1988). The input of coastal water normally causes sediment resuspension and consequently a release of inorganic nutrients like phosphate (Kantin and Baumgarten 1982; Fig. 8.1b). In general, intrusion of coastal water increases phytoplankton diversity in the estuary. If this process persists for a prolonged period, the residence time of estuarine water increases, and owing to reduced export of phytoplankton cells, the accumulation of phytoplankton biomass in the estuary is considerable. A subsequent change to northerly winds favours the discharge of estuarine waters, and large amounts of particulate organic carbon are exported as phytoplankton cells (Abreu 1987). Since much of the exported phytoplankton biomass is

U. Seeliger et al.
Subtropical Convergence Environments

composed of species like *Skeletonema costatum*, which enter the estuary during saltwater intrusion, part of the eutrophication process of coastal waters appears to depend on the addition of extra nitrate to the water column and on the prior seeding of the estuary with euryhaline neritic cells during saltwater input (Abreu et al. 1995a; Fig. 8.1c).

An increase in precipitation in southern Brazil is a consequence of El Niño-Southern Oscillation (ENSO) events, and thus long-term phytoplankton variability of the region might be related to this large-scale climatic phenomenon. Chlorophyll *a* variability in the Patos Lagoon estuary, with maximum values during years with high precipitation, shows a relationship between ENSO and the eutrophication process of coastal regions (Abreu et al., unpubl.). The freshwater flow from the Patos Lagoon, and especially from the La Plata River, is the main source of silicate and ammonia to the

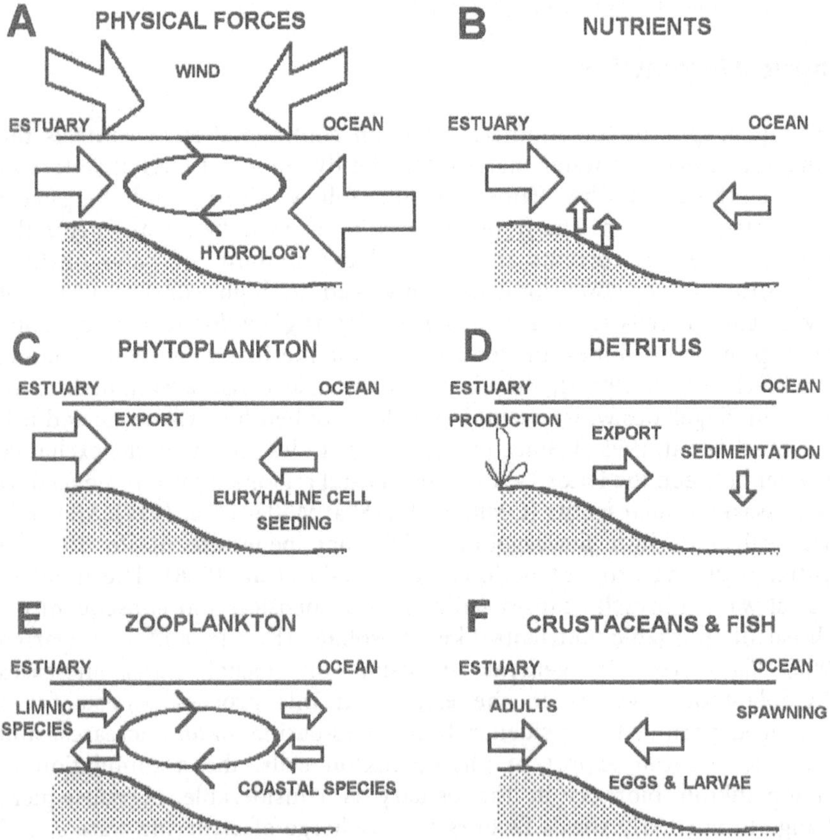

Fig. 8.1. Diagrammatic representation of the main estuarine-marine interactions

coastal euphotic zone, and thus high rainfall in the watershed should cause an increase of nutrients and primary production. A direct relation between ENSO events and a rise of chlorophyll *a* levels over the southern Brazilian continental shelf has been observed by Ciotti et al. (1995). Additionally, the large freshwater outflow increases the stability of the water column in coastal shelf regions, which causes higher spring chlorophyll *a* concentrations (Ciotti et al. 1995) than in years with reduced outflow. Similarly, water column stabilization favours the occurrence of large patches of anchovy *Engraulis anchoita* larvae, owing to probably both higher food availability and physical mechanisms which inhibit the dispersal of eggs and larvae, thus increasing their survival chance (Lima and Castello 1995).

During years of extreme precipitation, outflowing estuarine waters form a plume which may extend 30-50 km along the coast (Hartmann et al. 1980), and high levels of primary and secondary production at the confluence with coastal water masses are likely to influence the biological dynamics of the coastal region decisively (Abreu et al. 1995b). Together with natural and dredge removal of sediments, large amounts of plant detritus originated from marshes, seagrasses, and algae and animal debris are exported from the estuary (Castello 1985). These materials may form dense acoustic scattering layers (>100 mg l^{-1} suspended matter) at the bottom of the estuary inlet (Calliari 1980) and in the mid-water of the shelf. The sedimentation of detrital particles and of phytoplankton cells (Ciotti et al. 1995) and the presence of a rich benthic inner shelf community (Borzone and Griep 1991; Absalão 1991) suggest the significance of deposition processes in the coastal region (Fig. 8.1d).

Transport of Organisms

The meteorological processes which control water exchange between the Patos Lagoon and shelf regions are also responsible for the transport of zooplankton and fish and crustacean larvae into the estuary. Especially during the summer, the estuarine zooplankton community is dominated by neritic species which enter with salt water intrusion, whilst limnic species prevail during years with high precipitation (Montú 1980; Rieger and D'Incao 1991; Fig. 8.1e). Although fish eggs and larvae occur during the entire year in the estuary, peaks are associated with the spring and summer. Probably owing to elevated temperatures in the late spring and early summer, the spawning activities of most fish species are transferred to nearshore waters, which coincides with the period of maximum salt water intrusion into the estuary. The combined effect of these processes warrants the transport of eggs and larvae into the estuary, where food availability and sheltered conditions ensure growth (Muelbert and Weiss 1991; Ibagy and Sinque 1995; Fig. 8.1f).

The pink shrimp (*Peaneus paulensis)* is among the most prominent species which enter the estuary. The larvae of northern stocks (29° S) are transported by coastal currents to shelf regions near the Patos Lagoon inlet. The species is truly estuarine-dependent and migrates with varying success into the estuary in the spring and summer, trying to overcome the freshwater outflow. A negative relationship between the winter/fall precipitation and shrimp production during the following summer clearly indicates the impact freshwater runoff has on larval penetration (Castello and Möller 1978). After migration into the estuary, larvae settle in shallow embayments where they undergo metamorphosis and develop into adults (Fig. 8.1f).

Fish migration is perhaps one of the most common and best known manifestations of estuarine-coastal interactions, where migrations might be seen as an evolutionary solution to an optimization of population abundance. The Patos Lagoon estuary provides nursery, protection, and feeding grounds for several migratory fish species, like the marine catfish (*Netuma barba)* and the white croaker (*Micropogonias furnieri).* After spawning in coastal water during the late spring and summer, eggs and larvae of *Micropogonias furnieri* are transported into estuarine shoals where they are retained, owing to reduced water circulation and the presence of seagrass beds (Fig. 8.1f). However, *Micropogonias furnieri* is not an estuarine-dependent species as considered by Chao et al. (1985), since development and growth also occur in more haline adjacent coastal waters (Vieira and Castello, Sect. 4.13). Migration of fish in and out of the estuary exemplifies the import or export of biomass, and local fisheries benefit from this process. A clear understanding of this process by scientists, resource managers, and fishermen may often be limited by the unreliable quantifications of the biomasses involved, though issues of immediate financial profit may also adversely influence the decision-making policy. These conditions have largely inhibited the development of an integrated management of estuarine and coastal *Micropogonias furnieri* stocks which, together with *Netuma barba* stocks, eventually collapsed, leading to severe damage in the estuarine fisheries (Reis 1992).

In conclusion, the eutrophication processes produced by the land and sea, and the utilization of estuarine nursery and feeding grounds by organisms reflect two-way interactions which tightly couple the estuary to adjacent coastal waters and which significantly contribute to the region's considerable biological production.

9 Fisheries

M. Haimovici, J.P. Castello, and C.M. Vooren

For more than a century the Patos Lagoon estuary (von Ihering 1896) and the coastal sea have exemplified the fishing potential in the warm-temperate southwestern Atlantic (Yesaki 1973). Today about 6000 artisanal (Reis 1992) and 3000 industrial fishermen are temporarily or permanently involved in fishing activities. Despite the economic and social importance fisheries have historically assumed in this region, reliable records on fishing methods and landing statistics are lacking before 1945. Until then, artisanal trammel, gill, and channel net fishing with small wooden sail and row boats was largely restricted to the Patos Lagoon and estuary. The event of modern means of storage and transport and the introduction of a large number of synthetic fibre nets and motor-powered boats (< 10 m, 10-24 HP), which permitted artisanal trawl fishing, eventually led to the depletion of estuarine stocks. During the 1980s, artisanal fisheries quickly extended into shallow coastal waters where larger (12-15 m, 90-120 HP) wooden boats with up to 20 tons capacity used gill-nets, and occasionally purse seines and hooks. In the estuary and coastal waters artisanal fisheries follow a clearly defined seasonal pattern (Reis et al. 1994). The black drum (*Pogonias cromis)* and white croacker (*Micropogonias furnieri),* which actively feed in brackish waters, and the catfish (*Netuma barba),* which uses the estuary for reproduction, are fished in the spring, the pink shrimp (*Penaeus paulensis)* is caught in the summer and fall, the mullet (*Mugil platanus)* in the fall, and the bluefish (*Pomatomus saltatrix)*, castanha (*Umbrina canosai)*, and weakfish (*Cynoscion guatucupa)* in the winter. The mean annual artisanal fishery catch totals 21 500 t (Table 9.1; 58% demersal fishes, 15% pelagic fishes, 14% estuarine shrimp and crab, 12% freshwater fish) but, owing to often incomplete landing statistics, catch figures may be underestimated by as much as 25%. Industrial fisheries began in 1947 for demersal species (Yesaki and Bager 1975), and landings were mostly comprised of sciaenids, the red porgy *Pagrus pagrus* (Haimovici et al. 1989), and to a lesser extent elasmobranchs (Vooren et al. 1990). Pelagic fishery commenced in 1962 (Yesaki and Bager 1975) and intensified after 1977 when foreign vessels were hired and operated out of Rio Grande port (Zavala-Camin and Antero da Silva 1991; Antero da Silva 1994).

U. Seeliger et al.
Subtropical Convergence Environments

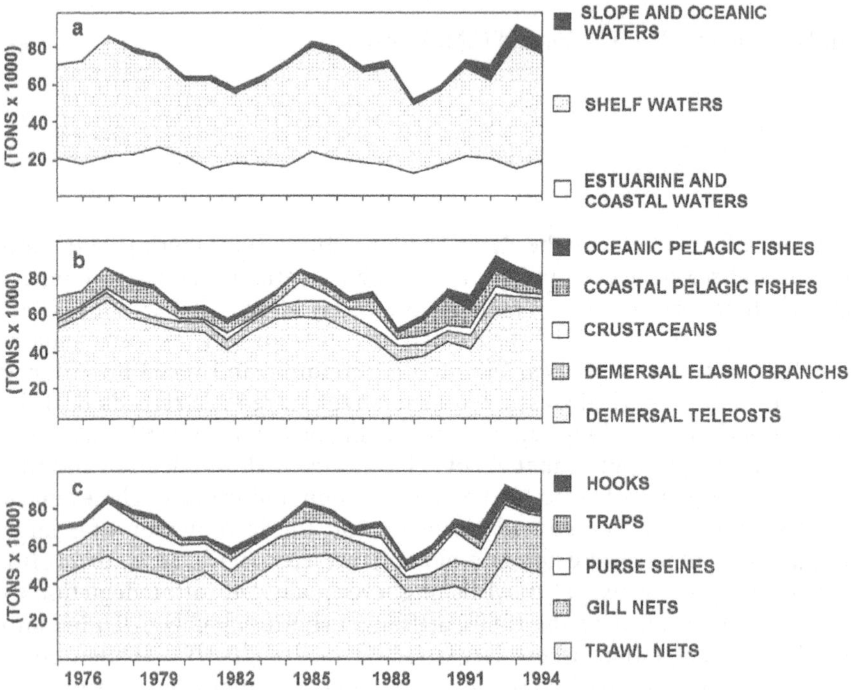

Fig. 9.1. Landed catches of estuarine and marine fisheries in the southwestern Atlantic between 29° and 35° S based on depth range (**a**), type of fishery resource (**b**), and fishing methods (**c**)

Seasonal changes in the composition and abundance of the necton (Vooren et al. 1990; Haimovici et al. 1994a; Lima and Castello 1995; Haimovici et al. 1996), and consequently also of fisheries, are induced by the alternate influence of cold subantarctic water and warm tropical water over the shelf. Shelf fisheries are more intense in the winter when a variety of commercially important species (castanha, weakfish, red porgy, bluefish, and the school shark *Galeorhinus galeus* immigrate from the south. Offshore fishing occurs all year roung, in the winter with long lines for southward migrating tunas and sailfishes and in the summer with live bait fishery for skipjack tuna (*Katsuwonus pelamis*). Most of the industrial fishery catches occur over the continental shelf (71%) and only 4% in oceanic and slope regions (Fig. 9.1a). About 80% of catches is landed in the fishing port of Rio Grande. Between 1975 and 1994 average annual catches approximated 74 000 metric tons but oscillated between a minima of 52 200 t in 1989 and a maxima of 91 800 t in 1993 (Haimovici et al. 1989; Vooren et al. 1990; IBAMA 1993, 1995). Demersal teleosts (71%) dominate among the catches, followed by pelagic shelf teleosts (10%), demersal elasmobranchs (8%), crustaceans (5.3%), and pelagic oceanic fish (4%; Fig. 9.1b). Trawl-net (61%) and gill-net (19%) methods account for most of the catch, whilst purse seines (10%), traps (4.5%), and hooks (4%) are less important (Fig. 9.1c).

Industrial Fishing Methods

In the early 1980s, when some stocks (*Pogonias cromis*, *Netuma barba*, *Pagrus pagrus*) collapsed and others (the royal weakfish *Macrodon ancylodon*, *Umbrina canosai*) showed signs of overfishing (Haimovici 1988a, b; Haimovici et al. 1989; Reis et al. 1994), demersal fishing diversified into double-rigg trawling, bottom gill-nets, bottom long-lines, and trap methods (Barcellos et al. 1991; Lima and Branco 1991) to exploit new resources.

Pair- and otter-trawling methods were introduced in 1947, but the trawler fleet first grew rapidly during the 1970s as a result of fiscal incentives (SUDEPE 1974; Yesaki and Bager 1975). Both methods use double sheet codends of 50-70 mm mesh size between opposite knots, which drastically reduces the chance of small fish escaping. Fishing vessels, which may employ either method depending on the season, are 22-35 m long and are powered by 250-650 HP engines. Since they lack refrigeration, the catch is stocked for up to 15 days in holds between layers of crushed ice (Haimovici et al. 1989). Although pair- and otter-trawling methods have changed little over the past few decades, fishing has become more efficient, owing to satellite navigation and more efficient use of echo-sounders. Pair-trawling, followed by otter-trawling, are the principal fishing methods for demersal sciaenids, like *Umbrina canosai*, *Cynoscion guatucupa*, *Macrodon ancylodon*, and *Micropogonias funieri*, which represent about 80% of total mean annual trawl landings. Otter-trawling over the inner shelf (20-100 m) landed 20 287 t in 1976, but this amount gradually decreased to 2708 t in the early 1990s. However, over the outer shelf (80-200 m) a few larger trawlers (49-52 m) landed an average of 4632 t in 1993 and 1994. Outer shelf otter-trawl catches include sharks and rays (31%) and sciaenids (20%), which are already intensely exploited, and the cutlassfish *Trichiurus lepturus* (24%), which is discarded by the inner shelf trawl fishery (Table 9.1).

Double-rigg trawling was first used in 1985 from 20-24 m long wooden vessels with 250-350 HP engines (Barcellos et al. 1991). Twin nets were employed for both shrimp fishing (*Artemesia longinaris*, *Pleoticus muelleri*) and demersal teleosts and elasmobranchs, though the length of the ground rope and the mesh size of the codend differed. Shrimps are fished in shallow waters in the spring and summer, whilst *Paralichthys patagonicus* is fished at 20-80 m depth between the fall and spring and *Squatina* spp. over the outer shelf at up to 140 m depth. Double-rigg trawlers landed about 30% of the total elasmobranch catches in 1989-1990, but their contribution decreased to 5% in 1992 (IBAMA 1995). Annual landings of double-rigg trawling totalled 3728 t after 1989 (IBAMA 1995; Haimovici and Mendonça, unpubl.).

Since trawl fishery is a non-selective method, small specimens of commercially valuable species and species without value are generally discarded from the catch. The estimated total annual discard ranges from 17 000 to 25 000 t. Onboard rejection by pair and otter-trawling of small demersal fishes and elasmobranchs (Haimovici and Palacios 1981; Haimovici and Habiaga 1982) may vary between 26 and 46%, though a larger (90 mm between opposite knots) mesh size would remedy this situation (Vooren 1983). Onboard discard of small elasmobranchs and teleosts during double-rigg trawling for fish exceeds 50% of the total catch. Discard during shrimp fishings may reach 20% (Haimovici and Mendonça 1996) which is significantly lower than the rejection rate reported for subtropical and tropical regions (Alverson et al. 1994).

Purse seine fishery for pelagic fishes has occurred in shallow shelf waters of less than 50 m depth since 1962. Seines (600-800 m length, 70-80 m height) are operated by vessels of 20-24 m length and 250-450 HP (Krug and Haimovici 1991). Mullets are caught in the fall and bluefish is the principal catch in the winter, though sometimes the jack mackerel (*Trachurus lathamis)*, mackerel (*Scomber japonicus)*, menhaden (*Brevoortia pectinata)*, and occasionally also demersal fish like white croaker and castanha represent an important portion of the catch which averages (1990-1994) 4261 t annually (Table 9.1).

Trap fishing methods at depths greater than 400 m were employed for the red crab (*Chaceon notialis)* by hired Japanese vessels in 1984/1985, and the catch totalled 1471 t over a 9-month period after which fishing was abandoned (Lima and Branco 1991). Trawl fishing for *Pagrus pagrus* dates back to 1973 (Yesaki and Barcellos 1974), and overfishing made exploitation uneconomical at the beginning of the 1980s (Haimovici et al. 1989). Between 1987 and 1992 *Pagrus pagrus* was caught by trap fishing methods, similar to those used in Argentina (Barcellos et al. 1991), over beachrock outcrops (70-80 m) and gravel bottoms (110-160 m). The average annual catch was 106 t and was composed 83% of *Pagrus pagrus* (Table 9.1).

Hand-line fishery, aimed at the wreckfish (*Polyprion americaus)*, was developed over hard bottoms of the upper slope in 1973 and was gradually replaced by vertical long-lines, bottom gill-nets, and bottom long-lines in recent years. Landing statistics of the commercially valuable wreckfish are incomplete but amount to several hundred tons annually (Table 9.1). Individual fishing grounds are abandoned after intense fishing for short periods and recover only after several years.

Bottom gill-nets replaced bottom long-lines, which were used for demersal shark fishing through 1989 (Barcellos et al. 1991). The gill-nets are pulled by 18-30 m long trawlers with 185 to 350 HP engines, and they may exceed several kilometers in length. Gill-netting is used at all depths (>50 m) of the continental shelf and has quickly become the main fishing method (IBAMA 1995). Between 1990 and 1994 mean annual catches equalled 5355 t

(Table 9.1). The increase of the total elasmobranch landings after 1990 clearly reflects the growth of the bottom gill-net fishery.

Japanese fishing vessels initiated surface long-line fishery for tunas, principally *Thunnus alalunga*, *Thunnus obesus*, and *Thunnus albacares*, and the swordfish *Xiphias gladius* over the southern shelf in 1961, followed by Brazilian long-liners in 1969 (Zavala-Camin and Tomas 1990; Zavala-Camin and Antero da Silva 1991), which dominated this fishery after 1977. Even so, a number (2-11) of large foreign vessels continue to fish for tunas and swordfish and land their catch in Rio Grande (Mello et al. 1993). During the 1977-1991 period, the average yield for this fishery was 728 kg/thousand hooks (Antero da Silva 1994). Pole and live bait fishery of the skipjack tuna (*Katsuwonus pelamis)* was introduced in 1979 (Castello and Habiaga 1988) and, owing to high initial yields, rapidly attracted a growing number (up to 97 in 1982) of 24-28 m long vessels (Matsuura 1982b). Four larger foreign vessels were hired in 1991. Almost 93% of the catch is composed of *Katsuwonus pelamis*, and annual landings average 3240 t (Table 9.1).

Table 9.1. Mean annual landed catches (tons) by fishing method in the southwestern Atlantic between 29° and 34° S from 1975 to 1994

Fishing method		1975-1979	1980-1984	1985-1989	1990-1994
Artisanal fishery		24926	20548	20617	20091
Pair-trawl		32056	33951	32051	29137
Otter-trawl (inner shelf)		15237	9160	10055	2708
Otter-trawl (outer shelf)	1993-1994	-	-	-	4362
Double-rigg trawl		-	-	-	3216
Bottom gill-net	Since 1989	-	-	435	5355
Fish trap	1988-1992	-	-	-	106
Hand line and vertical long-line		231	82	321	168
Purse seine		7105	1949	1833	4990
Offshore pelagic long-line	Since 1977	1295	1943	2260	3504
Pole and live bait fishery	Since 1991	-	-	-	3240

Demersal Teleost Fisheries

Demersal teleost catches (1975-1994) amount to 71% of all landings, and 70% are comprised of the sciaenids *Micropogonias furnieri*, *Umbrina canosai*, *Cynoscion guatucupa*, and *Macrodon ancylodon*. Annual landings do oscillate, but catch per unit effort (CPUE) for the first three species has decreased steadily (Fig. 9.2).

Regional *Micropogonias furnieri* stocks, represented by a spawning group near the Patos Lagoon inlet (32° S) and another further south (34° S; Haimovici and Umpierre 1996), have been heavily exploited by artisanal and industrial gill-net fisheries for more than two decades (Reis 1992), which has resulted in a drastic decrease in the density of juveniles in coastal waters between 1980 and 1990 (Ruffino and Castello 1992a). Although annual catches between 8000 and 17 000 t (mean 14 071 t) appear to be stable, the declining CPUE (Fig. 9.2) and changes in the age structure (Schwingel and Castello 1990; Haimovici and Umpierre 1996) indicate that the spawning stocks are overfished.

Umbrina canosai, a small species of low natural mortality (Haimovici and Reis 1984; Haimovici 1988b), is heavily fished along the southern Brazilian coast and to a lesser extent in Uruguay and Argentina (Fig. 9.2). This species represented the largest (19 000 t) part of landings during the 1970s, but gradually decreased to 10 000 t in recent years. Similarly, the biomass diminished from 75 000 t in 1976 to about 37 000 in 1983 (Haimovici 1988b). However, in contrast to *Micropogonias furnieri*, the age structure of *Umbrina canosai* has not changed in recent years. Therefore, in spite of overfishing, actual landings should persist if exploitation levels do not rise.

During the colder months adults of a single *Cynoscion guatucupa* stock migrate from fishing grounds in Uruguay and Argentina to more northern Brazilian shelf regions. Landings in southern Brazil, which represent about 30% of the total catch of the three countries, oscillate between 4000 t (1975 and 1988) and 14 000 t (1986 and 1994). These variations reflect either interannual differences in recruitment or availability to Brazilian trawlers, because the CPUE displays similar oscillations (Fig. 9.2).

A single stock of *Macrodon ancylodon* was found between 29° and 34° S (Yamaguti 1979) and has been exploited since the 1950s, with a maximum catch of 10 617 t in 1977. Landings gradually decreased to 3000 t in 1992 but showed an increase over the last 2 years (Fig. 9.2). Although *Macrodon ancylodon* may reach a maximum age of 12 years (Yamaguti 1967), since 1977 age classes above 6 years have been absent (Martins Juras 1980; Haimovici 1988a), which suggests that sustainable annual yields are below 4000 t. The slow growing and long-living *Pagrus pagrus* was overexploited by trawl fishing during the 1970s (Yesaki and Barcellos 1974; Haimovici et al. 1989), and stocks collapsed after a few years without showing signs of recuperation (Table 9.2). Landings of the flatfishes *Paralichthys* spp., the sea robin (*Prionotus punctatus*), and *Urophycis brasiliensis* averaged almost 3500 t annually over the last 5 years (Table 9.2). Landings of flatfishes alone reached a maximum of 2157 t in 1990 but decreased more recently. Since *Prionotus punctatus* and *Urophycis brasiliensis* are not abundant over the southern shelf (Haimovici et al. 1996), catches of these species probably cannot be sustained.

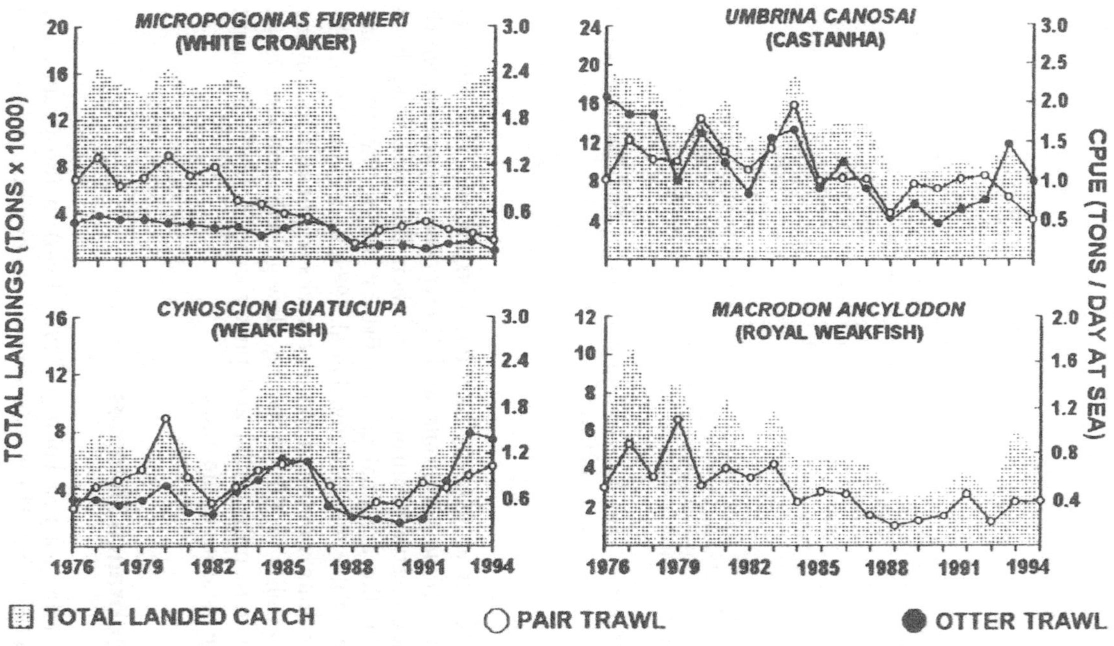

Fig. 9.2. Total landed catches and catch per unit effort (*CPUE*) of otter- and pair-trawls of the most abundant demersal teleosts between 29° and 35° S

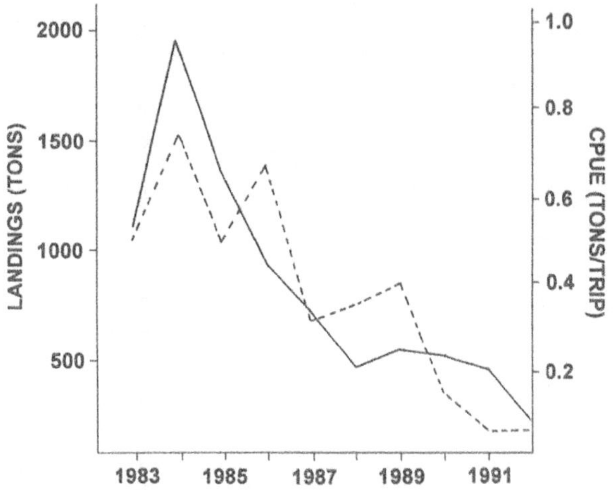

Fig. 9.3. Landings of guitar fish *Rhinobatos horkelii* at the Port of Rio Grande (*solid line*) and mean annual CPUE of the otter-trawl fleet (*dashed line*)

The stocks of demersal teleosts, like *Pogonias cromis*, the catfishes *Netuma barba* and *Netuma planifrons*, and *Micropogonias furnieri*, which use the Patos Lagoon estuary during part of their life-cycle, have been heavily exploited by the artisanal gill-net fishery during the 1970s and collapsed after 1980 (Table 9.2; Haimovici et al. 1989; Reis et al. 1994). The recuperation of *Pogonias cromis* and *Netuma* stocks will be difficult, owing to their high vulnerability to gill-net fishing, slow growth, and late sexual maturation. As a result, only mullet and pink shrimp actually represent economically viable fisheries in the estuary.

Demersal Elasmobranch Fisheries

About 10% of the total sea fish landings (1975-1994) in Rio Grande are comprised of elasmobranchs. Until 1988 almost 80% of all catches were by bottom-trawl fishing (Vooren et al. 1990) but in recent years most catches are being made by bottom gill-nets. Landing statistics are merely subdivided into sharks, angel sharks, guitar fish, and rays (Table 9.2).

Table 9.2. Mean annual landings (1975-1994) of the main fishery resources fished in the southwestern Atlantic between 29° and 34° S

		Mean annual landings (tons)			
		1975-1979	*1980-1984*	*1985-1989*	*1990-1994*
Estuary and shelf pelagic species					
White croaker	*Micropogonias furnieri*	14308	14904	12364	14709
Weakfesh	*Cynoscion guatucupa*	6439	7377	9572	8785
Royal weakfish	*Macrodon ancylodon*	7941	5865	3659	3966
Castanha	*Umbrina canosai*	16900	14877	11732	9629
Catfish	*Netuma spp.*	3983	1536	452	615
Black drum	*Pogonias cromis*	1044	359	246	63
Hake	*Merluccius hubbsi*	760	101	200	129
Flatfish	*Paralychthis spp.*	424	417	1486	1363
Red porgy	*Pagurs pagrus*	1419	327	294	238
Searobin	*Prionotus punctatus*	12	71	486	988
Abrótea	*Urophycis brasiliensis*	241	337	665	1186
Cutlassfish	*Trichiurus lepturus*	75	79	63	441
Demersal sharks[a]		1637	2486	3754	3748
Angel shark	*Squatina spp.*	947	1485	2607	2183
Guitar fish	*Rinobathos horketi*	1010	1253	901	460
Rays and skates	*various species*	116	461	718	746
Pink shrimp	*Peneus paulensis*	4143	1830	4354	2627
Marine shrimps[b]		-	-	1331	1148
Estuary and shelf demersal species					
Menhaden	*Brevoortia pectinata*	1549	343	646	1324
Bluefish	*Pomatomus saltatrix*	4290	2290	2456	3521
Mullet	*Mugil platanus*	2081	1709	1486	1524
Mackerel	*Scomber japonicus*	-	-	-	969
Jack mackerel	*Trachurus lathami*	-	-	-	1555
Demersal upper slope species					
Wreckfish	*Polyprion americanus*	152	68	145	176
Red crab	*Chaceon notialis*	-	-	-	-
Offshore pelagic species					
Bigeye tuna	*Thunus obesus*	312	419	640	500
Albacore	*Thunus alalunga*	269	446	494	1075
Yellowfin tuna	*Thunus albcarens*	212	336	325	684
Swordfish	*Xiphias gladius*	122	285	350	601
Skipjack tuna	*Katsuwonus pelamis*	-	-	-	2402
Pelagic sharks[c]		182	481	382	547

[a] Mostly *Galeorhinus galeus* and *Mustelus schmitti.*
[b] *Pleoticus muelleri* and *Artemesia longinaris.*
[c] *Isurus oxyrinchus, Prionace glauca, Sphrena* spp. among others.

The division "sharks" consists mostly of the school shark (*Galeorhinus galeus)*, with the narrownose smooth-hound (*Mustelus schmitti)* in second place. The latter species was discarded until about 1985 (Haimovici and Palacios 1981), but was subsequently landed in increasing quantities. Both species migrate from the Uruguayan and Argentinian shelf to Brazilian waters in the winter. Annual landings in Rio Grande rose gradually from 1414 t (1975) to 2482 t (1984), then increased sharply to about 3500 t (1985-1987), followed by a drop to 1247 t (1989). The decrease after 1987 reflects the reduced abundance of *Galeorhinus galeus* and possibly also of *Mustelus schmitti*. Owing to increased fishing efforts by bottom gill-netting, total shark landings increased to 2791 t in 1992. The otter-trawl CPUE showed a similar pattern, rising to a peak of about 5 t/trip in 1989, then dropping to 1.5 t/trip in 1992. The rise in CPUE until 1989 reflects increased targeting of shark and a decrease in discarding.

Catches of angel sharks consist mainly of *Squatina guggenheim*, which inhabits the inner shelf at depths down to 80 m, and *Squatina occulta* which is abundant on the outer shelf and the slope down to 350 m. *Squatina guggenheim* is a resident species, but the migratory status of *Squatina occulta* is not clear. Annual landings of angel shark increased from 820 t (1973) to a peak of 2442 t (1988) and then dropped sharply to 1031 t (1990). Similarly, otter-trawl CPUE of angel shark reached a maximum of 6.2 t/trip in 1988, followed by a decline to 0.7 t/trip in 1992, which reflects a major decrease of the abundance of both species of angel shark, the resident *Squatina guggenheim* probably being the most affected. However, while abundance decreased, landings increased to 1761 t in 1992, owing to increased fishing efforts by bottom gill-netting.

Adult females of the guitar fish (*Rhinobatos horkelii)* concentrate between December and February in shallow inshore waters to breed. During this period they are particularly vulnerable to intensive pair-trawl and beach seine fisheries. During the remainder of the year, adults of both sexes disperse over the inner shelf where catching is done by otter-trawling. A continuous decline of otter-trawl CPUE from 0.76 t in 1984 to 0.08 t in 1992 provides a measure of decreasing abundance (Fig. 9.3). Similarly, annual landings of this species decreased from a peak of 1927 t in 1984 to an all-time low of 178 t in 1992.

Pelagic Teleost Fisheries

The fishery for pelagic teleosts, like *Pomatomus saltatrix*, the cutlassfish (*Trichiurus lepturus)*, the anchovy (*Engraulis anchoita)*, and *Trachurus lathami*, is little developed and represents only about 10% of total landings.

Pomatomus saltatrix is a fast growing species which migrates from Argentina to southern Brazilian shelf waters. The species concentrates in surface waters warmed above 15^0C temperature (Krug and Haimovici 1989;

Haimovici and Krug 1992). In the winter mainly age classes between 2 and 5 years are fished (Krug and Haimovici 1991). Mean annual landings total 3528 t (1970-1992), with an exceptional peak of 12 126 t in 1971. Since climatic conditions regulate the abundance and migration cycles of *Pomatomus saltatrix*, the fishing potential of this species is highly unpredictable (Haimovici and Krug 1996). *Trichiurus lepturus* is fairly abundant, and the spring biomass may exceed 30 000 t (Haimovici et al. 1996). Since the commercial value of this species drastically declines when kept between ice layers, rather than being deep frozen onboard as is done by some hired Korean trawlers, the stock is underexploited. The small-sized *Engraulis anchoita* is the main pelagic species in southern Brazilian shelf waters. Since the species has no market value, it is unexploited, though the winter biomass may attain over one million tons. *Trachurus lathami* and *Scomber japonicus* may be caught by purse seines between 30 and 80 m depth in the winter and therefore may represent a potential resource for midwater trawling. The irregular occurrence of *Scomber japonicus* is probably related to the intensity of the subantartic water influence.

Large pelagic fishes are caught along the thermal front of the western margin of the Subtropical Convergence and landed in Rio Grande (Mello et al. 1993; Antero da Silva 1994). Landings (1977-1989) contain the yellowfin tuna (*Thunnus albacares;* 23%), the bigeye tuna (*Thunnus obesus;* 21.5%), the albacore tuna (*Thunnus alalunga;* 18%), the swordfish (*Xiphias gladius;* 16%), various pelagic sharks (i.e. *Prionace glauca, Sphyrna zygaena, Sphyrna lewini;* 18.5%), and billfishes (i.e. *Tetrapturus albidus, Makaira nigricans;* 3%; Mello et al. 1993). Although stock assessments are unreliable and catches fluctuate, the tuna fishery shows potential for expansion. The skipjack tuna (*Katsuwonus pelamis)* is caught during the summer and fall with pole and live-bait in waters between 80 and 1000 m depth and surface temperatures between 20 and 26°C (Castello and Habiaga 1988). Since the exploitation rates of skipjack tuna are moderate (Jablonski and Matsuura 1985; Vilela and Castello 1993), the catch could be increased. The availability of live bait appears to be the main limiting factor for the expansion of this fishery (Table 9.2).

Invertebrate Fisheries

Crustacean catches, mostly shrimp, blue crab, and red crab, amount to approximately 6% of the total landings (1975-1994). Subadults of the pink shrimp (*Penaeus paulensis)* migrate from northern coastal waters into the nursery grounds of the Patos Lagoon estuary and coastal lagoons in Uruguay, where the species is caught in the summer and fall (Valentin et al. 1991). Annual artisanal shrimp fishery landings change considerably as a consequence of variable recruitment conditions (Castello and Möller 1977; D'Incao 1991) and growth. Despite predictions that estuarine shrimp fisheries might collapse, owing to excessive catching of juveniles (Valentin et al. 1991), average annual landings between 1988 and 1994 were close to the 20-year annual mean of 3106 t. This leads to the conclusion that environmental variables rather than fishing pressure are the determinants for shrimp fishery catches. Juveniles and adults of the shrimps *Pleoticus muelleri* and *Artemesia longinaris* occur in coastal waters (Harán et al. 1992), and exploitation of *Artemesia longinaris* stocks appeared to have reached saturation in 1989 (Ruffino and Castello 1992b). However, the variability of catches during subsequent years coincided with a reduced catch of largely migratory demersal species, and thus oscillation in the abundance of *Artemesia longinaris* may depend on interannual variations in oceanographic conditions (Table 9.2).

The crab fishery is hardly developed. Small-scale trawl-net fishery for the blue crab *Callinectes sapidus* occurs regularly in the Patos Lagoon estuary. In 1984 the deep water red crab *Chaceon notialis* was experimentally fished over the continental slope (34° S; Lima and Branco 1991), but catches per trap-day decreased sharply in less than 6 months. Since this species has a wide low-density distribution, a biomass estimate (Delury method) of 2183 t of the unexploited stock in the fishing area (3000 km^2) may be conservative. The squids *Loligo sanpaulensis* and *Illex argentinus* frequently occur over the shelf between 30° and 34° S (Haimovici and Perez 1991a). Despite spring biomass estimates of 3500 t for *Loligo sanpaulensis* (Andriguetto and Haimovici 1991), landings of cephalopods are virtually non-existent.

Fisheries Interactions

Fishing activities interact with both sea birds and marine mammals. Between November and May flocks of several types of sea birds (*Puffinus gravis, Sterna hirundo, Procellaria aequinoctialis conspicillata, Diomedea chlororhynchos*) are reliable indicators of schools of the skipjack tuna *(Katsuwonus pelamis)* over the mid and outer shelf (Chiaradia 1991). A variety of sea birds (*Procellaria a. aequinoctialis, Fulmarus glacialoides, Diomedea melanophoris*) die on the hooks of long-line tuna fisheries along the western margin of the Subtropical Convergence, and the annual death toll may exceed 2650 birds (Vaske 1991). Fishery interactions may be the main cause for the decline of *Diomedea exulans* populations in the South Atlantic Ocean (Croxall and Prince 1990). Discard from trawl fisheries is composed of a large number of either small, floating teleosts or fast-sinking, small sharks and skates, and represents an important additional food source for sea birds like subantartic albatrosses and petrels (Vooren and Fernandes 1989) and the sea lion *Otaria flavescens*. Dead stranded specimens of *Otaria flavescens* are commonly found on southern Brazilian beaches. A large number have broken skulls or bullet holes inflicted by fishermen, while others may have died from entrapment in nets (Rosas et al. 1994). Gill-nets are the principal cause of small cetacean mortality in the southwestern Atlantic (Pinedo 1994a). The species most affected is the small dolphin *Pontoporias blainvillei*, but the bottle-nose dolphin (*Tursiops truncatus*) is also occasionally entangled and killed by gill-nets near the Patos Lagoon inlet (Pinedo 1994b).

Fishery Perspectives

Total annual catches in the southwestern Atlantic (38°-34°S), including a conservative estimate of 25% discard, range from 70 000 to 120 000 t, inter-annual differences being partly due to environmental factors. Changes of landing statistics with time are principally a result of the establishment of new fisheries (i.e. bottom gill-net fisheries for elasmobranchs and double-rigg trawling for flatfishes and shrimp) as the older stocks became overexploited. Today, most commercially important estuarine, coastal, and shelf stocks are either fully or overexploited, and landings are expected to decrease in coming years. Overfished stocks of demersal finfishes include *Netuma* spp., *Pogonias cromis*, and *Pagrus pagrus*. The abundance of *Rhinobathos horkelii*, *Galeorhinus galeus*, and other shark species have declined up to 90% in recent years, probably due to recruitment overfishing since these species are low-fecundity viviparous fishes. The assessment of the status of stocks of less abundant shark species and of many other species, which are simply grouped as rays and skates in the landing statistics, is at present impossible. Furthermore, about two-thirds of recent demersal catches have concentrated on migratory stocks (i.e. *Micropogonias furnieri*, *Umbrina canosai*, *Cynoscion guatucupa*, *Mustelus schmitti*) which are shared by Uruguayan and Argentinian fisheries. Estuarine and marine shrimp, menhaden, bluefish, and mullet catches are highly variable. These stocks do not clearly confirm overfishing but appear to lack potential for expansion. In addition, the growth potential of under- or unexploited stocks seems to be limited. Only *Engraulis anchoita* and *Trachurus lathami* stocks might have some fishery potential, and catches of offshore tunids may eventually increase.

Demersal teleost and elasmobranch stocks are affected as a whole, and thus their abundance and diversity are at risk under intensive and uncontrolled fishing. For example, if present trends continue, the populations of *Rhinobatos horkelii* and *Squatina guggenheim* will become extinct. The same may well be true for an unknown number of other species, whose decline remains unnoticed because they are not listed in the landing statistics. Although proposals to reduce estuarine and all demersal fishing efforts and to decrease onboard discard (Haimovici and Palacios 1981; Vooren 1983; Haimovici et al. 1989; IBAMA 1993, 1995) have provided an adequate legal basis, neither fishing efforts nor discard of undersized fish has diminished. Unquestionably, future fishery management strategies must not only involve both the fishermen and the fishing industry in creating effective legislation, but also encourage international management of shared stocks.

10 Natural and Human Impact

U. Seeliger and C.S.B. Costa

Early accounts of man's interference with the natural resources of coastal regions in the warm-temperate southwestern Atlantic date from the last century (von Ihering 1885). In particular, the Patos Lagoon and estuary have been subject to multiple human impacts during the past few decades. In the near future though, a gradual degradation of the ozone layer and increasing sea levels are likely to pose major ecological problems for the entire region.

Freshwater Flow

The water quality of the Patos Lagoon estuary and coastal nearshore environments is ultimately related to natural freshwater runoff. Because of the assumption that runoff is available for expanding agricultural and industrial activities and domestic needs without any limitation, the importance of freshwater flow into the estuary is often overlooked. The annual estimated streamflow through the Patos Lagoon inlet averaged 109 km^3 between 1965 and 1975 (Castello and Möller 1978) but varied significantly within (CV=80%) and between (CV=29%) years. During the drought years 1968-1969 (IPAGRO 1977), the water demand for irrigated rice cultivation (EPAGRI 1992) and for the population of the watershed approximated 6% of the expected (75 km^3) annual runoff. Since then, both the population and the area of rice cultivation have increased by more than 37% and 120%, respectively (IBGE 1993), and thus today as much as 13% of natural runoff may be diverted during drought periods (Fig. 10.1). Furthermore, the construction of locks (1977) in the São Gonçalo channel to prevent salinization of Mirim Lagoon waters used for irrigation has considerably reduced freshwater flow into the estuary. The growing encroachment on freshwater discharge into the estuary modifies seasonal variations in flow rates that may be essential for flushing and maintaining the balance of salinity and nutrients in the estuary. Many of the commercially important marine fish and crustaceans in the southwestern Atlantic have similar life-cycles and visit the estuary periodically to benefit from low salinity waters, physical protection, and food sources for either growth or reproduction. Therefore, any further intentions to control or divert freshwater flow must await the results of studies to establish minimum flow requirements for the estuary.

U. Seeliger et al.
Subtropical Convergence Environments

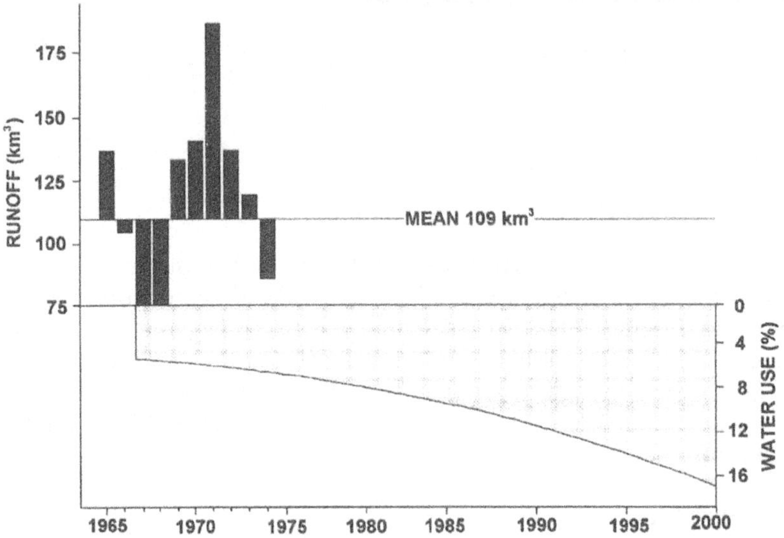

Fig. 10.1. Annual total freshwater runoff between 1965 and 1975 and estimated water use during drought years

Nutrient Enrichment and Eutrophication

Freshwater runoff not only influences estuarine salinity regimes but also affects the quality of surface waters through inputs of nutrients and toxic materials. Most urban centres around the Patos Lagoon (approx. 3.5 million inhabitants) either lack or have inadequate sewage treatment facilities which tend to overflow during heavy rains and discharge raw sewage directly into the lagoon or estuary. Nutrients are also added by the major tributaries from large-scale agricultural activities; phosphate (2-3 μM) and nitrogen (>40 μM) concentrations in particular tend to be seasonally elevated in the receiving waters of the Patos Lagoon (Vilas Boas 1990). In spite of a substantial reduction during transport, estuarine nutrient levels are also occasionally elevated (Kantin 1983; Niencheski and Windom 1994), owing to remobilization of bottom sediments in the lower estuary (Abreu et al. 1992), local discharge of industrial and domestic effluents, and agricultural runoff through the São Gonçalo channel. Excess nutrient loads in the estuary cause eutrophication with the development of blooms and changes in phytoplankton composition. Blooms of the potentially toxic bluegreen alga *Microcystis aeruginosa* (Yunes et al. 1994) form in limnic regions of the lagoon and, during years with high spring tributary discharge, are transported into the estuary (Odebrecht et al. 1987; Bergesch

1990). During the summer, estuarine embayments with prolonged water residence time favour blooms of other species (Persich 1993), which contributes to the decline of the submerged macrophyte biomass in the estuary, owing to attenuation of light penetration and epiphytic algae loading on *Ruppia* leaves (Cafruni 1983; Costa and Seeliger 1989).

Hazardous Materials

Since large extensions of coastline north and south of the inlet are devoid of industrial discharge, trace metals are added to the estuary either through freshwater runoff from the Patos Lagoon or through local discharge of effluents (Seeliger et al. 1988). Average metal concentrations in estuarine waters with high suspension loads correspond to natural background levels (Seeliger and Knak 1982a; Niencheski et al. 1994) reported for estuarine systems elsewhere (Windom 1975), though dissolved copper concentrations are occasionally slightly higher (Seeliger and Knak 1982b). Sporadic increases of suspended copper and lead concentrations in the estuary may reflect metal input from industrial effluents and mining activities to the upper limnic part of the Patos Lagoon (Baumgarten 1987; Vilas Boas 1990). In general, metal concentrations correlate significantly with suspended matter loads, and estuarine metal gradients are not well-defined (Seeliger and Cordazzo 1982) because physico-chemical processes add particulate metals through resuspension of estuarine bottom sediments by inflowing seawater, wind, and waves (Niencheski et al. 1994). The biological effects of trace metal pollution have not yet been documented for the Patos Lagoon estuary, probably because estuarine organisms tend to display elevated metal tolerance, and the concentrations are close to background levels. However, in shoals with reduced circulation, prolonged water residence time, and added domestic and industrial effluents even, low pollution levels appear to induce changes in the zooplankton community structure and anatomical anomalies (Montú and Gloeden 1982). Although data on the effect of agrotoxins on estuarine organisms are still lacking, the excessive application of pesticides over vast extensions of agricultural lands around the Patos Lagoon is likely to contribute to the estuary through runoff with elevated concentrations. Some intertidal estuarine species (i.e. *Chasmagnathus granulata*) which decisively influence energy transfer in submersed and emersed habitats (D'Incao et al. 1992) appear to be highly susceptible to pesticide concentrations in warmer waters during the summer (Monserrat and Bianchini 1995). Superficial coastal nearshore waters are espccially prone to pollution by different fractions of hydrocarbons discharged by vessels during the washing of tanks. These pollutants are likely to interfere with sea and shore birds and may occasionally cause high mortality among migrant penguins.

Sedimentation, Dredge, and Fill

The introduction of large amounts of suspended sediments from the Patos Lagoon and the presence of extensive shallow estuarine shoals with *Ruppia* beds and fringing marshes provide an ideal setting for natural deposition processes in the estuary. Over the last two centuries the water area of the lower estuary has decreased by approximately 11%, which emphasizes the depositional character of this environment. Natural erosion commonly occurs along the southern margins of the estuary and the inlet, with an estimated annual sediment removal of about 2 500 000 m^3 to coastal near-shore waters (Calliari 1980; Fig. 10.2.). Losses of lower estuarine water areas not only result from natural sedimentation but also from man-mediated processes. The expansion of Rio Grande port between 1909 and 1914 generated 8 776 000 m^3 of dredge material which served for the establishment of islands and was deposited along lower estuary margins. The construction of jetties in 1917 has probably intensified natural sedimentation at the northern side of the inlet. Furthermore, significant natural sediment deposition in the lower estuary has historically interfered with navigation, and dredging activities have concentrated in the navigation channel and the inlet since 1833. Modern vessels with increasing draft require deeper waterways, and today approximately 30 000 m^3 of channel and berth material needs to be dredged monthly from the main shipping channel. These human activities may have solved immediate problems but also resulted in often unexpected and long-term changes by modifying depth, circulation, and sediment deposition patterns (Calliari 1980). Since the accumulative effects of frequent dredging are comparable to natural geological processes (Nichols and Howard-Strobel 1991), the export of fine sediments to the adjacent coast is likely to have increased (Villwock and Martins 1972; Hartmann et al. 1980; Calliari and Abreu 1984; Borzone and Griep 1991; Calliari and Fachin 1993).

The filling of intertidal and shallow water flats in the lower estuary for port, residential, and more recently industrial development has decreased or destroyed vital seagrass and tidal marsh habitats. Filling along lower estuarine margins and over and around small islands has destroyed as much as 10% of the total salt marsh area. Seagrasses are especially susceptible to filling, and burial by dredged sediments results in their immediate eradication. Only if plants are lightly covered and the root-rhizome systems are not damaged may regrowth occur through the sediments. Since dredged sediments are readily resuspended, they represent a source of excess suspended matter for long periods of time. The gradually increasing water turbidity reduces seagrass production and depth distribution. Therefore, the extension of natural beds and their valuable functions as habitat and shelter decrease. Although the disposal of dredged material is clearly affecting estuarine resources, no disposal program has yet been implemented for the Patos Lagoon estuary.

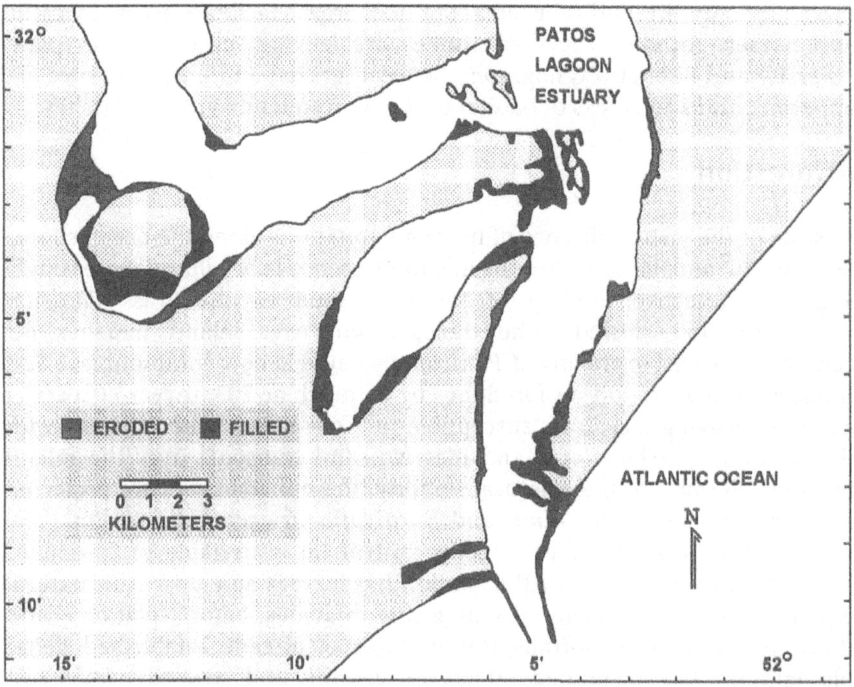

Fig. 10.2. Areas of the lower Patos Lagoon estuary which have suffered physical changes since 1776

Burning, Cutting, and Grazing

Several natural and man-induced events interact with salt marshes and their habitat function. Especially during drought periods, extensive upper salt marsh areas with large quantities of dead vegetative biomass become vulnerable to accidental burning. Although fire may cause high mortality of invertebrates and particularly of nesting birds, it opens the canopy of the dominant *Spartina densiflora* and *Juncus effusus* vegetation to invasion by secondary species and recycles large amounts of nutrients otherwise locked up in dead and live plant biomass. Rapid vegetative regrowth is stimulated by the nutrient enrichment, and the dominant species generally recover their aerial biomass densities within less than a year. Cutting of *Spartina alterniflora* and *Scirpus maritimus* to protect recently established summer vegetable crops against desiccation and to supply additional nutrients is a common practice, which represents only a small-scale disturbance (2-20 m^2). In contrast, a large-scale impact of grazing livestock on marginal marshes has existed since early colonization because the small tidal range, gentle marsh topography, and reduced net of tidal creeks facilitate access by cattle and horses. Although detailed studies are lacking, differences in plant diversity and habitat physiography between grazed and ungrazed sites are apparent, and a total loss of habitat occurs in frequently grazed marshes.

Afforestion

In spite of the virtual absence of human habitations along the ocean shores between 30^0 and 34^0 S, anthropogenic interference is becoming increasingly apparent. Effective binding of foredune sand by the grass *Panicum racemosum* has slowed dune transgression over inland marshes for centuries. Extensive grazing of *Panicum* by cattle is now commonplace and reduces sand accretion on foredunes by as much as 70 cm year^{-1} (Costa et al. 1991), leaving foredunes susceptible to erosion by onshore winds, with the formation of blow-outs and massive sand transport over the inland marshes. Because of their natural stability, these marshlands are preferred sites of afforestion with *Pinus* and *Eucalyptus* in excess of 100 km^2. The seasonally flooded marshes directly control the watertable of foredunes and associated vegetation. Afforestation has increasingly upset this natural equilibrium by artificially draining the marshes before planting and through the high evapotranspiration rates of already established dense plantations. The combined effects of grazing and hydrodynamic disturbances have reduced the sand stability of foredunes and increased transgression rates (Seeliger 1992a).

Sea Level Rise and Ozone Layer Depletion

Evidence suggests there is global warming due to the accumulation of atmospheric CO_2 and other gases (Schneider 1989) with a direct consequence of rising sea levels. Although the absence of conservative data does not allow us to trace secular trends of sea-level changes along coasts in the warm-temperate southwestern Atlantic (Schmack 1993), evidence of erosive processes along the Patos Lagoon margins and the coast (Villwock 1994) implies an estimated rate of sea-level rise of between 2 and 3 mm year^{-1}. If these rates are confirmed for the flat and wave-dominated coast (Costa et al. 1988a; Costa and Seeliger 1988a), storm overwash will more frequently impact areas further inland than in the past (Bernardi et al. 1987; Calliari and Klein 1993; Costa et al. 1995), coastal marshes and wetlands are likely to migrate further inland (Costa and Davy 1992), and salinization of the upper Patos Lagoon estuary will increase.

The gradual degradation of the ozone layer over Antartic regions is causing an increase in the transmission of ultraviolet B (UV-B) radiation and poses a threat to cold and warm-temperate latitudes in the southern hemisphere. Ozone-poor stratospheric air is spreading to northern

Argentina, Uruguay, and southern Brazil and has already led to a 13-20% decrease of ozone at 30^0 S during the spring (Kirchhoff et al. 1994; Santee et al. 1995). The potentially deleterious effect of increased UV-B exposure on marine plankton and benthic communities (Haeder and Worrest 1991; Karentz 1991) is likely to modify their distribution and abundance in the southwestern Atlantic with concomitant changes in productivity and trophic dynamics.

The coastal region of the warm-temperate southwestern Atlantic represents an enormous ecological and economic resource which has satisfied social and economic interests without management for more than 200 years. Because natural thresholds are being approached, vital habitats are being reduced or destroyed and fisheries will continue to decline if modern management strategies are not applied.

11 Management Considerations

M.L. Asmus and P.R.A. Tagliani

Among coastal regions of the southwestern Atlantic, the Patos Lagoon and surrounding lands, particularly the estuary and adjacent waters, stand out because of their ecological (zone of biological production and maintenance of biodiversity) and socio-economic significance (port, industrial, agricultural, fishing, tourist activities). Historically, socio-economic demands tend to collide with ecological preservation, and increasing anthropogenic activities are likely to jeopardize the health of the Patos Lagoon, estuary, and coast (Seeliger and Costa, Sect. 10). Therefore, the need to understand, plan, and manage these systems has become an issue of priority. During the 1980s a National Plan for Coastal Management was implemented by the Brazilian government, aiming to subsidize macrozonation, planning, management, and monitoring efforts. Thematic mapping (1:100 000) was expected to offer recommendations for the use of the coastal zone and the inner continental shelf, considering legal restraints, social needs, and the heterogeneity and potential of the region.

Initial management efforts were descriptive, but interpretative, simulative, and predictive phases soon followed (Asmus et al. 1991; Assis 1994). The coastal plain, with Patos Lagoon and estuary, reflects a considerable environmental heterogeneity, expressed by a mosaic of environmental units. Environmental units are characterized by small but distinct differences in, for example, topography, morphology, soil, vegetation, and climate (Cendrero and Diaz de Teran 1987). They can be defined by analytical or synthetic methods which correspond to descriptive and interpretative phases (Dyer and Vinogradov 1990), respectively. The analytical method, represented by the elaboration of thematic maps of the coastal plain, described geomorphological and hydrophysical aspects, fauna and flora, land use, and socio-economic attributes and furnished recommendations for use (Asmus et al. 1991).

The synthetic method, based on the interpretation of satellite imagery and ground truth verification, offers an integrated view of the different natural and anthropic components. A total of 33 environmental units have been identified in both emersed littoral, transitional, and lagoonal systems around the Patos Lagoon estuary and submersed areas within it. Units of emersed areas are principally comprised of marshes, lagoonal terraces, semifixed dunes, and marine deposits. The extensive monocultures of rice plantations and large-scale afforestation with *Pinus* sp. are the main

U. Seeliger et al.
Subtropical Convergence Environments

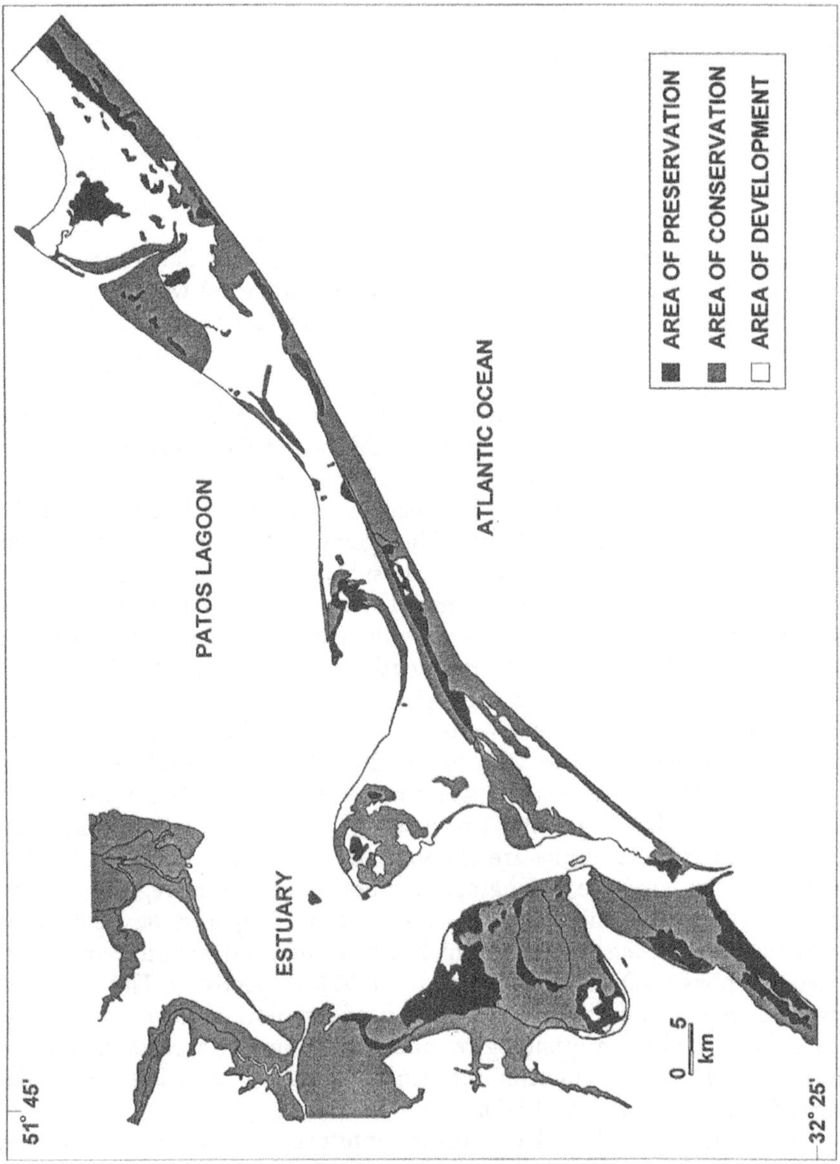

Fig. 11.1. Areas of preservation, conservation, and development of the Patos Lagoon estuary and adjacent coast

causes of their perturbation. These agricultural activities, owing to an increased equitability of the landscape pattern, induce low spatial heterogeneity (Tagliani 1995). Environmental units of submersed estuarine areas are dominated by protected shoals and channel zones. The ideal use of different environmental units is determined by their function within the coastal ecosystem (Tagliani 1995). A matrix of environmental functions can be constructed, based on the analysis of environmental characteristics, like biological production, support capacity, resilience, and socio-economic aspects (Groot 1986). This matrix proved to be an excellent tool to define and group environmental units into areas of preservation (not to be exploited), areas of conservation (limited and strictly controlled exploitation), and areas of development (Asmus et al. 1989, 1991; Fig. 11.1). Since the present use and exploitation by man of the estuary and the land around it conflict with management recommendations for some areas, mitigations will be necessary.

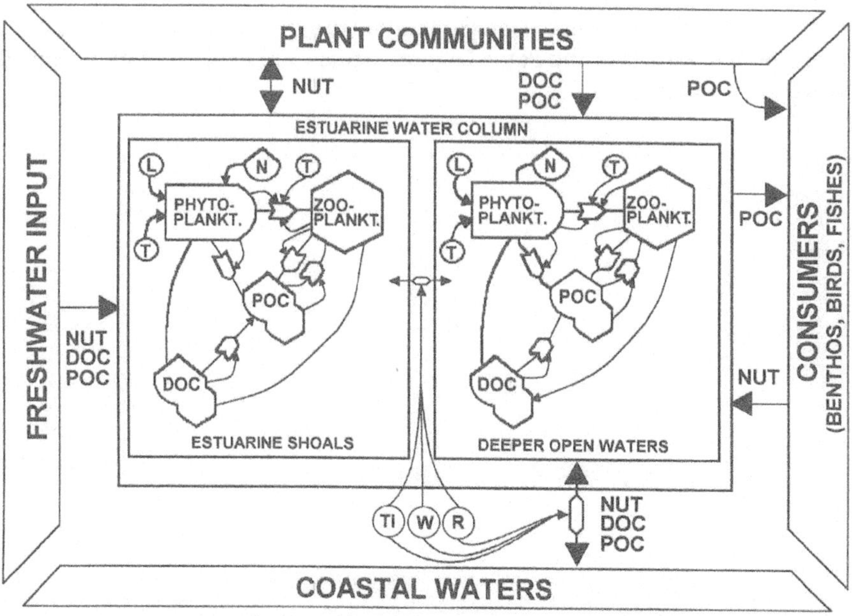

Fig. 11.2. Main modelled environments in the Patos Lagoon estuary. *NUT, N* nutrients; *POC* particulate organic carbon; *DOC* dissolved organic carbon; *L* light; *T* temperature; *Ti* tide; *W* wind; *R* rain

More recently, ecological modelling, representing simulative and predictive phases of management, has focused on dynamic aspects of the estuary and adjacent areas in order to characterize the dominant ecological, economic, and social processes of the region. Modelling of estuarine environments (Fig. 11.2) aims to determine dominant components and processes, the natural and anthropogenic control of processes, susceptibility of components/processes to management, and management recommendations (Faccini et al. 1995). All models emphasize the importance of metereology (wind, precipitation) and hydrodynamics (circulation) on critical ecological processes in the estuary (Bonilha and Asmus 1994; Rabelo da Cunha 1994; Teixeira da Silva 1995). Modelling efforts have identified gaps in our understanding of the estuary, which is an essential step for the improved comprehension of processes and indispensable for present and future management (Bonilha 1996).

The classification of national waters by the Brazilian government, based on use and quality, provided some legal support to administrate the use of waters in the Patos Lagoon estuary. However, environmental laws lack the desired detail to back urgently required management steps at municipal, state, and federal levels. Consequently, present and probably future management efforts of the Patos Lagoon estuary are hampered by technical and legal problems.

Appendix

Phylogenetic list of the 3193 coastal and marine taxa reported for the warm-temperate southwestern Atlantic since the beginning of this century.

KINGDOM PROKARYOTAE
PHYLUM CYANOBACTERIA (CYANOPHYTA)
CLASS CYANOPHYCEAE
Order Chroococcales
Family Chroococcaceae
Aphanothece castagnei (Bréb.) Rab.
A. stagnina (Spreng.) A.Braun
Chroococcus dispersus (Keissl.) Lemm.
C. limneticus Lemm.
C. membraninus (Menegh.) Näg.
C. minor Kütz.
C. pallidus Näg.
C. turgidus (Kütz.) Näg.
C. turicensis (Näg.) Hansg.
Gloeocapsa crepidinum Thuret
Gleothece confluens Näg.
Gomphosphaeria aponina Kütz.
G. lacustris Chodat
Merismopedia convoluta Bréb.
M. elegans A.Braun in Kütz.
M. glauca (Ehr.) Näg.
M. punctata Meyen
M. tenuissima Lemm.
Microcystis aeruginosa Kütz.
M. ichthyoblabe Kütz.
M. robusta (Clark) Nygard

Order Pleurocapsales
Family Hyellaceae
Xenococcus cladophorae (Ti.) Setch. & Gard.

Family Tubiellaceae

Order Nostocales
Johannesbaptista pellucida (Dickie) Taylor & Drouet

Family Oscillatoriaceae
Lyngbya confervoides C.Ag. ex Gom.
L. aestuarii Liebm. ex Gom.
L. epiphytica Hieron.
L. lutea (C.Ag.) Gom.
L. martensiana Menegh. ex Gom.
L. perelegans Lemm.
Microcoleus tenerrimus Gom.
M. chthonoplastes Thuret ex Gom.
Oscillaloria amphibia C.Ag. ex Gom.
O. animalis C.Ag. ex Gom.
O. earlei Gardner
O. obscura Bruhl & Biswas
O. limnetica Lemm.
O. limosa C.Ag. ex Gom.
O. meslini Fremy
O. nigroviridis Thw. ex Gom.
O. okeni C.Ag. ex Gom.
O. ornata Kütz. ex Gom.
O. princeps Vaucher ex Gom.
O. proboscidae Gom.
O. subbrevis Schmidle
Phormidium cf. *arcuatum* Skuja
P. crouani Gom.
P. fragile (Menegh.) Gom.
P. usterii Schmidle
Schizothrix lamyi Gom.
S. muelleri Näg. ex Gom.
Spirulina labyrinthiformis (Menegh.) Gom.
S. subsalsa Oersted ex Gom.
S. subtilissima Kütz.
Trichodesmium erythraeum (Ehr.) ex Gom.

Family Nostocaceae
Anabaena aphanizomenoides Forti.
A. doliolum Bharadw.
A. circinalis Kütz.
A. fertilissima Rao
A. cf. *iyengarii* Bharadw.
A. oscillarioides Bory ex Bornet & Flah.
A. variabilis Kütz.

Aphanizomenon gracile Lemm.
Cylindrospermopsis raciborskii (Wolos.) Seen. & Subba Raju
Cylindrospermum sp.

Nodularia harveyana (Thwaites) Thuret emend. Nordin & Stein
Nostoc commune Vaucher ex Born. & Flah.
N. paludosum Kütz.
Richelia intracellularis Schm. in Ostenf. & Schm.

Family Rivulariaceae
Calothrix parietina Thur ex Born. & Flah.

Order Stigonematales
Family Nostochopsidaceae
Mastigocoleus testarum Lagerh.

Family Stigonemataceae
Stigonema mesentericum Geitler

KINGDOM PROTOCTISTA
PHYLUM RHIZOPODA
CLASS LOBOSEA
Order Arcellinida
Family Arcellidae
Arcella vulgaris Ehr.

Family Centropyxidae
Centropyxis aculeata discoides Penard
C. (=*Cyclopyxis*) *arenata* (Cushman)
C. marsupiformis (Wall.)

Family Difflugiidae
Difflugia acuminata Ehr.
D. amphora Leidy
D. capreolata Penard
D. corona Wall.
D. curvicaulis Penard
D. elegans Penard
D. globularis Wall.
D. lageniformis Wall.
D. lebes Penard
D. lobostoma Leidy
D. mitriformis Wall.
D. pyriformis Perty
D. urceolata Carter
Pontigulasia compressa Rhumbler

Order Gromida
Family Euglyphidae
Euglypha alveolata Dujard.

PHYLUM RHODOPHYTA
CLASS BANGIOPHYCIDAE
Order Bangiales
Family Bangiaceae
Goniotrichum alsiddii (Zanard.) Howe
Erythrocladia subintegra Rosenv.
Erythrotrichia carnea (Dill.) J.Ag.
Bangia atropurpurea (Roth) C.Ag.
Porphyra pujalsii Coll & Oliveira F.
P.rizzini Coll & Oliveira F.

CLASS FLORIDEOPHYCIDAE
Order Nemalionales
Family Acrochaetiaceae
Kylinia crassipes (Börg.) Kylin
Acrochaetium flexuosum Vickers
A. globosum Börg.

Order Gelidiales
Family Gelidiaceae
Gelidium floridianum Taylor
G. crinale (Turn.) Lamour.
Pterocladia capillacea (Gmelin) Bornet & Thuret

Order Cryptonemiales
Family Hildebrandiaceae
Hildebrandia prototypus Nardo

Family Corallinaceae
Amphiroa beauvoisii Lamour.
Corallina officinalis L.
Jania rubens (L.) Lamour.

Family Grateloupiaceae
Grateloupia cuneifolia J.Ag.
G. filicina (Wulf.) J.Ag.

Order Gigartinales
Family Hypneaceae
Hypnea musciformis (Wulf.) Lamour.

Family Phyllophoraceae
Gymnogongrus griffithsiae (Turn.) Martius

Family Gigartinaceae
Gigartina teedii (Roth) Lamour.

Order Rhodymeniales
Family Rhodymeniaceae
Rhodymenia pseudopalmata (Lamour.) Silva

Order Ceramiales
Family Ceramiaceae
Ceramium brasiliensis Joly

C. brevizonatum Pet. & Börg.
Callithamnion uruguayense Taylor

Family Delesseriaceae
Cryptopleura ramosa (Hudson) Kylin ex Newton

Family Rhodomelaceae
Polysiphonia subtilissima Mont.
P. tepida Hollenb.
Polysiphonia sp.
Bryocladia thyrsigera (J.Ag.) Schmitz
Carradoria virgata (C.Ag.) Kylin

PHYLUM CROMOPHYTA
CLASS CRYPTOPHYCEAE
Order Cryptomonadales
Family Cryptomonadaceae
Rhodomonas Karst.
R. salina (Wisl.) Hill & Wether.

CLASS DYNOPHYCEAE
Order Ebriales
Family Ebriaceae
Ebria tripartita (Schm.) Lemm.
Hermesinum adriaticum Zach.

Order Oxyrrhinales
Family Oxyrrhinaceae
Oxyrrhis marina Dujard.

Order Prorocentrales
Family Prorocentraceae
Prorocentrum aporum (Schi.) Dodge
P. balticum (Lohm.) Loebl.
P. compressum (Bail.) Abé ex Dodge
P. dentatum Stein
P. gracile Schütt
P. lima (Ehr.) Dodge
P. micans Ehr.
P. maximum (Gourr.) Schi.
P. minimum (Pav.) Schi.
P. rostratum Stein
P. scutellum Schröd.

Order Dinophysiales
Family Citharistaceae
Citharistes apsteinii Schütt

Family Dinophysiaceae
Amphisolenia bidentata Schröd.
A. globifera Stein
A. lemmermanni Kof.
A. palaeotheroides Kof.
A. palmata Stein
A. retangulata Kof.
Dinophysis acuminata Clap. & Lach.
D. argus (Stein) Abé
D. caudata Sav.-Kent
D. cuneus (Schütt) Abé
D. diegensis Kof.
D. doryphora (Stein) Abé
D. exigua Kof. & Skogs.
D. hastata Stein
D. hindmarchii (Murr. & Whit.) Bal.
D. fortii Pav.
D. mitra (Schütt) Abé
D. odiosa (Pav.) Tai. & Skogs.
D. okamurai Kof. & Skogs.
D. ovum Schütt
D. parvula (Schütt) Bal.
D. paulsenii (Schi.) Bal.
D. porodictyum (Stein) Abé
D. punctata Jörg.
D. pusilla Jörg.
D. rapa (Stein) Abé
D. recurva Kof. & Skogs.
D. rotundata Clap. & Lach.
D. schroederi Pav.
D. schuettii Murr. & Whit.
D. tripos Gourr.
Histioneis dolon Murr. & Whit.
H. cymbalaria Stein
H. highleyi Murr. & Whit.
H. hyalina Kof. & Mich.
H. milneri Murr. & Whit.
H. para Murr. & Whit.
Ornithocercus cf. *heteroporus* Kof.
O. quadratus Schütt
O. magnificus (Stein) Kof. & Skogs.
O. splendidus Schütt
O. steinii Stein
O. thumii (Schm.) Kof. & Skogs.

Family Oxyphysaceae
Oxyphysis oxytoxoides Kof.

Order Actiniscales
Family Actiniscaceae
Achradina pulchra Lohm.
Actiniscus pentasterias (Ehr.) Ehr.
A. acutum Lohm.
A. crassum Lohm.
A. oceanicum Lohm.

Order Dinococcales
Family Thoracosphaeracea
Thoracosphaera sp.

Order Pyrocystales
Family Pyrocystaceae
Dissodinium pseudolunula Swift
Pyrocystis fusiformis (Wyv.Thoms.) Murr.
P. lunula (Schütt) Schütt
P. noctiluca Murr. ex Haeckel
P. obtusa Pav.
P. robusta Kof.
P. spinifera Lohm.

Order Brachydiniales
Family Brachydiniaceae
Asterodinium sp.
Brachidinium sp.

Order Noctilucales
Family Kofoidiniaceae
Cymbodinium sp.

Family Noctilucaceae
Noctiluca scintillans (Macartney) Kof. & Swezy

Family Protodiniferaceae
Pronoctiluca sp.

Order Gymnodiniales
Family Gymnodiniaceae
Amphidinium britannicum (C.Herdm.) Lebour
Cochlodinium sp.
Gymnodinium catenatum Grah.
G. exechegloutum Norris
G. splendens Lebour
Gyrodinium cf. *aureolum* Hulb.
G. fusiforme Kof. & Swezy
G. spirale (Bergh) Kof. & Swezy
Plectodinium nucleovelatum Biech.
Pseliodinium sp.
Torodinium robustum Kof. & Swezy
T. teredo (Pouch.) Kof. Swezy

Family Warnowiaceae
Erythropsidinium sp.
Warnowia sp.

Family Polykrikaceae
Polykrikos schwarzii Buetsch.

Order Peridiniales
Family Ceratiaceae
Ceratium arietinum Cl. f. *gracilentum* Jörg.
C. azoricum Cl.
C. belone Cl.
C. bucephalum (Cl.) Cl.
C. candelabrum (Ehr.) Stein
C. carriense (Gourr.)
C. concilians Jörg.
C. contortum (Gourr.) Cl.
C. contortum var. *longinum* (Kars.) Sour.
C. contrarium (Gourr.) Pav.
C. declinatum (Kars.) Jörg.
C. declinatum f. *brachiatum* Jörg.
C. declinatum f. *normale* Jörg.
C. declinatum var. *majus* Jörg.
C. deflexum (Kof.) Jörg.
C. dens Ostenf. & Schmidt
C. euarctuatum Jörg.
C. extensum (Gourr.) Cl.
C. falcatum (Kof.) Jörg.
C. furca (Ehr.) Clap. & Lach.
C. furca var. *eugrammum* (Ehr.) Jörg.
C. furca var. *hircus* (Schröd.)
C. fusus (Ehr.) Dujard.
C. fusus var. *seta* (Ehr.) Jörg.
C. gibberum Gourr.
C. gibberum f. *dispar* (Pouch.) Jörg.
C. gibberum f. *subaequale* Jörg.
C. hexacanthum Gourr.
C. hexacanthum f. *contortum* (Lemm.) Jörg.
C. horridum f. *denticulatum* (Jörg.) Böhm
C. horridum (Cl.) Gran *horridum*
C. horridum molle (Kof.) Jörg.
C. incisum Kars.
C. inflatum (Kof.) Jörg.
C. karstenii Pav.
C. karstenii f. *robustum* (Kars.) Jörg.
C. kofoidii Jörg.
C. limulus Gourr.
C. lineatum (Ehr.) Cl.
C. longirostrum Gourr.
C. macroceros (Ehr.) Vänh.
C. macroceros var. *gallicum* (Kof.) Jörg.
C. massiliense (Gourr.) Jörg.
C. massiliense f. *massiliense* Sourn.
C. massiliense f. *armatum* (Kars.) Jörg.
C. minutum Jörg.
C. paradoxides Cl.
C. pavillardii Jörg.
C. pentagonum pentagonum Gourr.
C. pentagonum var. *tenerum* Jörg.
C. praelongum Lemm.
C. pulchellum Schröd.
C. ranipes Cl.
C. setaceum Jörg.

C. strictum (Okam. & Nish.) Kof.
C. symmetricum Pav.
C. symmetricum var. *coarctatum* (Pav.) Grah. & Bronik.
C. symmetricum var. *symmetricum* Grah. & Bronik.
C. tenue (Ostenf. & Schm.)*tenue*
C. tenue var. *buceros* Bal.
C. tenue tenuissimum (Kof.)
C. teres Kof.
C. trichoceros (Ehr.) Kof.
C. tripos (O.Müll.) Nitz.
C. tripos breve Ostenf. & Schm.
C. tripos tripodioides Jörg.
C. tripos tripos (O.Müll.) Nitz.
C. vultur Cl.

Family Ceratocorythaceae
Ceratocorys gourretii Pauls.
C. horrida Stein

Family Cladopyxidaceae
Cladopyxis brachiolata Stein
Palaeophalacroma sp.

Family Goniodomataceae
Goniodoma polyedricum (Pouchet) Jörg.

Family Gonyaulacaceae
Alexandrium fraterculus (Bal.) Bal.
Alexandrium sp.
Gonyaulax birostris Stein
G. diegensis Kof.
G. digitalis (Pouch.) Kof.
G. fragilis (Schütt) Kof.
G. grindleyi Rein.
G. pacifica Kof.
G. polyedra Stein
G. polygramma Stein
G. cf. *spinifera* (Clap. & Lach.) Dies.
G. verior Sour.
Spiraulax kofoidii Grah.

Family Oxytoxaceae
Centrodinium sp.
Corythodinium sp.
C. constrictum (Stein) Tayl.
C. latum (Gaar.) Tayl.
C. milneri Murr. & Whit.
C. tesselatum (Stein)
Oxytoxum crassum Schi.
O. gracile Gran
O. longiceps Schi.
O. parvum Schi.
O. sceptrum (Stein) Schröd.
O. scolopax Stein
O. sphaeroideum Stein
O. turbo Kof.
Schuettiella mitra (Schütt) Bal.

Family Peridiniaceae
Diplopelta sp.
Oblea sp.
Peridinium quinquecorne Abé
Protoperidinium cf. *aspidiotum* (Bal.) Bal.
Protoperidinium cf. *claudicans* (Pauls.) Bal.
P. conicum (Gran) Bal.
P. depressum (Bail.) Bal.
P. divergens (Ehr.) Bal.
P. excentricum (Pauls.) Bal.
P. globulus (Stein) Bal.
P. granii (Ostenf.) Bal.
P. inclinatum (Bal.) Bal.
P. longipes (Kars.) Bal.
P. obtusum (Kars.) Bal.
P. oceanicum (Vänh.) Bal.
P. pentagonum (Gran) Bal.
P. punctulatum Pauls.
P. pyriforme (Pauls.) Bal.
P. quarnerense (Schröd.) Bal.
P. robustum (Kof.) Bal.
P. spiniferum (Schi.) Bal.
P. steinii (Jörg.) Bal.
Scrippsiella trochoidea (Stein) Fine & Loebl.III

Family Podolampadaceae
Podolampas bipes Stein
P. elegans Schütt
P. palmipes Stein
P. reticulata Kof.
P. spinifera Okam.

Family Pyrophacaceae
Pyrophacus horologium Stein
P. steinii Schi.

CLASS PRYMNESIOPHYCEAE
Order Isochrysidales
Family Gephyrocapsaceae
Emiliania huxleyi (Lohm.) Hay & Mohler
Gephyrocapsa oceanica Kampt.

Family Hymenomonadaceae
Cricosphaera sp.

Order Coccosphaerales
Family Calyptrosphaeraceae
Calyptrosphaera depressa Schi.
C. oblonga Lohm.
Helladosphaera cf. *cornifera* (Schi.) Kampt.
Zygosphaera sp.

Family Calciosoleniaceae
Anoplosolenia brasiliensis (Lohm.) Defl.
Calciosolenia murrayi Gran

Family Ceratolithaceae
Ceratolithus cristatus Kampt.

Family Coccolithaceae
Calcidiscus leptoporus (Murr. & Black.) Loebl.Jr. & Tap.
Coccolithus pelagicus (Wall.) Schi.
Oolithotus fragilis (Lohm.) Okada & McInt.
Umbilicosphaera sibogae (Weber-Van Bosse) Gaard.

Family Halopappaceae
Halopappus sp.
H. adriaticus Schi., emend. Manton, Bremer & Oates
Ophiaster hydroideus (Lohm.) Lohm.
O. formosus Gran

Family Pontosphaeraceae
Lohmannosphaera subclausa Schi.
Pontosphaera discopora Schi.
P. sessilis Lohm.
P. syracusana Lohm.
Scyphosphaera apsteinii Lohm.

Family Rhabdosphaeraceae
Acanthoica quattrospina Lohm.
Algirosphaera quadricornu (Schi.) Norris
Discosphaera tubifer (Murr. & Black.) Ostenf.
Rhabdosphaera claviger Murr. & Black.
R. hispida Lohm.

Family Syracosphaeraceae
Coronosphaera maxima (Hall. & Mark.) Gaard.
Deutschlandia anthos Lohm.
Syracosphaera dentata Lohm.
S. heimii Lohm.
S. histrica Kampt.
S. pirus Hall. & Mark.
S. pulchra Lohm.
S. spinosa Lohm.
S. tenuis Lohm.
Umbellosphaera tenuis (Kampt.) Paas.

Order Prymnesiales
Family Prymnesiaceae
Chrysochromulina sp.

Family Phaeocystaceae
Phaeocystis pouchetii (Hariot) Lager.

CLASS CHRYSOPHYCEAE
Order Ochromonadales
Family Chromulinaceae
Chromulina sp.

Family Chrysococcaceae
Chrysococcus cf. *ornatus* Paas.

Order Synurales
Family Mallomonadaceae
Mallomonas sp.

CLASS DICTYOCHOPHYCEAE
Order Dictyochales
Family Dictyochaceae
Dictyocha fibula Ehr.
D. speculum (Haeck.) Ehr.
Mesocena polymorpha Ehr.
Octactis pulchra Schi.

CLASS BACILLARIOPHYCEAE
Order Centrales
Family Asterolampraceae
Asterolampra marylandica Ehr.
Asteromphalus arachne (Bréb.) Ralfs
A. flabellatus (Bréb.) Grev.
A. heptactis (Bréb.) Ralfs ex Pritch.
A. hookeri Ehr.
A. robustus Castr.
A. sarcophagus Wall.

Family Biddulphiaceae
Biddulphia biddulphiana J.Sm. (Boy.)
B. tuomeyi (Bail.) Roper
Cerataulina pelagica (Cl.) Hend.
C. daemon (Grev.) Hasle
C. dentata Hasle
Climacodiun biconcavum Cl.
C. frauenfeldianum Grun.
Eucampia cornuta (Cl.) Grun.
E. zoodiacus Ehr.

E. zoodiacus cylindrocornis Syvert.
Eunotogramma dubium Hust.
E. leve Grun.
E. marinum (W.Sm.) Perag.
E. rostratum Hust.
Hemiaulus hauckii Grun. ex V.Heurck
H. membranaceus Cl.
H. sinensis Grev.
Hydrosera whampoensis (Schw.) Deby
Isthmia enervis Ehr.
Terpsinoe americana (Bail.) Ralfs
T. musica Ehr.

Family Chaetoceraceae
Bacteriastrum biconicum Pav.
B. elegans Pav.
B. elongatum Cl.
B. furcatum Shadb.
B. hyalinum Laud.
B. hyalinum var. *princeps* (Castr.) Ikari
B. mediterraneum Pav.
Chaetoceros affinis Laud.
C. atlanticus Cl.
C. atlanticus var. *neapolitanus* (Schoed.) Hust.
C. cf. *borealis* Bail.
C. brevis Schütt
C. coarctatus Laud.
C. compressus Laud. var. *compressus*
C. constrictus Gran
C. costatus Pav.
C. concavicornis Mangin
C. curvisetus Cl.
C. dadayi Pav.
C. danicus Cl.
C. debilis Cl.
C. decipiens Cl.
C. didymus Ehr.
C. didymus var. *anglica* Gran
C. didymus var. *protuberans* Gran
C. diversus Cl.
C. eibenii (Grun) Meun.
C. laciniosus Schütt
C. laevis Lend.-Fort.
C. lauderi Ralfs
C. lorenzianus Grun.
C. messanense Castr.
C. pelagicus Cl.
C. peruvianus Brightw.
C. peruvianus f. *gracilis* Schröd.
C. pseudocurvisetus Mangin
C. radians Schütt
C. radicans Shütt
C. rostratus Laud.
C. saltans Cl.
C. seriacanthus Gran
C. similis Cl.
C. socialis Laud.
C. subsecundus (Grun.) Hust.
C. subtilis Cl.
C. subtilis var. *abnormis* Prosch.-Lavr.
C. wighami Brightw.
C. willei Gran

Family Coscinodiscaceae
Coscinodiscus apiculatus Ehr.
C. apiculatus var. *ambigua* Grun.
C. argus Ehr.
C. asteromphalus Ehr.
C. asteromphalus var. *pabelanica* Grun.
C. brasiliensis M.Melch.
C. centralis Ehr.
C. concinnus W.Sm.
C. curvatulus Grun. ex Schm.
C. gigas Ehr.
C. gigas var. *praetextus* (Jan.) Hust.
C. granii Gough
C. janischii Schm.
C. jonesianus (Grev.) Ostenf.
C. jonesianus (Grev.) Ostenf. var. *jonesianus*
C. jonesianus var. *commutata* (Grun.) Hust.
C. kurzii Grun. & Schm.
C. marginatus Ehr.
C. obscurus A.Schm.
C. oculus-iridis Ehr.
C. oculus-iridis Ehr. var. *loculifera* Ratt.
C. oculus-iridis Ehr. var. *stelliger* Ratt.
C. pacificus Grun.
C. perforatus Ehr.
C. perforatus var. *cellulosa* Grun.
C. perforatus var. *pavillardii* (Forti) Hust.
C. radiatus Ehr.
C. radiatus var. *minor* A.Schm.
C. rothii (Ehr.) Grun.
C. sublineatus Grun.
C. thorii Pav.
C. variabilis Freng.
C. vidovichii M.Melch.
C. wailesii Gran & Angst.
Gossleriella tropica Schütt
Palmeria hardmanniana Grev.
Stellarima stellaris (Roper) Hasle & Sims

Family Psammodiscaceae
Psammodiscus nitidus (Greg.) Round & Mann

Family Cymatosiraceae
Campylosira cymbelliformis (A.Schm.)Grun. ex V.Heurck
Cymatosira adaroi Azpt. & Moros
C. atlantica Freng.
Plagiogrammopsis vanheurckii (Grun.) Hasle

Family Eupodiscaceae
Auliscus pruinosus Bail.
A. sculptus (W.Sm.) Ralfs ex Pritch.
A. caelatus Bail.
A. caelatus var. *rhipis* (A.Schm.) Perag.
Cerataulus smithii Ralfs
C. turgidus Ehr.
Eupodiscus antiguus (Cox) Hanna
E. radiatus Bail.
Odontella aurita (Lyng.) C.Ag.
O. aurita var. *obtusa* (Kütz.) Hust.
O. mobiliensis (Bail.) Grun.
O. regia (Schul.) Simon.
O. rhombus (Ehr.) Kütz.
O. longicruris Grev.
O. sinensis (Grev.) Grun.
Pleurosira laevis (Ehr.) Comp.
Triceratium alternans Bail.
T. antediluviana (Ehr.) V.Heurck
T. antediluviana var. *excavata* Freng.
T. dubium Brightw.
T. favus Ehr.
T. favus var. *quadrata* (Ehr.) Grun.
T. formosum f. *pentagonale* A.Schm.
T. patagonicum A.Schm.
T. pentacrinus (Ehr.) Wall.
T. reticulum (Ehr.) Boy.
T. robertsianum f. *inermis* Grev.
T. scitulum Brightw.
T. sculptum Shadb.

Family Heliopeltaceae
Actinoptychus campanulifer A.Schm.
A. cf. *concentricus* A.Schm.
A. senarius Ehr.
A. senarius cf. var. *tamanicus* Josúe
A. splendens (Shadb.) Ralfs ex Pritch.
A. vulgaris Schu.
A. vulgaris f. *octonaria* Freng.
A. vulgaris f. *polymera* Freng.
A. vulgaris f. *senaria* Freng.
Aulacodiscus petersii Ehr.

Family Hemidiscaceae
Actinocyclus barkly (Coat.) Grun.
A. brasiliensis M.Melch.
A. cholnokyi Van Land.
A. curvatulus Jan.
A. kuetzingii A.Schm.
A. normani (Greg.) Hust.
A. normani (Greg.) Hust. var. *normani*
A. octonarius Ehr.
A. octonarius var. *crassus* (W.Sm.) Hend.
A. octonarius var. *ralfsii* (W.Sm.) Hend.
A. octonarius var. *sparsa* (Greg.) Hend.
A. octonarius var. *tenella* (Bréb.) Hend.
A. platensis M.Melch.
A. subtilis (Greg.) Ralfs ex Pritch.
Hemidiscus cuneiformis Wall.
H. cuneiformis var. *ventricosa* (Castr.) Hust.
H. ovalis Lohm.
Roperia tessellata (Roper) Grun.

Family Lithodesmiaceae
Bellerochea sp.
Ditylum brightwellii (West) Grun. ex V.Heurck
Lithodesmium undulatum Ehr.
Streptotheca thamesis Shrubs.

Family Leptocylindraceae
Corethron criophilum Castr.
Leptocylindrus danicus Cl.
L. minimus Gran

Family Melosiraceae
Hyalodiscus laevis Ehr.
H. scoticus (Kütz.) Grun.
H. subtilis Bail.
Melosira juergensii C.Ag.
M. moniliformis (O.Müll.) C.Ag.
M. nummuloides (Dill.) C.Ag.
Paralia sulcata (Ehr.) Cl.
P. sulcata (Ehr.) Kütz. var. *sulcata*
P. sulcata var. *biseriata* Grun.
P. sulcata f. *coronata* (Ehr.) Grun.
P. sulcata f. *radiata* Grun.
Podosira montagnei Kütz.
P. stelliger (Bail.) Mann
Stephanopyxis palmeriana (Grev.) Grun.
S. turris (Grev.) Ralfs ex Pritch.

Family Rhizosoleniaceae
Dactyliosolen antarcticus Castr.
D. mediterraneus Perag.

Guinardia flaccida (Castr.) Perag.
Proboscia alata (Brightw.) Sund.
Proboscia indica (Perag.) Sund
Proboscia inermis (Castr.) Sund.
Pseudosolenia calcar-avis (Schul.) Sund.
Rhizosolenia acuminata (Perag.) Gran
R. bergonii Perag.
R. castracanei Perag.
R. delicatula Cl.
R. fragillissima Berg.
R. hebetata Bail.
R. hebetata f. *hiemalis* Gran
R. hebetata f. *semispina* (Hens.) Gran
R. hyalina Ostenf.
R. imbricata Brightw.
R. longiseta Zach.
R. phuketensis Sund.
R. pungens Cl.-Eu.
R. robusta Norm. ex Pritch.
R. setigera Brightw.
R. setigera f. *pungens* (Cl.) Brun.
R. setigera var. *daga* (Brightw.) M.Melch.
R. shrubsolei Cl.
R. stolterfothii Perag.
R. styliformis Brightw.

Family Sticodiscaceae
Ethmodiscus gazellae (Jan.) Hust.

Family Thalassiosiraceae
Aulacoseira ambigua (Grun.) Simon.
A. distans (Ehr.) Simon.
A. granulata (Ehr.) Simon.
A. islandica (O.Müll.) Simon.
A. islandica subsp. *helvetica* (O.Müll.) Simon.
A. italica (Ehr.) Simon.
A. nyassensis (O.Müller) Simon.
A. pseudogranulata (Cl.) Simon.
Cyclotella kuetzingiana Thw.
C. meneghiniana Kütz.
C. littoralis Lange & Syvert.
C. stelligera Cl. & Grun.
C. striata (Kütz.) Grun.
C. stylorum Brightw.
Cymatotheca weissflogii (Grun.) Hend.
Detonula pumila Castr.
Lauderia annulata Cl.
Planktoniella sol (Wall.) Schütt
Skeletonema costatum (Grev.) Cl.
S. subsalsum (Cl.) Bethge
S. cf. *tropicum* Cl.
Thalassiosira allenii Takano
T. angulata (Grun.) Jörg.
T. anguste-lineata (A.Schm.) Fryx. & Hasle
T. antarctica Comber
T. delicatula (chilensis) Ostenf.
T. eccentrica (Ehr.) Cl.
T. eccentrica var. *micropora* (Ehr.) Grun.
T. hendeyi Hasle & Fryx.
T. lentiginosa (Jan.) Fryx.
T. leptopus (Grun.) Hasle & Fryx.
T. levanderi Goor.
T. minuscula Krasske
T. nanolienata (Mann) Fryx. & Hasle
T. nodulineata (Hend.) Hasle & Fryx.
T. oestrupii (Ostenf.) Hasle var. *oestrupii*
T. oestrupii (Ostenf.) Hasle var. *venrickae*
T. pseudonana Hasle & Heimdal
T. punctigera (Castr.) Hasle
T. rotula Meun.
T. simonsenii Hasle & Fryx.
T. subtilis (Ostenf.) Grun.
T. tenera Prosch.-Lavr.
Trybliоptychus cocconeiformis (Cl.) Hend.

Order Pennales
Suborder Araphidineae
Family Fragilariaceae
Asterionellopsis glacialis (Castr.) Round
Climacosphaenia moniligera Ehr.
Delphineis surirella (Ehr.) G.W.Andr.
Diatoma anceps (Ehr.) Grun.
Dimeregramma hyalinum Hust.
D. cf. *marinum* (Greg.) Ralfs ex Pritch.
D. minor (Greg.) Ralfs in Pritch.
Fragilaria crotonenensis Kitt.
F. constricta Ehr.
F. construens Grun.
F. istvanffyi Pant.
F. lapponica Grun.
F. pinnata Ehr.
Glyphodesmis distans (Greg.) Grun.
Grammatophora angulosa Ehr.
G. angulosa Ehr. var. *angulosa*
G. hamulifera Kütz.
G. marina (Lyng.) Kütz.
G. marina (Lyng.) Kütz. var. *marina*
G. maxima Grun.
G. oceanica Ehr.
G. oceanica Ehr. var. *oceanica*
G. serpentina Ehr.

Licmophora abbreviata C.Ag.
L. argentescens C.Ag.
L. flabellata (Carm.) C.Ag.
L. gracilis (Ehr.) Grun.
L. lyngbyei (Kütz.) Grun ex V.Heurck
Opephora marina (Greg.) Petit
O. martyi Hérib.
Plagiogramma pulchellum Grev.
P. pulchellum Grev. var. *pygmaea* (Grev.) Perag. & Perag.
P. simplex Freng.
P. staurophorum (Greg.) Hérib.
Podocystis adriatica (Kütz.) Boyer
Rhabdonema adriaticum Kütz.
Rhaphoneis amphiceros (Ehr.) Ehr.
R. amphiceros var. *tetragona* Grun.
R. castracanei Grun.
R. discoides Subrahm.
R. surirella (Ehr.) Grun.
Striatella delicatula (Kütz.) Grun.
Synedra acus Kütz. var. *angustissima* Grun.
S. fasciculata (C.Ag.) Kütz.
S. fasciculata var. *truncata* (Grev.) Patr.
S. gaillonii (Bory) Ehr.
S. gaillonii (Bory) Ehr. var. *gaillonii*
S. goulardi (Bréb.) Grun.
S. reinboldii V.Heurck
S. rumpens Kütz.
S. ulna (Nitz.) Ehr.
S. ulna var. *contracta* Östr.
S. undulata Bail.
Tabellaria sp.
Thalassionema frauenfeldii (Grun.) Hallegr.
T. nitzschioides Hust.
Thalassiothrix delicatula Cupp
T. longissima Cl. & Grun.
T. mediterranea Pav.
T. mediterranea var. *pacifica* Cupp

Suborder Raphidineae
Family Eunotiaceae
Eunotia camelus Ehr.
E. camelus var. *denticulata* Grun.
E. camelus var. *didymodon* Grun.
E. diodon Ehr. var. *minor* Grun.
E. faba (Ehr.) Grun.
E. flexuosa (Bréb.) Kütz.
E. gracilis (Ehr.) Rab.
E. indica var. *biggiba* Freng.
E. indica f. *gracilis* Freng.
E. kocheliensis O.Müll.
E. maior (W.Sm.) Rab.
E. pectinalis (Kütz.) Rab.

Family Achanthaceae
Achnanthes brevipes C.Ag.
A. brevipes angustata (Grev.) Cl.
A. brevipes var. *intermedia* (Kütz.) Grun.
A. brevipes var. *parvula* (Kütz.) Cl.
A. curvirostrum Brun.
A. delicatula (Kütz.) Grun.
A. exigua var. *heterovalvata* Krasske
A. hauckiana Grun.
A. inflata (Kütz.) Grun.
A. javanica Grun. var. *rhombica* Grun
A. lanceolata (Bréb.) Grun.
A. longipes C.Ag.
A. longipes C.Ag. var. *longipes*
A. promonturii Giffen
A. saxonica Krasske
Anorthoneïs hyalina Hust.
Cocconeis costata Greg.
C. diminuta Pant.
C. dirupta Greg.
C. disculus (Schum.) Cl.
C. distans Greg.
C. heteroidea Hantzsch
C. pediculus Ehr.
C. pellucida Grun
C. placentula Ehr.
C. placentula var. *euglypta* (Ehr.) Cl.
C. placentula var. *lineata* (Ehr.) Cl.
C. pinnata Greg. ex Grev.
C. pseudomarginata Greg.
C. pseudomarginata Greg. var. *intermedia* Grun.
C. quarnerensis (Grun.) A.Schm.
C. scuttellum Ehr.
C. scuttellum var. *minutissima* Grun.
C. scutellum var. *ornata* Grun.
C. scuttelum var. *parva* Grun. in Cl.
C. scutellum Ehr. var. *scutellum*
C. scutellum var. *speciosa* (Greg.) Cl.
C. stauroneiformis (V.Heurck) Okuno

Family Naviculaceae
Amphipleura lindheimeri Grun.
Anomoeoneis serians (Bréb.) Cl.
Caloneis brevis (Greg.) Cl.
C. permagma (Bail.) Cl.
Diploneis bombus Ehr.
D. bombus var. *bombiformis* (Cl.) Hust.
D. crabro (Ehr.) Ehr.
D. didymus (Ehr.) Cl.
D. gruendleri (A.Schm.) Cl.

D. incurvata (Greg.) Cl.
D. littoralis (Donk.) Cl.
D. littoralis var. *clathrata* (Östrup) Cl.
D. oblongella (Näg. ex Kütz.) Ross
D. ovalis (Hilse) Cl.
D. ovalis var. *oblongella* (Näg.) Cl.
D. papula (A.Schm.) Cl.
D. papula (A.Schm.) Cl. var. *papula*
D. puella (Schum.) Cl.
D. smithii (Bréb.) Cl.
D. smithii (Bréb.) Cl. var. *smithii*
D. vacilans (A.Schm.) Cl.
D. vetula (A.Schm.) Cl. var. *vetula*
D. weissflogii (Schm.) Cl.
Donkinia carinata (Donk.) Ralfs
Frustulia interposta (Lew.) De-Toni
F. rhomboides (Ehr.) De Toni
Gomphonema constrictum Ehr.
G. gracile Ehr.
G. gracile Ehr. var. *gracile*
G. gracile var. *lanceolata* Kütz.
G. parvulum (Kütz.) Kütz.
G. turris Ehr.
Gyrosigma balticum (Ehr.) Rab.
G. balticum (Ehr.) Cl. var. *balticum*
G. eximium (Thw.) Boy.
G. exoticum Choln.
G. littorale (W.Sm.) Griff & Henfr.
G. rectum (Donk.) Cl.
G. scalproides (Rab.) Cl.
G. spencerii (W.Sm.) Cl.
Mastogloia apiculata W.Sm.
M. crucicula (Grun) Cl.
M. binotata (Grun.) Cl.
M. braunii Grun.
M. decussata Brun.
M. erythraea Grun.
M. exigua Lew.
M. exigua Lew. var. *exigua*
M. exilis Hust.
M. fimbriata (Brightw.) Cl.
M. jurgensii C.Ag.
M. meisteri Hust.
M. smithii Thw.
M. splendica (Greg.) Cl. & O.Müll.
Navicula anglica Ralfs
N. bacillum Ehr. var. *bacillum*
N. canalis Patrick
N. cancellata Donk.
N. capitata Ehr. var. *hungarica* (Grun.) Ross
N. cari Ehr.
N. cincta (Ehr.) Ralfs var. *cincta* in Pritch.
N. consentanea Hust.
N. cryptocephala Kütz.
N. crucicula (W.Sm.) Donk.
N. cuspidata (Kütz.) Kütz var. *cuspidata*
N. demerara Grun. ex Cl. var. *demerara*
N. digito-radiata (Greg.) Ralfs
N. exigua Greg. ex Grun. var. *capitata* Patr.
N. flagellifera Hust.
N. forcipata Grev.
N. gastrum (Ehr.) Kütz.
N. granulata Bail.
N. gregaria Donk.
N. hagelsteinii Hust.
N. humerosa Bréb. in W.Sm.
N. laevissima Kütz. var. *laevissima*
N. lorenzii (Grun.) Hust.
N. lyra Ehr.
N. lyroides Hend.
N. marina Ralfs ex Pritch.
N. membranacea (Cl.)
N. mutica Kütz.
N. notha Wall.
N. nummularia Grev.
N. pelagica Cl.
N. pennata A.Schm.
N. pennata A.Schm. var. *pennata*
N. pennata f. maxima Cl.
N. peregrina (Ehr.) Kütz.
N. pygmaea Kütz.
N. plagiostoma Grun.
N. porifera Hust.
N. praetexta Ehr.
N. punctulata W.Sm. var. *punctulata*
N. punctulata var. *marina* (Ralfs) Cl.
N. pupula Kütz.
N. radiosa var. *tenella* (Bréb.) Grun.
N. ramosissima (C.Ag.) Cl.
N. rostellata Kütz.
N. salinaroides Choln.
N. scopulorum Bréb. ex Kütz.
N. serena Freng.
N. sovereigne Hust.
N. soodensis Krasske
N. spectabilis Greg.
N. transfuga var. *plagiostoma* (Grun.) Cl.
N. cf. *tripuncata* (O.Müll.) Bory
N. tubulosa Brun.
N. yarrensis Grun. in Schm.
N. zostereti Grun.
Neidium affine var. *amphyrhynchus* (Ehr.) Cl.
N. iridis (Ehr.) Cl.

Pinnularia borealis Ehr.
P. cruciformis (Donk.) Cl.
P. divergens W.Sm.
P. gibba Ehr.
P. hemiptera (Kütz.) Cl.
P. intermedia (Lagers.) Cl.
P. major (Kütz.) Rab.
P. microstauron (Ehr.) Cl.
P. neuquina Freng.
P. trevelyana (Donk.) Rab.
Pleurosigma angulatum (Quek.) W.Sm.
P. atlanticum Freng.
P. brasiliana M.Melch.
P. diverse-striatum Meister
P. elongatum W.Sm.
P. hamuliferum Brun. & Temp.
P. intermedium W.Sm.
P. intermedium var. *nubecula* (W. Sm.) Grun. ex Van Heurck
P. lanceolatum Donk.
P. marinum Donk.
P. naviculaceum Bréb.
P. normanii Ralfs.
P. strigosum W.Sm.
Rhoicosphaenia curvata (Kütz.) Grun. ex Rab.
Stauroneis salina W.Sm
S. phoenicenterum (Nitz.) Ehr.
Trachyneis antillarum Cl.
T. aspera (Ehr.) Cl.
T. aspera var. *intermedia* (Grun.) Cl.
Tropidoneis lepidoptera (Greg.) Cl.
T. lepidoptera Greg. var. *minor* Cl.

Family Epithemiaceae
Rhopalodia gibberula (Ehr.) O.Müll.
R. musculus (Kütz.) O.Müll.

Family Nitzschiaceae
Bacillaria paxillifer (O.Müll.) Hend.
Cylindrotheca closterium (Ehr.) Reim. & Lew.
C. fusiformis var. *fusiformis* Reim. & Lew.
C. signata Reim. & Lew.
Cymatonitzschia marina (Lew.) Simon.
Hantzschia amphioxus (Ehr.) Grun.
H. amphioxus var. *vivax* (Hantzsch)
H. distinctepunctata (Hust.) Hust.
H. longiareolata Garc.-Bapt.
H. psammicola Garc. -Bapt.
H. virgata (Roper) Grun.
H. virgata var. *capitellata* Hust.
H. virgata var. *wittii* (Grun.) Grun.
Nitzschia acuminata (W.Sm.) Cl.
N. angularis var. *affinis* Grun.
N. apiculata (Greg.) Grun. in Cl. & Grun.
N. archibaldii Lange-Bertalot
N. arenosa Garc.-Bapt.
N. bilobata W.Sm.
N. circumsuta (Bail.) Grun. in Cl. & Grun.
N. coarctata var. *oceanica* (Grun.) Freng.
N. compressa (Bail.) Boy.
N. debilis (Arn.) Grun.
N. dissipata Kütz.
N. epithemioides Grun.
N. filiformis (W.Sm.) Schütt
N. gandersheimiensis Krasske
N. granulata Grun. in Cl. & Grun.
N. hantzschiana Rab.
N. insignis Greg.
N. insignis var. *spathulifera* Grun.
N. lanceolata W.Sm.
N. cf. *lanceolata* var. *pygmaea* Cl.
N. linearis Wm.Sm. var. *linearis*
N. longissima (Bréb. in Kütz.) Ralfs
N. lorenziana Grun.
N. marginata Hust.
N. obtusa W.Sm.
N. pacifica Cupp
N. palea (Kütz.) W.Sm.
N. panduriformis Greg.
N. panduriformis var. *minor* Grun.
N. punctata (W.Sm.) Grun. var. *punctata*
N. scalaris (Ehr.) W.Sm. var. *scalaris*
N. sigma (Kütz.) W.Sm.
N. sigma var. *intercedens* Grun.
N. sigma var. *sigmatella* Grun.
N. socialis Ralfs
N. spathulata W.Sm
N. tryblionella Hantzch
N. tryblionella var. *victoriae* Grun.
N. ventricosa (Palmer.) Kitt.
N. vidovichii (Grun.) Grun.
N. vitrea Norm.
Pseudoeunotia doliolus (Wall.) Grun.
Pseudo-nitzschia australis Freng.
P. cf. *cuspidata* Hasle
P. delicatissima complex
P. pseudodelicatissima (Hasle) Hasle
P. pungens Grun. f. *pungens*
P. seriata complex
Tryblionella coarctata (Grun.) Mann

Family Cymbellaceae
Amphora acutiuscula Kütz.
A. angusta Grev. var. *angusta*
A. bigibba Grun.
A. binodis Greg. var. *bigibba* Grun.
A. bongrainii M.Perag. var. *bongrainii*
A. castellata Giffen
A. coffeaeformis (C.Ag.) Kütz.
A. granulata Greg.
A. laevis Greg. var. *laevissima* (Greg.) Cl.
A. marina (W.Sm.) V.Heurck
A. ovalis (Kütz.) Kütz.
A. proteoides Hust.
A. sabiniana Reim.
A. turgida Greg.
A. ventricosa Greg.
Cymbella affinis Kütz.
C. amphicephala Näg.
C. cuspidata Kütz.
C. cymbiformis C.Ag.
C. tumida (Bréb.) V.Heurck
C. tumida var. *tumida* (Bréb.) V.Heurck
C. turgida (Greg.) Cl.
C. ventricosa C.Ag.

Family Surirellaceae
Campylodiscus daemelianus Grun.
C. decorus Bréb.
C. echineis Ehr. var. *echineis*
C. ecclesianus Grev.
Petrodictyon gemma (Ehr.) Mann
Surirella bisseriata Bréb.
S. fastuosa Ehr.
S. febigerii Lewis
S. guatimalensis Ehr.
S. linearis var. *constricta* (Ehr.) Grun.
S. recendes A.Schm.
S. rorata Freng.
S. tenera Greg.
S. cf. *mulleri* Hust.
Stenopterobia intermedia (Lew.) Van Heurck

Family Entomoneidaceae
Entomoneis alata Ehr.

CLASS RAPHIDOPHYCEAE
Order Chattonellales
Family Chattonellacea
Chattonella sp.
Fibrocapsa japonica Toriumi & Takano

CLASS TRIBOPHYCEAE (XANTHOPHYCEAE)

Order Chloramoebales
Family Chloramoebaceae
Chloromeson parva Carter

Order Vaucheriales
Family Vaucheriaceae
Vaucheria longicaulis Hoppaugh

CLASS PHAEOPHYCEAE
Order Ectocarpales
Family Ectocarpaceae
Ectocarpus siliculosus (Dillwyn) Lyngb.
Giffordia mitchellae Hamel

Order Punctariales
Family Punctariaceae
Petalonia fascia (Müll.) Kuntze

PHYLUM CHLOROPHYTA
CLASS EUGLENOPHYCEAE
Order Euglenales
Family Eutreptiaceae
Eutreptia sp.
Eutreptiella sp.

Family Euglenaceae
Euglena viridis Ehr.

CLASS PRASYNOPHYCEAE
Order Chlorodendrales
Family Halosphaeraceae
Halosphaera viridis Schmitz
Pterosperma sp.
Pyramimonas sp.

CLASS CHLOROPHYCEAE
Order Volvocales
Family Chlamydomonadaceae
Brachiomonas sp.
Chlamydomonas sp.

Order Chlorococcales
Family Oocystaceae
Koliella longiseta Hindak
Lagerheimia citriformis (Snow) Collins

Family Scenedesmaceae
Crucigenia tetrapedia (Kirch.) W. & S.West
Scenedesmus acuminatus (Lager.) Chodat
S. acutus Meyen

S. quadricauda (Turp.) Bréb.

Order Ulotrichales
Family Ulotrichaceae
Planctonema lauterbornii Schmidle
Ulothrix flacca (Dillwyn) Thuret

Family Chaetophoraceae
Phaeophila viridis (Reinke) Burrows

Family Gomontiaceae
Gomontia lignicola Moorex

Order Ulvales
Family Ulvaceae
Enteromorpha bulbosa (Suhr) Kütz.
E. clathrata (Roth) J.Ag.
E. compressa (L.) Grev.
E. flexuosa (Wulf) J.Ag.
E. intestinalis (L.) Link
E. linza (L.) J.Ag.
E. micrococca Kütz.
E. ramulosa (Smith) Hooker
Ulvaria oxysperma (Kütz.) Bliding
Ulva lactuca L.
U. fasciata Delile

Order Cladophorales
Family Cladophoraceae
Chaetophora aerea (Dillwyn) Kütz.
Cladophora crispata (Roth) J.Ag.
C. fascicularis (Mert.) Kütz.
C. flexuosa (Dill.) Harv.
C. prolifera (Roth) Kütz.
C. scitula (Suhr) Kütz.
Rhizoclonium kerneri Stockm.
R. ripanum (Roth) Harv.
R. tortuosum Kütz.

Order Siphonales
Family Bryopsidaceae
Bryopsis plumosa (Huds.) C.Ag.
Bryopsis sp.

Order Zygnematales
Family Zygnemataceae
Mougeotia sp.
Spirogyra sp.

Family Desmidiaceae
Cosmarium sp.
Mesotaenium macrococcum var. *micrococcum* (Kütz.) West & West

PHYLUM GRANULORETICULOSA
CLASS FORAMINIFEREA
Order Textulariida
Family Rzehakinidae
Miliammina earlandi Loeblich & Tappan
M. fusca (Brady)
Trilocularena patensis Closs

Family Hormosinidae
Protoschista findens (Parker)
Reophax arctica Brady
R. nana Rhumbler

Family Lituolidae
Ammoastuta inepta (Cushman & McCulloch)
Ammobaculites exiguus Cushman & Bronnimann
Ammoscalaria pseudospiralis (Williamson)
Ammotium cassis (Parker)
A. planissimum (Cushman)
A. pseudocassis Cushman & Bronnimann
A. salsum (Cushman & Bronnimann)
Haplophragmoides wilberti Andersen

Family Trochamminidae
Arenoparrella mexicana (Kornfeld) emend. Andersen
Trochammina inflata (Montagu)
T. macrescens Brady
T. ochracea (Williamson)
Trochamminita salsa (Cushman & Bronnimann)

Family Textulariidae
Textularia candeiana Orbigny
T. earlandi Parker
T. gramen Orbigny
Spiroplectammina biformis (Parker & Jones)

Order Miliolida
Family Miliolidae
Massilina secans (Orbigny)
Pyrgo nasuta Cushman
Pyrgo sp.
Quinqueloculina atlantica Boltovskoy
Q. aff. *frigida* Parker
Q. isabellei Orbigny
Q. lamarckiana Orbigny

Q. cf. *patagonica* Orbigny
Q. polygona Orbigny
Q. seminulum (Linnaeus)
Q. aff. *seminulum* (Linnaeus)
Triloculina trigonula (Lamarck)

Order Lagenida
Family Nodosariidae
Nodosaria catesbyi Orbigny

Family Polymorphinidae
Globulina caribaea Orbigny
Guttulina problema Orbigny

Family Lagenidae
Lagena laevis (Montagu)
L. laevis f. *perlucida* (Montagu)
L. sulcata f. *lyellii* (Seguenza)
Oolina costata Williamson

Order Globigerinida
Family Globigerinidae
Globigerina bulloides Orbigny
G. digitata Brady
G. falconensis Blow
G. rubescens Hofker
Globigerinella aequilateralis (Brady)
G. calida (Parker)
Globigerinoides conglobatus (Brady)
G. elongatus (Orbigny)
G. ruber (Orbigny)
G. ruber f. *pyramidalis* (van den Broeck)
G. ruber f. *triloba* (Reuss)
G. sacculifer (Brady)
G. trilobus (Reuss)
G. trilobus f. *sacculifera* (Brady)
G. trilobus (Reuss)
G. tenellus Parker
Orbulina universa Orbigny
Sphaeroidinella dehiscens (Parker & Jones)

Family Candeinidae
Candeina nitida Orbigny

Family Hantkeninidae
Hastigerina pelagica (Orbigny)
Hastigerinella digitata (Rhumbler)

Family Globorotaliidae
Globoquadrina dutertrei (Orbigny)
G. pachyderma (Ehrenberg)
Globorotalia anfracta Parker
G. crassaformis (Galloway & Wissler)
G. hirsuta (Orbigny)
G. inflata (Orbigny)
G. menardii (Orbigny)
G. pumilio Parker
G. scitula (Brady)
G. tumida (Brady)
G. truncatulinoides (Orbigny)
G. ungulata Bermúdez
Pulleniatina obliquiloculata (Parker & Jones)

Family Catapsydracidae
Globigerinita clarkei (Rögl & Bolli)
G. glutinata (Egger)
G. humilis (Brady)
G. iota Parker

Order Rotaliida
Family Buliminidae
Bolivina striatula Cushman
B. pusilla Schwager
Bulimina marginata Orbigny
B. marginata f. *aculeata* Orbigny
B. marginata f. *echinata* Orbigny
B. marginata f. *subalata* Cushman & Parker
B. patagonica Orbigny
B. patagonica f. *glabra* Cushman & Wickenden
B. pseudoaffinis Kleinpell
Buliminella elegantissima Orbigny
Uvigerina peregrina f. *parvula* Cushman
Virgulina cf. *pauciloculata* H.Brady

Family Cibicididae
Cibicides bertheloti (Orbigny)
C. bertheloti f. *boueana* (Orbigny)
C. umbonatus Phleger & Parker

Family Nonionidae
Nonion grateloupe (Orbigny)
N. tisburyensis Butcher
Noniella atlantica Cushman

Family Rotaliidae
Bucella frigida (Cushman)
B. cf. *inusitata* Anderson
B. peruviana campsi (Boltovskoy)
Discorbis williamsoni (Chapman & Parr)
Poroeponides lateralis (Terquem)
Rolshausenia rolshauseni (Cushman & Bermúdez)
Rotalia beccarii ex gr. *parkinsoniana* (Orbigny)

Family Elphidiidae
Cancris sagra (Orbigny)
Elphidium discoidale (Orbigny)
E. aff. *discoidale* (Orbigny)
E. excavatum (Terquem)
E. galvestonense Kornfeld
E. gunteri Cole
E. incertum (Williamson)

PHYLUM CILIOPHORA
CLASS KINETOFRAGMINOPHORA
Order Haptorida
Family Didiniidae
Mesodinium rubrum (Lohm.) Hamb. & Budd.

CLASS POLYHYMENOPHORA
Order Oligotrichida
Suborder Oligotrichina
Family Strombidiidae
Strombidium spp.

Family Strobilidiidae
Lohmaniella sp.
Strobilidium sp.

Suborder Tintinnina
Family Codonellidae
Codonaria fimbriata (Meun.) Bal.
Codonella sp.
Leprotintinnus sp.
Tintinnidium sp.
Tintinnopsis sp.
T. beroidea Stein
T. campanula Ehr.
T. radix (Imhof) Brandt
T. levigata Kof. & Camp.

Family Codonellopsidae
Codonellopsis morchella (Cl.) Jörg.
Stenosemella sp.

Family Cyttarocylidae
Favella sp.

Family Tintinnidae
Salpingacantha ampla Kof. & Camp.
Salpingella secata Brandt

Family Dictyocystidae
Dictyocysta lata Kof. & Camp.
D. mitra Haeckel
D. mülleri (Imhof) Jörg.

Family Rhabdonellidae
Rhabdonellopsis sp.

Family Xystonellidae
Parundella inflata Kof. & Camp.
Xystonella sp.

KINGDOM PLANTAE
PHYLUM PTERIDOPHYTA
Family Lycopodianaceae
Lycopodium alopecuroides L.

Family Equisetaceae
Equisetum giganteum L.

Family Polypodiaceae
Acrosticum aureum L.

Family Sphagnaceae
Sphagnum sp.

PHYLUM ANGIOSPERMOPHYTA
CLASS MONOCOTYLEDONEAE
Order Pandanales
Family Typhaceae
Typha domingensis Pers.

Order Helobiae
Family Potamogatonaceae
Potamogeton striatus Ruiz & Pav.
Ruppia maritima L.
Zannichellia palustris L.

Family Juncaginaceae
Triglochin striata Ruiz & Pav.
Triglochin sp.

Order Graminales
Family Gramineae
Andropogon arenarius Hack.
A. selloanus Kunth.
Axonopus compressus (Sw.) Beauv.
Briza erecta Lam.
B. minor L.
Cenchrus echinatus L.
Cynodon dactylon (L.) Pers.
Digitaria aequiglumes (Hack. & Arech.) Parodi
Elyonurus rostratus Nees
Eragrostis airoides Nees
E. bahiensis (Schrad. ex. Schult.)
E. cataclasta Nicora. ex Burk.
E. tnchocolea Hack. & Arech.

Erianthus angustifolius Nees
Eriochloa punctata (L.)Desv.
Imperata brasiliensis Trin.
Ischaemum minus Presl.
I. urvilleanum Kunth.
Panicum racemosum (Beauv.) Spreng.
P. repens Berg.
Paspalum distichum L.
P. nicorae Parodi
P. vaginatum Sw.
Poidium rufum (Presl.) Matthei
Polypogon maritimus Willd.
P. monspeliensis (L.) Desf.
Setaria geniculata (Lam.) Beauv.
Spartina alterniflora Lois.
S. ciliata Brongn.
S. densiflora Brongn.
Sporobulus indicus (L.) Brongn.
Stenotaphrum secundatum (Walt.) O.Ktze

Family Cyperaceae
Androtrichum trigynum (Spreng.) Pfeiffer
Cladium jamaiscensis Crantz.
Cyperus brevifolius (Rottb.) Hassk.
C. cayennensis (Lam.) Britt.
C. giganteus Vahl
C. obtusatus (Presl.) Mattf. & Kükenth.
C. polystachyus Rottb.
C. reflexus Vahl
C. sesquiflorus (Torr.) Mattf. & Kükenth.
C. tener (Clarke.) Ostenf.
Eleocharis montevidensis Kunth.
E. nana Kunth.
E. nodulosa (Roth.) Schult.
E. obtusa (Willd.) Schult.
E. sellowiana Kunth.
Fimbristylis autumnalis (L.) Roem. & Schult.
F. diphylla (Retz.) Vahl
F. spadicea (L.) Vahl
Scirpus olneyi Gray
S. americanus Pers.
S. cernuus Vahl
S. maritimus L.
Scleria hirtella Sw.

Order Xyridales (Farinosae)
Family Eriocaulaceae
Eriocaulon modestum Kunth.

Family Pontederiaceae
Heteranthera reniformis R. & P.
Eichhornia crassipes (Mart.) Solms.

Family Bromeliaceae
Bromelia antiacatha Bert.

Family Xyrydaceae
Xyris caroliniana Walt.

Order Liliales
Family Juncaceae
Juncus acutus L.
J. capillaceus Lam.
J. dombeyanus Gay.
J. effusus L.
J. microcephalus H.B.K.
Juncus spp.

Family Iridaceae
Sisynnchium sp.
Sisyrinchium graminifolium Lindl.

Order Orchidales
Family Orchidaceae
Habenaria parviflora Lindl.

CLASS DICOTYLEDONEAE
Order Polygonales
Family Polygonaceae
Polygonum negletum Bess.

Order Chenopodiales
Family Aizoaceae
Sesuvium portulacastrum L.

Family Portulacaceae
Portulaca grandiflora Hook.

Family Caryophyllaceae
Cardionema ramosissima (Weinm.) A.Nels. & Macbr.
Spergularia cf. *marina* (L.) Griseb.

Family Amaranthaceae
Alternanthera phyloxeroides (Mart.) Griseb.
Blutaparon portulacoides (St.-Hil.) Mears.

Family Chenopodiaceae
Atriplex cf. *hastata* L.
A. cf. *patula* L.
Chenopodium album L.
C. retusum Juss. ex Moq.
Salicornia ambigua Michx.

S. gaudichaudiana Moq.

Order Ranales
Family Ceratophyllaceae
Ceratophyllum demersum L.

Order Papaverales
Family Cruciferae
Cakile maritima Scop.

Order Sarraceniales
Family Droseraceae
Drosera brevifolia Pursh.

Order Rosales
Family Rosaceae
Margyricarpus pinnatus (Lam.) O.Ktze.

Family Leguminoseae
Aeschynomene sensitiva Sw.
Indigofera sabulicola Benth.
Vigna luteola (Jacq.) Benth.

Order Geraniales
Family Polygalaceae
Polygala cyparissias St.-Hil.

Order Parietales
Family Guttiferae
Hypencum connatum Lam.

Order Myrtales
Family Onagraceae
Oenothera affinis Camb.

Family Gunneraceae
Gunnera herteri Osten

Family Haloragaceae
Myriophyllum brasiliense Camb.

Order Umbelliflorae
Family Umbelliferae
Apium graveolens L.
Centella hirtella Nannf.
Eryngium nudicaule Lam.
Eryngium sp.
Hydrocotyle bonarensis Lam.
Lilaeopsis attenuata (Hook. & Arn.) Fern.

Order Primulales
Family Myrsinaceae
Rapanea parviflora Mez.

Family Primulaceae
Anagallis arvensis L.

Order Plumbaginales
Family Plumbaginaceae
Limonium brasiliensis (Boiss.) O.Ktze.

Order Gentianales
Family Gentianaceae
Zigostigma australe (Cham. & Schlecht) Griseb.

Family Apocynaceae
Nymphoides indica (L.) O.Ktze.

Family Asclepiadaceae
Asclepias mellodora St.-Hil.

Order Tubiflorae
Family Convolvulaceae
Calystegia sepium (L.) R.Br.
C. soldanella (L.) Roem. & Schult.
Cressa truxillensis H.B.K.

Family Verbanaceae
Phyla canescens (H.B.K.) Greene

Family Solanaceae
Solanum cf. *americanum* (Mill.)

Family Scrophulariaceae
Agalinis communis (Cham. & Schlecht.) D'Arcy
Bacopa monnieri (L.) Penn.
Lindernia dubia (L.) Penn.
Stemodia hyptoides Cham. & Schlecht.

Family Lentibulariaceae
Utricularia tricolor St.-Hil.

Order Plantaginales
Family Plantaginaceae
Plantago australis Lam.
P. paralias Dcne.
P. cf. *lanceolata* L.

Order Campanulales
Family Compositae
Achyrocline satureoides (Lam.) DC.
Aster squamatus (Spreng.) Hieron.
Baccharis patens Baker
B. spicata (Lam.) Baill.
B. trimera DC.
Conyza blakei Cabrera

C. bonariensis (L.) Cronq.
C. floribunda H.B.K.
C. pampeana (Parodi) Cabrera
Cotula coronopifolia L.
Eclipta prostrata L.
Gamochaeta americana (Mill.) Wedd.
G. spicata (Lam.) Cabrera
Gnaphalium filaginea (DC.) Cabrera
Picrosia longifolia D.Don.
Pluchea sagittalis (Lam.) Cabrera
Porophyllum cf. *americanum* DC.
P. obscurum (Spreng.) DC.
Pterocaulon angustifolium DC.
P. bakeri Malme
P. purpurascens Malme
Senecio crassiflorus (Poir.) DC.
S. selloi (Spreng.) DC.
S. tweediei Hook & Arn.
Spilanthes sp.

Family Calyceraceae
Acicarpha spathulata R.Br.
Calycera crassifolia (Miers) Hick.

Order Campanulales
Family Campanulaceae
Pratia hederacea (Cham.) G.Don.

KINGDOM ANIMALIA
PHYLUM PORIFERA
CLASS HEXATINELLIDA
Order Dyctyonina (= Hexasteropoda)
Family Euretidae
Dactylocalix pumiceus Stutchubury, 1841
Margaritella sp.

CLASS DEMOSPONGIAE
Order Astrophorida
Family Ancorinidae
Penares sp.
Stelletta sp.

Family Geodiidae
Cydonium glariosus Sollas, 1886
Erylus oxyaster Lendenfeld, 1910
E. formosus Sollas, 1886
Geodia eosaster (Sollas, 1886)
Geodia sp.

Family Pachastrellidae
Pachastrella monilifera Schmidt, 1870
Sphinctrella cribrifera (Sollas, 1886)

Order Epipolasida
Family Jaspidae
Jaspis johnstoni (Schmidt, 1868)

Order Lithistida
Family Corallistidae
Corallistes typus Schmidt, 1870

Family Pleromidae
Pleroma sp.
Siphonidium sp.

Order Hadromerida
Family Tethydae
Halicometes minuta Sarà & Barbosa, 1995

Family Clionidae
Cliona sp.

Family Suberitidae
Pseudosuberites melanos Laubenfels, 1934
Suberites caminatus Ridley & Dendy, 1886

Order Axinellida (= Halicondrida)
Family Axinellidae
Axinella sp.
Bubaris sp.
Phakelia sp.

Order Poecilosclerida
Family Microcionidae
Clathria sp.
Microciona sp.

Family Raspailiidae
Raspailia sp.

Family Rhabderemiidae
Rhabderemia uruguaiensis (Van Soest & Hooper, 1993)

Family Iophonidae
Iophon sp.

Family Tedaniidae
Tedania spinata (Ridley, 1881)
Tedania sp.

Family Desmacellidae
Biemna sp.

Order Haplosclerida
Family Chalinidae
Acervochalina sp.

Family Niphatidae
Isodictya sp.

Order Dictyoceratida
Family Thorectidae
Ircinia sp.

Family Dysideidae
Dysidea sp.

PHYLUM CNIDARIA
CLASS HYDROZOA
Order Trachylina
Family Geryonidae
Geryonia proboscidalis (Forskal, 1775)
Liriope tetraphylla (Chamisso & Eysenhart, 1821)

Family Halicreidae
Halitrephes maasi Bigelow, 1909

Family Rhopalonematidae
Aglantha digitale (Müller, 1776)
Aglaura hemistoma Péron & LeSueur, 1809
Persa incolorata McCrady, 1857
Rhopalonema velatum Gegenbaur, 1856
Sminthea erygaster Gegenbaur, 1856

Family Solmariidae
Pegantha clara Bigelow, 1909
P. laevis Bigelow, 1909
P. martagon Haeckel, 1879
P. rubiginosa (Kölliker, 1853)
P. triloba Haeckel, 1879
Solmaris corona (Keferstein & Ehlers, 1861)

Family Cuninidae
Cunina duplicata Maas, 1893
C. frugifera Kramp, 1948
C. octonaria McCrady, 1857
C. peregrina Bigelow, 1909

Family Aeginidae
Aegina citrea Eschscholtz, 1829
Solmundella bitentaculata Quoy & Gaimard, 1833

Order Hydroida
Suborder Limnomedusae
Family Olindiidae
Olindias sambaquiensis (Delle Chiaje, 1841)
Vallentinia gabriellae Mendes, 1948

Family Proboscidactylidae
Proboscidactyla ornata (McCrady, 1857)

Suborder Anthomedusae
Family Zancleidae
Zanclea costata Gegenbaur, 1856

Family Cytaeidae
Cytaeis tetrastyla Eschscholtz, 1829

Family Bougainvilliidae
Bougainvillia niove Mayer, 1894
B. platygaster (Haeckel, 1879)

Family Clavidae
Cordylophora sp.
Turritopsis nutricula McCrady, 1856

Family Corynidae
Dipurena reesi Vannucci, 1956
Stauridiosarsia producta (Writh, 1858)

Family Tubulariidae
Ectopleura dumortieri (Van Beneden, 1844)
Euphysora furcata Kramp, 1948
E. gracilis (Brooks, 1882)

Family Pandeidae
Amphinema dinema (Péron & LeSueur, 1809)
Halitholus intermedius (Browne, 1902)
Pandea conica (Quouy & Gaimard, 1827)

Family Calycopsidae
Calycopsis borchgrevinki (Browne, 1910)
Heterotiara anonyma Maas, 1905

Suborder Leptomedusae
Family Laodiceidae
Laodicea undulata (Forbes & Goodsir, 1853)

Family Mitrocomidae
Cosmetira pilosella Forbes, 1848

Family Eurenidae
Phialopsis diegensis Torrey, 1909

Family Phialuciidae
Phialucium carolinae (Mayer, 1900)

Family Lovenellidae
Eucheilota ventricularis McCrady, 1857

Family Campanullaridae
Obelia spp.
Phialidium discoidum (Mayer, 1900)
P. ovale (Mayer, 1900)
P. simplex Browne, 1902

Family Plumulariidae
Kirchenpaueria sp.

Order Siphonophora
Suborder Cystonectae
Family Physalidae
Physalia physalis (Linnaeus, 1759)

Suborder Physonectae
Family Agalmidae
Marrus antarticus Totton, 1954
Stephanomia bijuga (Delle Chiaje, 1841)

Suborder Calycophorae
Family Prayidae
Amphicaryon ernesti Totton, 1954

Family Hippopodiidae
Hippopodius hippopus (Forskal, 1776)
Vogtia glabra Bigelow, 1918

Family Diphyidae
Chelophyes appendiculata (Eschscholtz, 1829)
C. contorta (Lens & van Riemsdijk, 1908)
Dimophyes arctica (Chun, 1897)
Diphyes bojani (Eschscholtz, 1829)
D. dispar Chamisso & Eisenhardt, 1821
Diphyopsis mitra (Huxley, 1859)
Eudoxoides spiralis (Bigelow, 1911)
Lensia campanella (Moser, 1925)
L. conoidea (Keferstein & Ehlers, 1860)
L. cossack Totton, 1941
L. grimaldi (Leloup, 1933)
L. hardy Totton, 1941
L. meteori (Leloup, 1934)
L. subtilis (Chun, 1886)
Muggiaea kochi (Wild, 1844)
Sulculeolaria biloba (Sars, 1846)
S. chuni (Lens & van Riemsdijk, 1908)
S. quadrivalvis Blainville, 1834
S. turgida (Gegenbaur, 1853)

Family Abylidae
Abyla trigona Quoy & Gaimard, 1827
Abylopsis eschscholtzi (Huxley, 1859)
A. tetragona (Otto, 1823)
Bassia bassensis (Quoy & Gaimard, 1834)
Ceratocymba leuckarti Huxley, 1859
C. sagittata Quoy & Gaimard, 1827
Enneagonum hyalinum (Quoy & Gaimard, 1827)

CLASS SCYPHOZOA
Order Semaeostomeae
Family Pelagiidae
Chrysaora hysoscella (Linnaeus, 1766)
C. lactea Eschscholtz, 1829
C. quinquecirrha (Desor, 1848)

Family Ulmaridae
Aurelia aurita (Linnaeus, 1746)

Family Cyaneidae
Drymonema gorgo F.Müller, 1883

Order Rhizostomeae
Family Lychnorhizidae
Lychnorhiza lucerna Haeckel, 1880

CLASS ANTHOZOA
SUBCLASS OCTOCORALLIA
Order Gorgonacea (=Alcyonaria)
Family Anthothelidae
Tripalea clavaria (Studer, 1878)

Family Plexauridae
Muricea midas (Bager, 1959)
Muricea sp.
Muriceides hirtus (Pourtalés, 1867)
Muriceides sp.

Family Elliselidae
Elisella barbadensis (Duchassaing & Michelotti, 1864)
E. elongata (Pallas, 1766)

Family Primnoidae
Primnoella biserialis Wright & Studer, 1889
P. compressa Kukenthal, 1908
P. distans Studer, 1878
P. magelhaenica Studer, 1878
P. philippii Aurivillius, 1931

Family Acanthogorgiidae
Acanthogorgia scharammi(Duchassaing & Michelotti, 1864)

Family Paramuriceidae
Bebryce cinerea (Deichamann, 1936)

Family Gorgoniidae
Leptogorgia setacea (Pallas, 1766)

Family Isididae
Acanella eburnea (Pourtalés, 1868)

Order Pennatulacea
Family Renillidae
Renilla kollikeri
R. mulleri Kolliker, 1872
R. musaica
R. reniformis (Pallas, 1766)
R. tentaculata

SUBCLASS HEXACORALLIA (= ZOANTHARIA)
Order Zoanthidea
Family Zoanthidae
Epizoanthus sp.

Order Actiniaria
Family Hormathiidae
Actinauge longicornis Verril, 1882
Phelliactis lophohelia Riemann-Zürneck, 1973
P. pelophyla Riemann-Zürneck, 1973

Family Actiniidae
Bunodosoma caicarum
B. cangicum
Phlyctenanthus australis Carlgren, 1949

Family Edwardsiidae
Edwardsia sp.

Order Scleractinia
Family Faviidae
Cladocora debilis Milne-Edwards & Haime, 1849

Family Rhizangiidae
Astrangia rathbuni Vauhan, 1906

Family Caryophylliidae
Caryophyllia smithii Stokes & Broderip, 1828
Dasmosmilia lymani Pourtalés, 1880
Deltocyathus conicus Zibrowius, 1980
Lophelia pertusa (Linnaeus, 1758)

Family Flabellidae
Flabellum brasiliensis
Flabellum spp.
Javania cailleti (Duchassaing & Michelotti, 1846)

Family Dendrophylliidae
Rhizopsammia manuelensis Chevalier, 1966

Order Antipatharia
Family Antipathidae
Antipathes sp.

PHYLUM NEMERTINA
Family Ototyphlonemertidae
Ototyphlonemertes sp.

PHYLUM PLATYHELMINTHES
CLASS TURBELLARIA
Order Polycladida
Family Stylochidae
Stylochoplana sp.
Stylochus sp.

PHYLUM SIPUNCULIDA
Family Sipunculidae
Sipunculus nudus

Family Golfingiidae
Golfingia sp.
Nephasomma sp.
Themistes sp.

PHYLUM ECHIURIDA
Family Echiuridae
Echiurus sp.

Family Bonelliidae
Bonellia sp.

PHYLUM PRIAPULIDA
Priapulus sp.

PHYLUM ANNELIDA
CLASS POLYCHAETA
Order Orbiniida
Family Orbiniidae
Phylo felix Kinberg, 1866
Scoloplos (= Leodomas) ohlini (Ehlers, 1901)

Family Paraonidae
Aedicira belgicae (Fauvel, 1936)

Order Cossurida
Family Cossuridae
Cossura candida Hartman

Order Spionida
Family Spionidae
Apoprionospio pygmaea (Hartman, 1961)
Boccardia hamata (Webster, 1879)
Dispio sp.
Laonice cirrata (Sars, 1851)
Paraprionospio pinnata (Ehlers, 1901)
Polydora ligni Webster, 1879
Prionospio steenstrupi Malmgren, 1867
Spio gaucha Orensanz & Gianuca, 1974
Spiophanes bombyx (Claparède, 1870)
S. kroeyeri (Grube, 1860)
S. soederstroemi Hartman, 1953

Family Magelonidae
Magelona riojai Jones, 1963

Family Poecilochaetidae
Poecilochaetus australis Nonato, 1963

Family Chaetopteridae
Phyllochaetopterus socialis
Spiochaetopterus spp.

Family Cirratulidae
Dodecaceria sp.

Order Capitellida
Family Capitellidae
Heteromastus similis Southern, 1921
Notomastus lobatus Hartman, 1947

Family Maldanidae
Asychis amoena Kinberg
A. atlanticus Kinberg
A. brasiliensis Kinberg
Euclymene sp.
Nicomache sp.
Rhodine sp.

Order Opheliida
Family Opheliidae
Ammotrypane trigintae Hartman, 1953
Ammotrypane sp.
Euzonus furciferus (Ehlers, 1897)
Travisia forbesii Johnston, 1840

Order Phyllodocida
Family Phyllodocidae
Anaitides mucosa (Oersted)
Anaitides sp.
Dodecaceria spp.
Eulalia spp.
Phyllodoce mucosa
Polynoe sp.

Family Alciopidae
Alciopa reynaudii Audouin & Milne-Edwards, 1833
Alciopina parassitica Claparède & Panceri, 1867
Plotohelmis capitata (Greeff, 1876)
Rhynchonereella gracilis Costa, 1862
R. petersii (Langerhans, 1880)
Vanadis crystallina Greeff, 1876
V. formosa Claparède, 1870
V. longissima Levinsen, 1885
V. minuta Treadwell, 1906
V. studeri Apstein, 1893

Family Aphroditidae
Aphrodita longicornis Kinberg, 1855
Aphroditella alta (Kinberg, 1855)
Laetmonice parva Amaral & Nonato, 1982
Lamelleulepethus orensanzi1 Pettibone, 1986

Family Polynoidae
Euphionella besnardi Amaral & Nonato, 1982
Halosydnella australis (Kinberg, 1855)
H. brasiliensis (Kinberg, 1858)
Lagisca lepida Amaral & Nonato, 1982
Lepidastenia esbelta Amaral & Nonato, 1982
Lepidastenia sp.

Family Polyodontidae
Eupanthalis kinbergi McIntosh, 1876
E. rudipalpa Amaral & Nonato, 1984
Panthalis bicolor Grube, 1877
Polyodontes oculea (Treadwell, 1901)
P. oerstedi Kinberg, 1855

Family Pholoididae
Pholoe minuta (Fabricius, 1780)

Family Eulepethidae
Grubeulepis bracteata Nonato

Pareulepis multibranchiata Amaral & Nonato, 1984

Family Sigalionidae
Ehlersileanira incisa (Grube, 1877)
Psammolyce flava Kinberg, 1855
Sigalion cirriferum Orensanz & Gianuca, 1974
Sthenelais limicola (Ehlers, 1901)
Sthenolepis oculata (Hartman, 1942)
Thalenessa edwardsi (Kinberg, 1855)

Family Chrysopetalidae
Dysponetus sp.
Paleonotus heteroseta Hartman, 1945

Family Pilargiidae
Parandalia americana (Hartman, 1947)
Sigambra grubii Fritz Müller, 1858

Family Syllidae
Odontosyllis hetero-falcata Temperine, 1981
Syllis prolixa Ehlers, 1901
Typosyllis ehlersioides Merenzeller, 1890

Family Nereidae
Eunereis patagonica (McIntosh, 1885)
Laeonereis acuta (Treadwell, 1923)
Lycastopsis pontica (Bobretzky, 1872)
Neanthes bruaca Lana & Sovierzoski, 1987
N. succinea (Frey & Leuckart, 1847)
Nereis falsa Quatrefages, 1865

Family Glyceridae
Glycera americana Leidy, 1855
Hemipodus olivieri Orensanz & Gianuca, 1974

Family Goniadidae
Glycinde multidens Fritz Müller, 1858
Goniada littorea Hartman, 1950
G. maculata Oersted
Ophioglycera eximia (Ehlers, 1900)

Family Iospilidae
Iospilus affinis (Viguier, 1911)
Phalacrophorus pictus Greeff, 1879
P. uniformis Reibisch, 1893

Family Nephtyidae
Aglaophamus praetiosus (Kinberg, 1866)
A. uruguayi Hartman, 1953
Nephtys acrochaeta Hartman, 1950
N. fluviatilis Monro, 1937
N. simoni Perkins, 1980
N. squamosa Ehlers

Family Tomopteridae
Tomopteris elegans Chun, 1888
T. kefersteini Greeff, 1879
T. planktonis Apstein, 1900
T. rolasi Greeff, 1885
T. septentrionalis Quatrefages, 1865

Family Typhloscolecidae
Sagitella kowalevskii N.Wagner, 1872
Travisiopsis dubia Stop-Bowitz, 1948
T. levinseni Southern, 1910
T. lobifera Levinsen, 1885
Typhloscolex muelleri Busch, 1851

Order Amphinomida
Family Amphinomidae
Amphinome rostrata (Pallas, 1766)
Hipponoa gaudichaudi agulhana Day, 1967

Order Eunicida
Family Onuphidae
Americonuphis casamiquelorum Orensanz, 1974
Diopatra cuprea (Bosc, 1802)
D. tridentata Hartman, 1944
D. viridis Kinberg, 1865
Hyalonecia juvenalis Moore, 1911
H. tubicola (Müller, 1788)
Kinbergonuphis difficilis (Fauchauld, 1982)
Nothria benthophyla Temperini, 1981
Onuphis eremita Audouin & Milne-Edwards, 1833
O. eremita oculata Hartman, 1951
O. setosa Orensanz, 1974
O. tenuis Hansen, 1881
Ramphobrachium verngreni (Kinberg, 1865)

Family Eunicidae
Eunice argentinensis (Treadwell, 1929)
E. frauenfeldi Grube, 1868
E. rubra Grube, 1856
E. vittata (Delle Chiaje, 1828)
Marphysa conferta Moore
M. disjuncta Hartman
M. kinbergi McIntosh, 1910
M. sanguinea (Montagu, 1815)

Family Lumbrineridae
Aglaurides fulgida (Savigny)
Arabellonereis janeiroensis (Augener, 1934)
Lumbrinerides gesae Orensanz, 1973
Lumbrineriopsis nucronata (Ehlers, 1908)
Lumbrineris atlantica Kinberg, 1865
L. cingulata (Ehlers, 1897)
L. latreilli Audouin & Milne-Edwards, 1834
L. magalhaensis Kinberg, 1865
L. tetraura (Schmarda, 1861)
Ninoe brasiliensis Kinberg, 1865
N. gayheadia Hartman, 1965
Ninoe sp.
Paraninoe antarctica (Monro, 1930)

Family Arabellidae
Arabella iricolor (Montagu, 1804)
Drilonereis filum (Claparède, 1868)
D. simplex Orensanz, 1974
D. tenuis (Ehlers, 1900)
Notocirrus lorum Ehlers, 1897

Family Lysaretidae
Lysarete brasiliensis Kinberg, 1865

Family Dorvilleidae
Pettiboneia sanmatiensis Orensaz, 1973

Order Sternaspida
Family Sternaspidae
Sternaspis sp.

Order Oweniida
Family Oweniidae
Owenia fusiformis Delle Chiaje, 1844
Owenia spp.

Order Flabelligerida
Family Flabelligeridae
Pherusa capitata Nonato, 1966
P. scutigera Ehlers

Order Terebellida
Family Sabellariidae
Phragmatopoma sp.
Sabellaria bellis Hansen, 1882

Family Pectinariidae
Cystenides sp.
Pectinaria laelia Nonato, 1981

Family Ampharetidae
Amphicteis latibranchiata Nonato
Melinna uruguayi Hessle, 1917

Family Terebellidae
Artacama benedeni Kinberg, 1867
Telephus setosus (Quatrefages, 1865)

Family Trichobranchidae
Terebellides stroemi Sars
Tricobranchus lobiungens Hessle, 1917

Order Sabellida
Family Sabellidae
Chone infundibiliformis Kroyer, 1856
Megalomma sp.
Sabellaria bella Grube, 1870

Family Serpulidae
Filograna implexa Berkeley, 1827
Hydroides parvus Treadwell, 1901
Omphalopoma sp.
Protula sp.

Family Protodrilidae
Protodrilus sp.

PHYLUM MOLLUSCA
CLASS POLYPLACOPHORA
Order Neoloricata
Family Ischnochitonidae
Chaetopleura angulata (Splenger, 1797)
C. isabellei (Orbigny, 1841)

CLASS GASTROPODA
SUBCLASS PROSOBRANCHIA
Order Archeogastropoda
Family Pleurotomariidae
Mikadotrochus notialis Leme & Penna, 1969
Perotrochus amabilis Bayer, 1963
P. atlanticus Rios & Matthews, 1968

Family Scissurellidae
Anatoma aedonia (Watson, 1886)

Family Fissurellidae
Diodora patagonica (Orbigny, 1847)
Emarginula tuberculosa Libassi, 1859
Fissurellidea megatrema (Orbigny, 1841)
Lucapinella henseli (Martens, 1900)
L. limatula (Reeve, 1850)
Puncturella granulata Seguenza, 1863

Family Acmaeidae
Collisella subrugosa (Orbigny, 1846)

Family Trochidae
Calliostoma axelolssoni Quinn, 1992
C. carcellesi Clench & Aguayo, 1939
C. coppingeri (E.A.Smith, 1880)
C. jucundum (Gould, 1849)
C. militaris (Ihering, 1907)
Calliotropis actinophora (Dall, 1890)
C. calatha (Dall, 1927)
Halistylus columna Dall, 1890
Mirachelus clinocnemus Quinn, 1979
Solariella lubrica Dall, 1881
S. patriae Carcelles, 1953
Tegula patagonica (Orbigny, 1840)

Family Turbinidae
Arene microforis (Dall, 1889)
A. variabilis (Dall, 1889)

Family Seguenziidae
Ancistrobasis costulata (Watson, 1879)

Order Mesogastropoda
Family Hidrobiidae
Heleobia australis (Marcus & Marcus, 1963)
H. parchapei (Orbigny, 1835)

Family Littorinidae
Littorina ziczac (Gmelin, 1791)

Family Barleeidae
Barleeia rubrooperculata (Castellanos, 1927)

Family Caecidae
Caecum achironum (Folin, 1867)
C. antillarum Carpenter, 1857
Mioceras nitidum (Stimpson, 1851)

Family Vitrinellidae
Solariorbis shimeri (Clapp, 1914)

Family Tornidae
Macromphalina argentina Castellanos, 1975

Family Fossaridae
Fossarus perplexa (Dall, 1927)

Family Calyptraeidae
Calyptraea centralis (Conrad, 1841)
Crepidula aculeata (Gmelin, 1791)
C. protea Orbigny, 1835

Family Xenophoridae
Xenophora caribaea Petit, 1856

Family Ovulidae
Primovula vanhyningi (M.Smith, 1940)

Family Lamellariidae
Lamellaria patagonica E.A.Smith, 1881

Family Naticidae
Natica limbata Orbigny, 1840
Polinices hepaticus (Roding, 1798)
Polinice sp.

Family Tonnidae
Tonna galea Brunnich, 1772

Family Cassidae
Phalium granulatum (Born, 1778)
P. labiatum iheringi Carcelles, 1953

Family Ranellidae
Cabestana felipponei (Ihering, 1907)
Cymatium parthenopeum (von Salis, 1793)
Fusitriton magellanicus (Roding, 1798)

Family Triphoridae
Triphora pulchella (C.B.Adams, 1850)

Family Epitoniidae
Cylindriscala andrewsii (Verril, 1882)
C. watsoni (de Boury, 1911)
Epitonium albidum (Orbigny, 1842)
E. *angulatum* (Say, 1830)
E. candeanum (Orbigny, 1842)
E. celesti (Aradas, 1854)
E. dallianum Verril & Smith, 1880
E. georgettina (Kiener, 1839)
E. humpheysii (Kiener, 1838)
E. magellanicum (Philippi, 1845)
E. pourtalesii Verril & Smith, 1880
E. tenuistriatum (Orbigny, 1840)
E. unifasciatum (Sowerby, 1844)

Family Eulimidae
Eulima auricincta Abbot, 1959
E. bifasciata (Orbigny, 1842)
Hemiaclis aff. *georgiana* (Dall, 1927)
Melanella arcuata (C.B.Adams, 1850)
M. conoidea (Kurtz & Simpson, 1851)

Niso aeglees Bush, 1895

Order Neogastropoda
Family Muricidae
Chicoreus beauii (Fischer & Bernardi, 1857)
Murexiella hidalgoi (Crosse, 1869)
Muricopsis necocheana (Pilsbry, 1900)
Orania fusulus (Brocchi, 1814)
Trophon acanthodes Watson, 1882
T. pelseneeri E.A.Smith, 1915
T. plicatus (Lightfoot, 1786)
Typhis cleryi (Petit, 1842)
T. riosi Bertsch & d'Attilio, 1980
Urosalpix cala (Pilsbry, 1897)
U. haneti (Petit, 1856)

Family Thaididae
Thais haemastoma (Linnaeus, 1767)

Family Coralliophilidae
Coralliophila aedonius (Watson, 1886)

Familiy Columbellidae
Aesopus metcalfei (Reeve, 1858)
Amphissa acuminata (E.A.Smith, 1915)
A. cancellata (Castellanos, 1982)
Anachis isabellei (Orbigny, 1841)
A. sertulariarum (Orbigny, 1841)

Family Nassariidae
Buccinanops deformis (King, 1831)
B. duartei Klappenbach, 1961
B. *gradatum* (Deshayes, 1844)
B. lamarckii (Kiener, 1834)
B. uruguayensis (Pilsbry, 1897)
Dorsanum moniliferum (Valenciennes, 1834)
Nassarius scissuratus (Dall, 1889)

Family Fasciolariidae
Fusinus frenguelli (Carcelles, 1953)
Metula anfractura Matthews & Rios, 1968

Family Vasidae
Fulgurofusus coronatum (Penna-Neme & Leme, 1978)

Family Olividae
Ancilla dimidiata (Sowerby, 1850)
Olivancillaria auricularia (Lamarck, 1810)
O. carcellesi Klappenbach, 1965
O. deshayesiana (Duclos, 1857)
O. urceus (Roding, 1798)
O. uretai Klappenbach, 1965
Olivella formicacorsii Klappenbach, 1962
O. orejasmirandai Klappenbach, 1986
O. plata (Ihering, 1909)
O. puelcha (Duclos, 1840)
O. riosi Klappenbach, 1991
O. tehuelcha (Duclos, 1840)

Family Marginellidae
Granulina ovuliformis Orbigny, 1841
Prunum martini (Petit, 1853)
P. rubens (Martens, 1881)

Family Mitridae
Mitra larrangai (Carcelles, 1947)

Family Conidae
Conus clerii Reeve, 1844
C. mazei Deshayes, 1874
C. tostesi Petuch, 1986
C. villepinii Fisher & Bernardi, 1867

Family Volutidae
Adelomelon ancilla (Lighfoot, 1786)
A. beckii (Broderip, 1836)
A. brasiliana (Lamarck, 1811)
A. riosi Clench & Turner, 1964
Nanomelon vossi Leal & Rios, 1990
Odontocymbiola americana (Reeve, 1856)
O. corderoi (Carcelles, 1953)
Zidona dufresnei (Donovan, 1823)

Family Turridae
Benthomangelia macra (Watson, 1881)
Brachytoma rioensis (E.A.Smith, 1915)
Carinodrillia braziliensis (E.A.Smith, 1915)
Cochlespira radiata (Dall, 1889)
Cryoturris adamsi (E.A.Smith, 1884)
Fusiturricula lavinoides limonensis (Olsson, 1922)
F. taurina (Olsson, 1922)
Kurtziella dorvillae (Reeve, 1845)
K. padolina (Fargo, 1953)
Leucosyrinx dalli (Busch, 1893)
L. hemimeres(Watson, 1881)
Oenopota magellanica (Martens, 1881)
Pleurotomella aguayoi (Carcelles, 1953)
P. cala (Watson, 1886)
P. ipara (Dall, 1881)

P. meriacum (Locard, 1896)
Polystira formosissima (E.A.Smith, 1915)
Spirotropis lithocolleta (Watson, 1881)
S. patagonica (Orbigny, 1841)
S. stirophora (Watson, 1881)
Splendrillia carolinae (Bartsch, 1934)
Veprecula morra (Dall, 1881)

Family Terebridae
Terebra doello-juradoi Carcelles, 1953
T. gemmulata Kiener, 1839

SUBCLASS HETEROBRANCHIA
Order Heterostropha
Family Architectonicidae
Heliacus bisulcatus (Orbigny, 1845)
Philippia uruguaya (Carcelles, 1953)
Pseudomalaxis nobilis Verril, 1885

Family Pyramidellidae
Eulimastoma engonium surinamense Altena, 1975
Odostomia canaliculata C.B.Adams, 1850
O. jadisi Olsson & McGinty, 1958
O. seminuda (C.B.Adams, 1837)
Turbonilla americana (Orbigny, 1840)
T. atypha Bush, 1900
T. dispar Pilsbry, 1897
T. fasciata (Orbigny, 1840)
T. interrupta (Totten, 1835)
T. rushii Bush, 1899
T. uruguayensis Pilsbry, 1897

Family Amathinidae
Iselica anomala (C.B.Adams, 1850)

SUBCLASS OPISTHOBRANCHIA
Order Cephalaspidea
Family Acteonidae
Acteon pelecais Marcus, 1981
Mysouffa cumingii (A.Adams, 1854)

Family Cylichnidae
Acteocina bidentata (Orbigny, 1841)
A. candei (Orbigny, 1842)
A. inconspicua Olsson & McGinty, 1958
Cylichna aff. *crispula* Watson, 1883
C. discus Watson, 1883
C. verrillii Dall, 1889
Scafander watsoni Dall, 1881

Family Philinidae
Philine falklandica Powell, 1951

Family Bullidae
Bulla striata Bruguiere, 1792

Family Retusidae
Pyrunculus caelatus (Bush, 1885)
Volvulella paupercula (Watson, 1883)
V. persimilis (Morch, 1875)

Order Thecosomata
Family Limacinidae
Limacina bulimoides (Orbigny, 1836)
L. inflata (Orbigny, 1836)
L. lesueuri (Orbigny, 1836)
L. retroversa f. *retroversa* (Fleming, 1823)
L. trochiformis (Orbigny, 1836)

Family Cavoliniidae
Cavolinia gibbosa f. *flava* (Orbigny, 1836)
C. inflexa (LeSueur, 1813)
C. inflexa f. *inflexa* (LeSueur, 1813)
C. inflexa f. *imitans* (Pfeffer, 1880)
C. inflexa f. *labiata* (Orbigny, 1836)
C. *longirostris* f. *longirostris* (Blainville, 1821)
C. longirostris f. *limbata* (Orbigny, 1836)
C. longirostris f. *strangulata* (Deshayes, 1823)
C. tridentata (Niebuhr, 1775)
C. uncinata (Rang, 1829)
C. uncinata f. *uncinata* (Rang, 1829)
Clio cuspidata (Bosc, 1802)
C. pyramidata f. *antartica* (Dall, 1908)
C. pyramidata f. *lanceolata* (LeSueur, 1813)
Creseis. acicula Rang, 1828
C. acicula f. *acicula* (Rang, 1828)
C. acicula f. *clava* (Rang, 1828)
C. virgula f. *conica* Eschscholtz, 1929
C. *virgula* f. *virgula* (Rang, 1828)
Cuvierina columnella f. *atlantica* Van der Spoel, 1970
Diacria rampali Dupont, 1979
D. trispinosa f. *trispinosa* (Blainville, 1821)
Hyalocylis striata (Rang, 1828)
Styliola subula (Quoy & Gaimard, 1827)

Order Nudibranchia
Family Tritoniidae
Marionia cucullata (Gould, 1852)

Tritonia eriosi Ev.Marcus, 1983

Order Basommatophora
Family Siphonariidae
Siphonaria lessoni (Blainville, 1824)

CLASS PELECYPODA
SUBCLASS PROTOBRANCHIA
Order Solemyoida
Family Solemyidae
Solemya patagonica E.A.Smith, 1885

Order Nuculoidea
Family Nuculidae
Nucula puelcha Orbigny, 1846
N. semiornata Orbigny, 1846

Family Tindariidae
Tindaria striata (King, 1831)

Family Nuculanidae
Adrana electa (A.Adams, 1846)
A. patagonica (Orbigny, 1846)
Nuculana fortiana Esteves, 1984
N. larranagai Klappenbach & Scarabino, 1968

Family Yoldiidae
Yoldia riograndensis Esteves, 1984

Family Malletiidae
Malletia cumingii (Hanley, 1860)

SUBCLASS PTERIOMORPHA
Order Arcoida
Family Arcidae
Anadara chemnitzi (Philippi, 1851)
A. ovalis (Bruguìere, 1789)

Family Noetiidae
Noetia bisulcata (Lamarck, 1819)

Family Limopsidae
Limopsis davinae Esteves, 1984
L. janeiroensis E.A.Smith, 1915
L. minuta (Philippi, 1836)

Family Glycymerididae
Glycymeris longior (Sowerby, 1833)

Order Mytiloida
Family Mytilidae
Amygdalum dendriticum Muhlfels, 1811
A. sagittatum Rehder, 1934
Brachidontes exustus (Linnaeus, 1758)
B. rodriguezi (Orbigny, 1846)
B. solisianus (Orbigny, 1846)
Crenella divaricata (Orbigny, 1846)
Lithophaga patagonica (Orbigny, 1846)
Modiolus carvalhoi Klappenbach, 1966
Musculus viator (Orbigny, 1846)
Mytilus edulis platensis (Orbigny, 1846)
Perna perna (Linnaeus, 1758)

Order Pterioida
Family Pteriidae
Pteria hirundo (Linnaeus, 1758)

Family Pinnidae
Atrina seminuda (Orbigny, 1846)

Order Limoida
Family Limidae
Limatula pigmaea (Philippi, 1845)

Order Ostreoida
Family Ostreidae
Ostrea equestris Say, 1834
O. puelchana (Orbigny, 1841)

Family Pectinidae
Chlamys felipponei (Dall, 1922)
C. patagonica (King, 1832)
C. tehuelchus (Orbigny, 1846)
Cyclopecten leptaleus (Verrill, 1884)
Leptopecten bavayi (Dautzenberg, 1900)

Family Propeamussiidae
Propeamussium pourtalesianum (Dall, 1886)

Family Anomiidae
Anomia ephippium Linnaeus, 1758
Pododesmus leloiri Carcelles, 1941
P. rudis (Broderip, 1834)

SUBCLASS HETERODONTA
Order Veneroida
Family Lucinidae
Myrtea lens (Verrill & Smith, 1880)

Family Thyasiridae
Thyasira croulinensis Jeffreys, 1874

Family Ungulinidae
Diplodonta patagonica (Orbigny, 1842)
Felaniella vilardeboana (Orbigny, 1846)

Phlyctiderma semiaspera (Philippi, 1836)

Family Carditidae
Carditamera floridana Conrad, 1838
C. plata (Ihering, 1907)
Cyclocardia moniliata (Dall, 1902)
Pleuromeris sanmartini Klappenbach, 1971)

Family Condylocardiidae
Americuna besnardi Klappenbach, 1962
Carditopsis smithii (Dall, 1896)

Family Chamidae
Arcinella brasiliana (Nicol, 1953)

Family Crassatellidae
Crassatella riograndensis Vokes, 1973
Crassinella lunulata (Conrad, 1834)
C. marplatensis Castellanos, 1970

Family Cardiidae
Trachycardium muricatum (Linnaeus, 1758)

Family Mactridae
Mactra isabelleana Orbigny, 1846
M. janeiroensis Smith, 1915
M. marplatensis Doello-Jurado, 1918
M. patagonica Orbigny, 1846
M. petiti Orbigny, 1846
Mulinia aff. *branneri* Dall, 1901
Raeta plicatella (Lamarck, 1818)

Family Mesodesmatidae
Mesodesma mactroides Deshayes, 1854

Family Solenidae
Solen tehuelchus Orbigny, 1843

Family Tellinidae
Macoma uruguayensis (E.A.Smith, 1885)
Strigilla carnaria (Linnaeus, 1758)
Tellina alerta Boss, 1964
T. gibber Ihering, 1907
T. petitiana Orbigny, 1846
T. sandix Boss, 1968

Family Semelidae
Abra lioica (Dall, 1881)
A. uruguayensis (Pilsbry, 1897)
Semele casali Doello-Jurado, 1949
S. proficua (Pulteney, 1799)

Family Psammobiidae
Tagelus plebeius (Lightfoot, 1786)

Family Donacidae
Donax gemmula Morrison, 1971
D. hanleyanus Philippi, 1842

Family Veneridae
Amiantis purpuratus (Lamarck, 1818)
Callista eucymata (Dall, 1889)
Chione paphia (Linnaeus, 1767)
Clausinella gayi (Hupé, 1854)
Eurhomalea exalbida (Dill. 1817)
Pitar rostratus (Koch, 1844)
Protothaca antiqua (King & Broderip, 1835)
Tivela isabelleana (Orbigny, 1846)
T. ventricosa (Gray, 1838)
Transenpitar americana (Doello-Jurado, 1951)

Family Ptricolidae
Petricola lapicida (Gmelin, 1791)
P. stellae Narchi, 1975

Order Myoida
Family Corbulidae
Corbula caribaea Orbigny, 1842
C. lyoni Pilsbry, 1897
C. patagonica Orbigny, 1846
C. tryoni E.A.Smith, 1880

Family Erodonidae
Erodona mactroides Bosc, 1802

Family Hiatellidae
Panopea abbreviata Valenciennes, 1839

SUBCLASS ANOMALODESMATA
Order Pholadina
Family Pholadidae
Barnea lamellosa (Orbigny, 1846)
Cyrtopleura lanceolata (Orbigny, 1846)
Pholas campechiensis Gmelin, 1791

Order Pholadomyoida
Family Thraciidae
Asthenothaerus rushii (Pilsbry, 1987)
Thracia distorta (Montagu, 1808)
T. similis Couthony, 1839

Family Periplomatidae
Periploma compressa Orbigny, 1846
P. ovata Orbigny, 1846

Family Lyonsiidae
Entodesma alvarezi Orbigny, 1846
E. patagonica (Orbigny, 1846)

Family Pandoridae
Pandora braziliensis Sowerby, 1874
P. bushiana Dall, 1886

Family Cuspidariidae
Cardiomya cleryana (Orbigny, 1846)
Cuspidaria platensis E.A.Smith, 1915

Family Verticordiidae
Euciroa grandis (E.A.Smith, 1885)
Verticordia acuticostata (Philippi, 1844)
V. fisheriana Dall, 1881
V. ornata (Orbigny, 1842)

CLASS SCAPHOPODA
Order Dentaliida
Family Dentaliidae
Antalis ceratum (Dall, 1881)
A. infractum (Odhner, 1931)
Fissidentalium amphialum (Watson, 1879)
F. carduum (Dall, 1889)

Order Gadilida
Family Entalinidae
Pertusiconcha callithrix (Dall, 1889)

Family Gadilidae
Cadulus braziliensis Henderson, 1920
Polyschides tetraschistus (Watson, 1879)

CLASS CEPHALOPODA
Order Sepioidea
Family Spirulidae
Spirula spirula Lamarck, 1801

Family Sepiolidae
Heteroteuthis atlantis Voss, 1955
Semirossia tenera (Steenstrup, 1887)

Order Teuthoidea
Family Loliginidae
Loligo plei (Blainville, 1823)
L. sanpaulensis Brakoniecki, 1984

Family Ctenopterygidae
Ctenopteryx siculus

Family Ommastrephidae
Illex argentinus (Castellanos, 1960)
Ornithoteuthis antillarum Adam, 1957
O. volatilis Sasaki, 1915
Ommastrephes bartramii (LeSueur, 1821)
Todarodes angolensis Adam, 1962

Family Thysanoteuthidae
Thysanoteuthis rhombus Troschel, 1857

Family Onychoteuthidae
Moroteuthis ingens (Smith, 1881)
M. robsoni Adam, 1962
Onychia caribaea LeSueur, 1821
Onychoteuthis banksi (Leach, 1817)

Family Enoploteuthidae
Abralia redfieldi (Voss, 1955)
A. veranyi (Ruppel, 1844)
Abraliopsis pfefferi Joubin, 1896
Ancistrocheirus leusueri (Vúrany, 1851)
Pterygioteuthis giardi Fischer, 1896

Family Histioteuthidae
Histioteuthis atlantica (Hoyle, 1885)
H. dofleini (Pfeffer, 1912)

Family Cranchiidae
Liocranchia reinhardti (Steenstrup, 1856)
Megalocranchia sp.

Family Chiroteuthidae
Chiroteuthis veranyi (Fúrussac, 1835)

Family Lycoteuthidae
Lycoteuthis diadema (Chun, 1900)

Order Octopoda
Family Octopodidae
Eledone gaucha Haimovici, 1988
E. massyae Voss, 1964
Octopus tehuelchus Orbigny, 1834
O. vulgaris (Cuvier, 1797)
Pareledone charcoti (Joubin, 1905)
P. turqueti (Joubin, 1905)
Pteroctopus tetracirrhus Delle Chiaje, 1830
Scaeurgus unicirrhus (Orbigny, 1840)
Vosseledone charrua Palacio, 1978

Family Bolitaenidae
Eledonella pigmaea Verril, 1884
Japetella diaphana Hoyle, 1885

Family Allopsidae
Allopsus mollis Verril, 1880

Family Argonautidae
Argonauta argo Linnaeus, 1758
A. hians Solander, 1786
A. nodosa Lightfoot, 1786

Family Tremoctopodidae
Tremoctopus violaceous Della Chiaje, 1830

Family Ocythoidae
Ocythoe tuberculata Rafinesque, 1814

PHYLUM ARTHROPODA
SUBPHYLUM CHELICERATA
CLASS ARACHNIDA
Order Araneae
Family Licosidae
Glieschiella (=*Moenkhausiana*) *halophila* (Mello-Leitão, 1932)

CLASS PYCNOGONIDA
Family Tanystylidae
Tanystylum orbiculare Wilson, 1878

Family Colossendeidae
Colossendeis geoffreyi Mañe-Garzón, 1944

SUBPHYLUM CRUSTACEA
CLASS BRANCHIOPODA
Order Cladocera
Family Polyphemidae
Evadne nordmanni Loven, 1835
E. spinifera Müller, 1868
Pleopis polyphemoides (Leuckart, 1859)
Podon intermedius Lilljeborg, 1859
P. schmackeri (Poppe)
Pseudoevadne tergestina (Claus, 1877)

Family Sididae
Diaphanosoma brachyurum (Lievin, 1848)
D. brevireme Sars, 1901
D. fluviatile Hansen, 1899
D. sarsi Richard, 1895
Penilia avirostris Dana, 1852
Pseudosida bidentata Herrick, 1884

Family Moinidae
Moina micrura Kurz, 1874

Family Daphniidae
Ceriodaphnia cornuta cornuta (Sars, 1886)
C. cornuta rigaudi (Richard, 1894)
C. dubia Richard, 1894
C. reticulata (Jurine, 1820)
Daphnia ambigua Scourfield, 1946
Scapholeberis spinifera (Nicolet, 1879)
Simocephalus exspinosus (Koch, 1841)
S. serrulatus Schoedler, 1858
S. vetulus (O.F.Müller, 1776)

Family Bosminidae
Bosminopsis deitersi Richard, 1895
Eubosmina tubicen (Brehm, 1953)

Family Macrothricidae
Echinisca triserialis Brady, 1886
Ilyocryptus spinifer Herrick, 1884
I. verrucosus Daday, 1905
Macrothrix laticornis (Jurine, 1820)

Family Chydoridae
Acroperus harpae (Baird, 1834)
Alona costata Sars, 1862
A. davidi davidi Richard, 1895
A. monacantha monacantha Sars, 1901
Alonella dentifera Sars, 1901
Biapertura affinis affinis (Leydig, 1860)
B. karua (King, 1852)
B. pseudoverrucosa verrucosa Sars, 1901
Camptocercus australis Sars, 1896
Chydorus ciliatus Poggenpol, 1874
C. eurynotus eurynotus Sars, 1901
C. faviformis Birge, 1893
C. sphaericus sphaericus (O.F.Müller, 1785)
Dunhevedia odontoplax Sars, 1901
Euryalona orientalis Daday, 1898
Eurycercus lamellatus (O.F.Müller, 1785)
Graptoleberis testudinaria (Fisher, 1851)
Kurzia latissima (Kurz, 1875)
Pleuroxus denticulatus Birge
P. similis Vavra, 1900
P. striatus Schoedler, 1858
P. truncatus truncatus (O.F.Müller, 1785)

CLASS OSTRACODA
Family Cypridinidae
Gigantocypris muelleri Skogsberg, 1920
Macrocypridina castanea (Brady, 1897)

Family Asteropidae
Leuroleberis poulseni (Moguilevsky & Ramirez, 1970)

Family Archiconchoecidae
Archiconchoecia cucullata (Brady, 1902)
A. striata Müller, 1894

Family Euconchoecidae
Euconchoecia chierchiae Müller, 1891

Family Halocyprididae
Halocypris inflata (Dana, 1852)
H. globosa Claus, 1874
Fellia cornuta (Müller, 1906)

Family Conchoecidae
Conchoecia acuminata (Claus, 1890)
C. acuticosta Müller, 1906
C. aequiseta Müller, 1906
C. allotherium Müller, 1906
C. amblypostha Müller, 1906
C. ametra Müller, 1906
C. antipoda Müller, 1906
C. atlantica (Lubbock, 1856)
C. bispinosa Claus, 1890
C. brachyaskos Müller, 1906
C. chuni Müller, 1906
C. ctenophora Müller, 1906
C. curta Lubbock, 1860
C. daphnoides (Claus, 1890)
C. dentata Müller, 1906
C. dichotoma Müller, 1906
C. discophora Müller, 1906
C. echinata Müller, 1906
C. edentata Müller, 1906
C. elegans Sars, 1866
C. gaussae Müller, 1908
C. glandulosa Müller, 1906
C. hadoni Brady & Norman, 1896
C. hettacra Müller, 1906
C. hirsuta Müller, 1906
C. hyalophyllum Claus, 1890
C. imbricata (Brady, 1880)
C. incisa Müller, 1906
C. isocheira Müller, 1906
C. kyrtophora Müller, 1906
C. lophura Müller, 1906
C. loricata Claus, 1894
C. macrocheira Müller, 1906
C. macromma Müller, 1906
C. magna Claus, 1874
C. major Müller, 1906
C. mamillata Müller, 1906
C. nasotuberculata Müller, 1906
C. oblonga (Claus, 1890)
C. obtusata Sars, 1866
C. parthenoda Müller, 1906
C. parvidentata Müller, 1906
C. plactolycos Müller, 1906
C. plinthina Müller, 1906
C. porrecta Claus, 1890
C. procera Müller, 1890
C. pseudoparthenoda Angel, 1972
C. pusilla Müller, 1906
C. reticulata Müller, 1906
C. rhynchena Müller, 1906
C. secernenda Vavra 1906
C. serrulata serrulata Claus, 1874
C. spinifera (Claus, 1890)
C. spinirostris Claus, 1874
C. stigmatica Müller, 1906
C. subarcuata Claus, 1890
C. symmetrica Müller, 1906
C. teretivalvata Iles, 1953
C. valdiviae Müller, 1906

CLASS MAXILLOPODA
SUBCLASS COPEPODA
Order Calanoida
Family Calanidae
Nannocalanus minor (Claus, 1863)
Calanus tenuicornis (Dana, 1849)
Neocalanus gracilis (Dana, 1849)
Calanus australis Brodskii, 1959
Undinula vulgaris (Dana, 1849)
Calanoides carinatus Krøyer, 1848

Family Eucalanidae
Eucalanus elongatus (Dana, 1849)
E. longiceps Matthews, 1925
E. crassus Giesbrecht, 1888
E. monachus Giesbrecht, 1888
E. subtenuis Giesbrecht, 1888
E. pileatus Giesbrecht, 1888
Rhincalanus cornutus (Dana, 1849)

Family Paracalanidae
Paracalanus aculeatus Giesbrecht, 1888
P. campaneri Björnberg, 1979
P. crassirostris F.Dahl, 1894
P. parvus (Claus, 1863)
P. quasimodo Bowman, 1971
P. indicus Wolfenden, 1905
Delius sewelli Björnberg, 1979

Family Calocalanidae
Calocalanus pavo (Dana, 1849)

C. pavoninus Farran, 1936
C. styliremis Giesbrecht, 1888
Ischnocalanus plumulosus Claus, 1963
Dolichocerea tenuis (Farran, 1926)

Family Mecynoceridae
Mecynocera clausi J.C.Thompson, 1888

Family Clausocalanidae
Clausocalanus furcatus (Brady, 1883)
C. pergens Farran, 1926
C. parapergens Frost & Fleminger, 1968
C. brevipes Frost & Fleminger, 1968
Ctenocalanus vanus Giesbrecht, 1888
C. citer Heron & Bowman, 1971

Family Aetideidae
Aetideus armatus (Boeck, 1872)
Euaetideus bradyi (A.Scott, 1909)
E. giesbrechti (Cleve, 1904)
E. acutus (Farran, 1929)
Paracomantena magalyae Campaner, 1978

Family Euchaetidae
Euchaeta spinosa Giesbrecht, 1892
E. acuta Giesbrecht, 1888
Paraeuchaeta barbata Brady, 1883
P. scotti Farran, 1909

Family Phaennidae
Phaenna spinifera Claus, 1863
Xanthocalanus marlyae Campaner, 1978
X. minor Giesbrecht, 1892
X. agilis Giesbrecht, 1892
Brachycalanus bjornbergae Campaner, 1978

Family Scolecithricidae
Scottocalanus securifrons (J.Scott, 1894)
Scaphocalanus magnus (T.Scott, 1894)
S. curtus (Farran, 1926)
S. echinatus (Farran, 1905)
S. impar (Wolfenden, 1911)
Lophotrix latipes (T.Scott, 1893)
Scolecithrix danae (Lubbock, 1856)
Scolecithricella bradyi (Giesbrecht, 1888)
S. dentata (Giesbrecht, 1892)

Family Temoridae
Temora stylifera (Dana, 1848)
T. turbinata (Dana, 1849)
Temeropia mayumbaensis T.Scott, 1893

Family Metridinidae
Pleuromamma xiphias Giesbrecht, 1889
P. piseki Farran, 1929
P. gracilis Claus, 1863

Family Centropagidae
Centropages velificatus (de Oliveira, 1947)
C. brachiatus (Dana, 1849)
C. violaceus (Claus, 1863)
C. bradyi Wheeler, 1899

Family Lucicutiidae
Lucicutia wolfendeni Sewell, 1932
L. ovalis (Giesbrecht, 1889)
L. gaussae Grice, 1963
L. longicornis (Giesbrecht, 1889)
L. flavicornis (Claus, 1863)

Family Heterorhabdidae
Heterorhabdus spinifrons (Claus, 1863)

Family Augaptilidae
Euaugaptilus hecticus (Giesbrecht, 1889)
E. filiger (Claus, 1863)

Family Diaptomidae
Notodiaptomus incompositus (Brian, 1925)
N. amazonicus (Wright, 1935)
N. carteri (Lowndes, 1934)
Argyrodiaptomus denticulatus Pesta, 1927

Family Pseudodiaptomidae
Pseudodiaptomus acutus (F.Dahl, 1894)
P. richardi F.Dahl, 1894

Family Candaciidae
Paracandacia simplex (Giesbrecht, 1892)
Candacia longimana Claus, 1863
C. bipinnata (Giesbrecht, 1888)
C. pachydactyla (Dana, 1848)
C. curta (Dana, 1849)
C. armata Boeck, 1872

Family Pontellidae
Calanopia americana F.Dahl, 1894
Labidocera fluviatilis F.Dahl, 1894
Pontellopsis brevis (Giesbrecht, 1889)
P. villosa Brady, 1883
Pontella marplatensis Ramírez, 1966
Pontellina plumata (Dana, 1849)

Family Acartiidae
Acartia longiremis (Lilljeborg, 1853)
A. lilljeborgi Giesbrecht, 1892
A. danae Giesbrecht, 1889
A. negligens (Dana, 1848)
A. tonsa Dana, 1848

Order Cyclopoida
Family Oithonidae
Oithona plumifera Baird, 1843
O. similis Claus, 1863
O. setigera (Dana, 1849)
O. oculata Farran, 1913
O. nana Giesbrecht, 1892
O. hebes Santos, 1973
O. oswaldocruzi Oliveira, 1945
O. atlantica Farran, 1908

Family Cyclopidae
Macrocyclops albidus (Jurine, 1820)
M. ater (Herrick, 1882)
Mesocyclops annulatus (Wierzejski, 1892)
M. meridianus (Kiefer, 1926)
M. longisetus (Thiebaud, 1914)
Microcyclops mendocinus (Wierzejski, 1902)
Eucyclops ensifer Kiefer, 1936

Family Clausidiidae
Hemicyclops subadhaerens Gooding, 1960

Order Poecilostomatoida
Family Oncaeidae
Oncaea conifera Giesbrecht, 1891
O. venusta Philippi, 1843
O. venusta var.*venella* Farran, 1929
O. media Giesbrecht, 1891
O. mediterranea (Claus, 1863)
O. curta G.O.Sars, 1916
O. minuta Giesbrecht, 1892
O. dentipes Giesbrecht, 1891
Pachos punctatum (Claus, 1863)
Lubbockia squillimana Claus, 1863
L. aculeata Giesbrecht, 1892
Conea rapax Giesbrecht, 1891

Family Sapphirinidae
Sapphirina angusta Dana, 1849
S. auronitens-sinuicauda Brady, 1883
S. nigromaculata Claus, 1863
S. ovatolanceolata Dana, 1849
Copilia mirabilis Dana, 1849

Family Corycaeidae
Corycaeus typicus (Kroyer, 1849)
C. speciosus Dana, 1849
C. furcifer Claus, 1863
C. limbatus Brady, 1888
C. flaccus Giesbrecht, 1891
C. amazonicus F.Dahl, 1894
C. giesbrechti F.Dahl, 1894
C. latus (Dana, 1849)
C. lautus Dana, 1852
C. ovalis Claus, 1863
Farranulla rostrata (Claus, 1863)
F. gracilis (Dana, 1853)

Family Ergasilidae
Ergasilus euripedesi Montú, 1980

Order Harpacticoida
Family Miraciidae
Macrosetella gracilis (Dana, 1848)
Miracia efferata (Dana, 1852)

Family Ectinosomatidae
Microsetella rosea (Dana, 1848)
M. norvegica (Boeck, 1864)

Family Pseudopeltiidae
Clytemnestra rostrata Brady, 1883
C. scutellata Dana, 1848

Family Tachydiidae
Euterpina acutifrons (Dana, 1852)

SUBCLASS CIRRIPEDIA
Order Thoracica
Suborder Lepadomorpha
Family Scalpellidae
Arcoscalpellum boubaloceros Young, 1992
A. portoricanum (Pilsbry, 1907)
A. triangulare Hock, 1883
Hamatoscalpellum rathbunae (Pilsbry, 1907)
Litoscalpellum henriquecostai (Weber, 1960)
L. regina (Pilsbry, 1907)
Heterolepas cornuta (Darwin, 1851)
Ornatoscalpellum gibberum (Aurivillius, 1892)

Suborder Verrucomorpha
Family Verrucidae
Altiverruca sp.
Rostratoverruca nexa (Darwin)

Verruca caribea Pilsbry, 1916

Suborder Balanomorpha
Family Balanidae
Balanus amphitrite Darwin, 1854
B. improvisus (Darwin, 1854)
B. venustus Darwin, 1854
Megabalanus coccopoma (Darwin, 1854)

Family Chthamalidae
Chthamalus bisinuatus Pilsbry, 1916

Family Tetraclitidae
Tetraclita stalactifera (Lamarck, 1818)

CLASS MALACOSTRACA
Superorder Hoplocarida
Order Stomapoda
Family Squillidae
Squilla brasiliensis Claman, 1917

Family Hemisquillidae
Hemisquilla brasiliensis (Moreira, 1903)

Superorder Peracarida
Order Mysidacea
Family Mysidae
Bowmaniella (=*Coifmanniella*) *brasiliensis* Bacescu, 1968
Bowmaniella sp.
Mysidopsis coelhoi Bacescu, 1968
M. tortonesi Bacescu, 1968
Metamysidopsis elongata Holmes, 1900
Neomysis americana (S.I.Smith, 1873)
Promysis atlantica Tattersall, 1923
Brasilomysis castroi Bacescu, 1968

Order Tanaidacea
Family Tanaidae
Tanais stanfordi (Richardson, 1901)

Family Kalliapseudidae
Hemikalliapseudes cavooreni Bacescu & Absalão, 1985
Kalliapseudes gianucai Bacescu, 1979
K. *schubarti* Mañe-Garzón, 1949

Order Cumacea
Family Pseudocumatidae
Leptocuma kinbergii Sars, 1873

Family Diastylidae
Diastylis sympterigiae Bacescu & Queiroz, 1985

Order Isopoda
Family Idoteidae
Cleantoides sp.
Edotea sp.
Idotea balthica (Pallas, 1772)
Macrochiridotea giambiagiae Torti & Bastida, 1969
M. lilianae Moreira, 1973
M. robusta Bastida & Torti, 1967
Synidotea marplatensis Giambiagi, 1922

Family Munnidae
Munna peterseni (Vanin, 1985)

Family Sphaeromatidae
Ancinus gaucho Pires, 1987
Dies fluminensis (Mañe-Garzón, 1944)
Pseudosphaeroma mourei Loyola, 1960

Family Serolidae
Serolis bonaerensis Bastida & Torti, 1967
S. elliptica Sheppard, 1933
S. exigua Nordenstam, 1933
S. foresti Bastida & Torti, 1970
S. marplatensis Bastida & Torti, 1970
S. polaris Richardson, 1911
S. uaperta Moreira, 1971
Serolis sp.

Family Aegidae
Aega sp.

Family Cymothoidae
Anilocra sp.

Family Cirolanidae
Bathynomus giganteus (Milne-Edwards)
Cirolana spp.
Excirolana armata (Dana, 1852)
E. braziliensis Richardson, 1912

Order Amphipoda
Family Ampeliscidae
Ampelisca spp.
Byblis sp.
Haploops sp.

Family Amphithoidae
Amphitoe ramondi Audouin, 1826

Family Atylidae
Atylus minikoi (Walker, 1916)

Family Cheidae
Cheus sp.

Family Corophiidae
Corophium acherusicum Costa, 1857
Leptocheirus sp.
Paracorophium sp.

Family Gammaridae
Ceradocus sp.
Maera spp.
Melita appendiculata
Melita mangrovi Oliveira, 1953
Melita spp.
Netamelita sp.

Family Isaeidae
Cheiriphotis sp.
Photis sp.

Family Lysianassidae
Wecomedon sp.

Family Oedicerotidae
Bathyporeiapus bisetosus Escofet, 1970
B. ruffoi Escofet, 1971

Family Phoxocephalidae
Heterophoxus sp.
Metharpinia sp.
Paraphoxus sp.
Phoxocephalopsis sp.
P. zimmeri Schellenberg, 1931
Pseudoharpinia sp.
Puelche orensanzi Barnard & Clark, 1982
Urothoe sp.

Family Platyischnopidae
Platyischnopus sp.

Family Talitridae
Orchestia platensis Kroyer, 1845
Orchestoidea brasiliensis Dana, 1853

Family Urohaustoriidae
Huarpe escofeti Barnard & Clark, 1982

Family Caprellidae
Caprella dilatata Kroyer, 1842

Superorder Eucarida
Order Euphausiacea
Family Euphausiidae
Euphausia americana Hansen, 1911
E. tenera Hansen, 1905
E. brevis Hansen, 1905
E. gibboides Ortmann, 1893
E. hemigibba Hansen, 1910
E. longirostris Hansen, 1908
E. lucens Hansen, 1905
E. recurva Hansen, 1905
E. similis G.O.Sars, 1883
E. spinifera G.O.Sars, 1883
E. vallentini Stebbing, 1900
Nematobrachion flexipes (Ortmann, 1893)
Nematoscelis megalops G.O.Sars, 1883
N. microps G.O.Sars, 1883
N. tenella G.O.Sars, 1883
Stylocheiron abbreviatum G.O.Sars, 1883
S. affine Hansen, 1910
S. carinatum G.O.Sars, 1883
S. elongatum G.O.Sars, 1883
S. longicorne G.O.Sars, 1883
S. maximum Hansen, 1908
S. suhmii G.O.Sars, 1883
Thysanoessa gregaria G.O.Sars, 1883
Thysanopoda aequalis Hansen, 1905
T. monacantha Ortmann, 1893
T. obtusifrons G.O.Sars,1883
T. orientalis Hansen, 1910
T. acutifrons Holt & Tattersall, 1905
T. tricuspidata Milne-Edwards, 1830

Order Decapoda
Suborder Dendrobranchiata
Infraorder Penaeidea
Family Aristeidae
Aristeomorpha foliacea (Risso, 1827)

Family Penaeidae
Artemesia longinaris Bate, 1888
Funchalia villosa (Bouvier, 1905)
Parapenaeus americanus Rathbun, 1901
Penaeopsis serrata Bate, 1881
Penaeus brasiliensis Latreille, 1817
P. paulensis Pérez-Farfante, 1967
P. schmitti Burkenroad, 1936
Xiphopenaeus kroyeri (Heller, 1862)

Family Sicyoniidae
Sicyonia laevigata Stimpson, 1871
S. typica (Boeck, 1864)

Family Solenoceridae
Mesopenaeus tropicalis (Bouvier, 1905)

Pleoticus muelleri (Bate, 1888)
Solenocera necopina Burkenroad, 1939

Family Sergestidae
Lucifer faxoni Borradaile, 1915
Lucifer sp.

Suborder Pleocyemata
Infraorder Caridea
Family Pasiphaeidae
Leptochela serratorbita Bate, 1888

Family Palaemonidae
Palaemonetes argentinus Nobili, 1901

Family Hippolytidae
Latreutes sp.

Family Pandalidae
Dichelopandalus sp.
Hexhippolysmata oplophoroides (Holthius, 1948)
Parapandalus sp.
Pleistonika sp.
Pontocaris boschi Christoffersen, 1988

Infraorder Astracidea
Family Nephropidae
Metanephrops rubellus Moreira, 1903

Infraorder Thalassinidea
Family Axiidae
Meticonaxius sp.

Family Callianassidae
Callichirus mirim (Rodrigues, 1971)

Infraorder Palinura
Family Polychelidae
Stereomastis sp.

Family Scyllaridae
Scyllarides deceptor Holthius, 1963

Infraorder Anomura
Family Diogenidae
Dardanus arrosor insignis (De Saussure, 1858)
Loxopagurus loxochelis (Moreira, 1901)
Paguristes robustus Forest & Saint Laurent, 1967

Family Paguridae
Pagurus criniticornis (Dana, 1852)
P. exilis Benedict, 1892
P. trichocerus Forest & Laurent, 1967
Rodochirus rosaceus (Milne-Edwards & Bouvier, 1893)

Family Galatheidae
Munida flinti Benedict, 1819
M. forceps Milne-Edwards, 1880
M. irrasa Milne-Edwards, 1880
M. longipes Milne-Edwards, 1880
M. valida Smith, 1883

Family Porcellanidae
Pachycheles chubutiensis Boschi, 1963
P. haigae Stimpson, 1858
Porcellana sayana Leach, 1820

Family Hippidae
Emerita brasiliensis Schmitt, 1935

Infraorder Brachyura
Family Homolidae
Homola barbata Fabricius, 1793

Family Latreilliidae
Latreillia sp.

Family Calappidae
Acanthocarpus alexandri Stimpson, 1871
Hepatus pudibundus (Herbst, 1785)

Family Leucosiidae
Myropsis quinquespinosa Stimpson, 1871
Persephona punctata Stimpson, 1859

Family Majidae
Collodes rostratus Milne-Edwards, 1878
Leurocyclus tuberculosus (Milne-Edwards & Lucas, 1842)
Libinia spinosa Milne-Edwards, 1834
Pelia rotunda Milne-Edwards, 1875
Persephona punctata
Podochela sp.
Rochinia gracilipes Milne-Edwards, 1875
Stenocionops sp.

Family Parthenopidae
Parthenope pourtalesi (Stimpson, 1871)

Family Atelecyclidae
Corystoides chilensis Milne-Edwards & Lucas, 1844

Family Geryonidae
Chaceon notialis Manning & Holtius, 1989

Family Portunidae
Arenaeus cribarius (Lamarck, 1818)
Callinectes danae Smith, 1869
C. ornatus Ordway
C. sapidus Rathbun, 1895
Ovalipes trimaculatus (de Haan, 1833)
Portunus spinicarpus (Stimpson, 1815)
P. spinimanus Latreille, 1919

Family Goneplacidae
Batyplax typhla Milne-Edwards, 1880
Chasmocarcinus rathbuni Bouvier, 1923
C. typicus Rathbun, 1898
Cyrtoplax spinidentata (Benedict)
Frevillea hirsuta (Borradaile, 1916)
Speocarcinus meloi D'Incao & Silva, 1991

Family Xanthidae
Allactaea lithostrata Williams, 1974
Hexapanopeus paulensis Rathbun, 1930
Panopeus sp.
Pilumnoides hassleri Milne-Edwards, 1880
Pilumnus reticulatus Stimpson, 1860
Platyxanthus patagonicus Milne-Edwards, 1879
Rhithropanopeus harrissii Gould, 1841
Tetraxanthus rathbunae Chace, 1939

Family Grapsidae
Chasmagnathus granulata Dana, 1851
Cyrtograpsus affinis (Dana, 1851)
C. altimanus Rathbun, 1914
C. angulatus Dana, 1851
Hepatus pudibundus (Herbst, 1785)
Metasesarma rubripes (Rathbun, 1897)

Family Pinnotheridae
Dissodactylus crinitichelis Moreira, 1901
Favia emiliae (de Melo, 1971)
Pinnixa brevipolex Rathbun, 1898
P. chaetopterana Stimpson
P. patagoniensis Rathbun, 1918
P. rapax Bouvier
Pinnotheres maculatus Say

Family Ocypodidae
Ocypode quadrata (Fabricius, 1787)
Uca uruguayensis Nobili, 1901

SUBPHYLUM UNIRAMIA
CLASS INSECTA
SUBCLASS PTERYGOTA
Order Orthoptera
Family Gryllotalpidae
Scapteriscus riograndensis Corseuil, 1992

Family Tridactylidae
Neotridactylus carbonelli Gunther, 1972

Order Dermaptera
Family Labiduridae
Labidura riparia Pallas, 1773

Order Homoptera
Family Cicadidae
Proarna uruguayensis Berg, 1882
Proarna sp.

Family Delphacidae
Delphacodes kuscheli Fennah, 1955

Order Coleoptera
Family Cicindelidae
Cicindela nivea conspersa Dejean, 1825
C. patagonica bergiana W. Horn, 1895
Megacephala (=*Tetracha*) *brasiliensis* Kirby, 1912

Family Carabidae
Bembidion (=*Notaphus*) sp.
Scarites spp.
Schizogenius costiceps Steinheil, 1869
Tetragonoderus undatus Dejean, 1829
T. variegatus Dejean, 1829

Family Hydrophilidae
Paracimus rufocinctus Bruch, 1915

Family Staphylinidae
Bledius bonariensis Bernhauer, 1912
B. fernandezi Bernhauer, 1939
B. microcephalus Fauvel, 1901

Family Heteroceridae
Efflagitatus freudei Pacheco, 1973

Family Anthicidae
Lagrioida nortoni Costa & Vanin, 1995

Family Oedemeridae
Hypasclera dorsalis (Melsheimer, 1846)

Family Tenebrionidae
Phaleria testacea Say, 1824

Family Scarabaeidae
Plectris bonariensis Frey, 1966
Psammodius sp.

Family Dinastidae
Ligyrus gianucai Dechambre & Lumaret, 1985
Thronistes rouxi Burmeister, 1847

Family Geotrupidae
Athyreus chalibeatus Fairmaire, 1892

Family Curculionidae
Listroderes uruguayensis Kuschel, 1952

Order Lepidoptera
Family Arctiidae
Ecpanteria indecisa Walker, 1855

Order Diptera
Family Agromyzidae
Aplopeudes vogti Steysical, 1980

Family Asilidae
Eccritosia rubriventris (Macquart, 1850)

Family Tabanidae
Scepsis nivalis Walker, 1850

Order Hymenoptera
Family Cynipidae
Eucoila sp.

Family Formicidae
Subfamily Dolichoderinae
Conomyrma spp.

Subfamily Formicinae
Brachymyrmex sp.
Camponotus (=*Tanaemyrmex*) *punctulatus* Mayer, 1868

Subfamily Myrmicinae
Monomorium sp.
Mycetophylax simplex Emery, 1887
Pheidole spininodis Emery, 1887
Solenopsis geminata Fabricius, 1804

Family Larridae
Tachytes ornatipes Cameron, 1889

Family Pompilidae
Anoplius (=*Arachnophroctonus*) *bilunulatus* Saussure, 1867

Family Scoliidae
Campsomeris (*Pygodasis*) *cineraria* (Sichel, 1864)

PHYLUM PHORONIDA
Phoronis psammophila Cori

PHYLUM ECHINODERMATA
CLASS CRINOIDEA
Order Isocrinida
Family Pentacrinidae
Diplocrinus maclearanus (W.Thomson, 1877)

Order Millecrinida
Family Bourgueticrinidae
Rhizocrinus lofotensis Sars, 1864

Order Comatulida
Family Charitometridae
Crinometra brevispina brasiliensis(Pourtalès, 1868)

Family Antedonidae
Isometra angustipinna (P.H.Carpenter, 1888)
I. challengeri (A.H.Clark, 1907)
I. vivipara Mortensen, 1918
Phrixometra longipinna (P.H.Carpenter, 1888)

CLASS STELLEROIDEA
SUBCLASS OPHIURIDEA
Order Phrynophiurida
Family Gorgonocephalidae
Astrocyclus caecilia Lútken, 1856
Gorgonocephalus chilensis Philippi, 1858

Order Ophiurida
Family Ophiuridae
Ophiomastus satellitae Tommasi & Abreu, 1976
Ophioplocus januarii (Lútken, 1856)
Ophiura sp.

Family Ophiacantidae
Ophiacantha vivipara Ljungman, 1870

Family Amphiuridae
Amphilimna olivacea (Lyman, 1869)

Amphiodia pulchella (Lyman, 1869)
Amphioplus lucyae Tommasi, 1971
Amphiura complanata Ljungman, 1867
A. crassipes Ljungman, 1866
A. flexuosa Ljungman, 1867
A. joubini Koehler, 1912
A. kinberi Ljungman, 1871
A. latispina Ljungman, 1867
A. stimpsonii Lútken, 1859
Axiognathus squamatus (Delle Chiaje, 1829)
Diamphiodia planispina (Martens, 1873)
Hemipholis elongata (Say, 1825)
Micropholis subtilis Ljungman, 1871
Nudamphiura carvalhoi Tommasi, 1965
Ophiocnida sp.
Ophiurolepis sp.

Family Ophiactidae
Ophiactis lymani Ljungman, 1871

Family Ophiotrichidae
Ophiotrix angulata (Say, 1825)
O. rathbuni Ludwig, 1882

Family Ophiodermatidae
Ophioderma januarii Lútken, 1856

Family Ophiochitonidae
Ophioplax clarimundae Tommasi, 1970

Family Ophiolepididae
Ophioceramis januarii (Lútken, 1856)
Ophiomisidium pulchellum (Wyville & Thompson, 1877)

SUBCLASS ASTEROIDEA
Order Paxillosida
Family Astropectinidae
Astropecten armatus brasiliensis (Müll & Troschel, 1842)
A. articulatus (Say, 1825)
A. cingulatus Sladen, 1883
A. marginatus Gray, 1840
Bathybiaster herwigi Bernasconi, 1971
B. loripes Sladen, 1889

Family Goniasteridae
Anthenoides brasiliensis Bernasconi, 1956
Hippasteria argentinensis Bernasconi, 1961
Mediaster pedicellaris (Perrier, 1881)

Order Platyasterida
Family Luidiidae
Luidia clathrata (Say, 1825)
L. quequenensis Bernasconi, 1943
L. scotti (Bell, 1917)

Order Valvatida
Family Ophidiasteridae
Calliophidiaster psicodelica Tommasi, 1970

Order Spinulosida
Family Asterinidae
Enoplopatiria stellifera (Möbius, 1859)

Family Echinasteridae
Allostichaster hartii (Rathbun, 1879)
Echinaster brasiliensis Müller & Troschel, 1842

CLASS ECHINOIDEA
Order Cidaroida
Family Cidaridae
Australocidaris canaliculata (A.Agassiz, 1836)
Stylocidaris affinis (Philippi, 1845)

Order Spatangoida
Family Brissidae
Brissopsis atlantica Mortensen, 1907

Order Clypeasteoidea
Family Scutellidae
Mellita quinquiesperforata (Leske, 1778)
Encope emarginata (Leske, 1778)

CLASS HOLOTUROIDEA
Order Dendrochirotida
Family Psolidae
Psolus patagonicus Ekman, 1925

Order Molpadiida
Family Molpadiidae
Molpadia sp.

PHYLUM CHAETOGNATHA
Eukrohnia hamata (Möbius, 1875)
E. bathypelagica Alvariño, 1962
Krohnitta pacifica (Aida, 1897)
K. subtilis (Grassi, 1881)
Pterosagitta draco (Krohn, 1853)
Sagitta bipunctata Quoy & Gaimard, 1827
S. enflata Grassi, 1881

S. tenuis Conant, 1896
S. helenae Ritter-Zahóny, 1910
S. hexaptera Orbigny, 1843
S. hispida Conant, 1895
S. lyra Krohn, 1853
S. macrocephala Fowler, 1905
S. maxima (Conant, 1896)
S. minima Grassi, 1881
S. planctonis Steinhaus, 1896
S. serratodentata Krohn, 1853
S. tasmanica Thomson, 1947
S. decipiens Fowler, 1905
S. gazellae Ritter-Zahony, 1909

PHYLUM HEMICHORDATA
CLASS PTEROBRANCHIATA
Order Cephalodiscida
Cephalodiscus sp.

PHYLUM CHORDATA
SUBPHYLUM UROCHORDATA
CLASS THALIACEA
Family Salpidae
Cyclosalpa bakeri Ritter, 1905
C. polae Sigl, 1912
Brooksia rostrata (Traustedt, 1893)
Ihlea magalhanica (Apstein, 1894)
I. punctata (Forskål, 1775)
Ritteriella retracta (Ritter, 1906)
Salpa aspera Chamisso, 1819
S. fusiformis Cuvier, 1804
S. thompsoni Foxton, 1961
S. maxima Forskal, 1775
Thalia cicar van Soest, 1973
T. democratica (Forskal, 1775)
T. orientalis Tokioka, 1937
T. longicauda (Quoy & Gaimard, 1824)
Weelia cylindrica (Cuvier, 1804)
Traustedtia multitentaculata (Quoy & Gaimard, 1824)

CLASS APPENDICULARIA
Family Fritillaridae
Appendicularia sicula Fol, 1874
Fritillaria aberrans Lohmann, 1896
F. aequatorialis Lohmann, 1896
F. borealis Lohmann, 1896
F. formica Fol, 1872
F. fraudax Lohmann, 1896
F. haplostoma Fol, 1872
F. megachile Fol, 1872
F. messanensis Lohmann, 1899
F. pellucida (Busch, 1851)
F. tenella Lohmann, 1896
F. venusta Lohmann, 1896

Family Oikopleuridae
Folia gracilis Lohmann, 1892
F. mediterranea (Lohmann, 1899)
Oikopleura albicans (Leuckart, 1854)
O. cophocerca (Gegenbaur, 1855)
O. drygalskii Lohmann, 1926
O. fusiformis Fol, 1872
O. gracilis Lohmann, 1896
O. longicauda (Vogt, 1854)
O. parva Lohmann, 1896
O. rufescens Fol, 1872
Stegosoma magnum (Langerhans, 1880)

SUBPHYLUM GNASTHOTOMATA
SUPERCLASS PISCES
CLASS CHONDRICHTHYES
Order Squatiniformes
Family Squatinidae
Squatina argentina (Marini, 1930)
S. guggenheim Marini, 1936
S. occulta Vooren & Silva, 1991

Order Carcharhiniformes
Family Triakidae
Galeorhinus galeus (Linnaeus, 1758)
Mustelus canis (Mitchill, 1815)
M. fasciatus Garman, 1913
M. schmitti (Springer, 1940)

Family Sphyrnidae
Sphyrna lewini (Griffith & Smith, 1834)
S. zygaena (Linnaeus, 1758)

Family Carcharhinidae
Carcharhinus acronotus (Poey, 1860)
C. brachyurus (Günther, 1870)
C. brevipinna (Müller & Henle, 1839)
C. obscurus (LeSueur, 1818)
C. plumbeus (Nardo, 1827)
C. signatus (Poey, 1868)
Galeocerdo cuvier (Peron & LeSueur, 1822)
Rhizoprionodon lalandii (Valenciennes, 1839)

Family Scyliorhinidae
Schroederichthys bivius (Smith, 1838)
Scyliorhinus besnardi Springer & Sadowsky, 1970

Order Lamniformes
Family Odontaspididae

Eugomphodus taurus (Rafinesque, 1810)

Family Alopiidae
Alopias vulpinus (Bonnaterre, 1788)

Order Squaliformes
Family Saqualidae
Centroscymnus cryptacanthus Regan, 1906
Etmopterus hillianus (Poey, 1861)
E. pusillus (Lowe, 1839)
Squaliolus laticaudus Smith & Radcliffe, 1912
Squalus acanthias Smith & Radcliffe, 1912
S. cubensis Howell-Rivero, 1936
S. megalops (Macleay, 1881)
S. mitsukurii Jordan & Snyder, 1903

Family Echinorhinidae
Echinorhinus brucus (Bonnaterre, 1788)

Order Hexanchiformes
Family Hexanchidae
Heptranchias perlo (Bonnaterre, 1788)
Hexanchus griseus (Bonnaterre, 1788)
Notorhynchus cepedianus (Peron, 1807)

Order Rhinobatiformes
Family Rhinobatidae
Rhinobatos horkelii (Müller & Henle, 1841)
Zapteryx brevirostris (Müller & Henle, 1841)

Order Rajiformes
Family Rajidae
Gurgesiella dorsalifera McEachran & Compagno, 1980
Psammobatis bergi Marini, 1932
P. glansdissimilis McEachran, 1983
P. lentiginosa McEachran, 1983
P. rutrum Jordan, 1890
Raja agassizi (Müller & Henle, 1841)
R. castelnaui Ribeiro, 1907
R. cyclophora Regan, 1903
R. leptocauda Krefft & Stehmann, 1975
R. platana Günther, 1880
Sympterigia acuta Garman, 1877
S. bonapartei Müller & Henle, 1841

Order Myliobatiformes
Family Myliobatidae
Myliobatis freminvillei LeSueur, 1824
M. goodei Garman, 1885
M. NT (Vooren, Sect. 6.16)
M. BT (Vooren, Sect. 6.16)

Family Dasyatidae
Dasyatis centroura (Mitchill, 1815)
D. say (LeSueur, 1817)

Family Gymnuridae
Gymnura altavela (Linnaeus, 1758)

Family Rhinopteridae
Rhinoptera bonasus (Mitchill, 1815)

Order Torpediniformes
Family Narcinidae
Benthobatis sp.
Discopyge tschudii (Heckel, 1845)
Narcine brasiliensis (Olfers, 1831)

Family Torpedinidae
Torpedo puelcha Lahille, 1928

Family Callorhynchidae
Callorhynchus callorhynchus (Linnaeus, 1758)

CLASS OSTEICHTHYES
Family Megalopidae
Tarpon atlanticus (Valenciennes, 1846)

Family Xenocongridae
Chlopsis bicolor Rafinesque, 1810

Family Muraenidae
Gymnothorax conspersus Poey, 1867
G. ocellatus Agassiz, 1831

Family Muraenesocidae
Hoplunnis tenuis Ginsburg, 1951

Family Congridae
Ariosoma opisthophthalma (Ranzani, 1838)
Conger orbignyanus Valenciennes, 1847
Rechias dubius (Breder, 1927)

Family Ophichthidae
Myrophis frio Jordan & Davis, 1892
Ophichthus gomesii (Castelnau, 1855)

Family Clupeidae
Brevoortia pectinata (Jenyns, 1842)

Chirocentrodon bleekerianus (Poey, 1867)
Opisthonema oglinum (LeSueur, 1818)
Pellona harroweri (Fowler, 1917)
Ramnogaster arcuata (Jenyns, 1842)
Sardinella brasiliensis (Steindachner, 1789)

Family Engraulididae
Anchoa marinii Hildebrand, 1943
A. tricolor (Agassiz, 1829)
Engraulis anchoita Hubbs & Marini, 1935
Lycengraulis grossidens (Agassiz, 1829)

Family Gonostomatidae
Maurolicus muelleri (Gmelin, 1788)
Vicinguerria sp.

Family Argentinidae
Argentina striata Goode & Bean, 1896

Family Ariidae
Genidens genidens (Valenciennes, 1839)
Netuma barba (Lacépède, 1803)
N. planifrons Higuchi Reis & Araújo, 1982

Family Pimelodidae
Parapimelodus nigribarbis (Boulenger, 1889)
Pimelodus maculatus (Lacépède, 1803)

Family Synodontidae
Saurida brasiliensis Norman, 1935
S. caribbaea Breder, 1927
Synodus foetens (Linnaeus, 1766)

Family Chlorophthalmidae
Chlorophthalmus agassizi Bonaparte, 1840
Parasudis truculenta (Goode & Bean, 1895)

Family Paralepididae
Lestidium atlanticum Boradin, 1928

Family Alepisauridae
Alepisaurus brevirostris Gibbs, 1960

Family Myctophidae
Diaphus brachycephalus Taning, 1928
D. dumerilii (Bleeker, 1856)
D. effulgens (Goode & Bean, 1896)
Gymnoscopelus piabilis (Whitley, 1931)
Lepidophanes guentheri (Goode & Bean, 1896)
Symbolophorus barnardi (Taning, 1932)
S. veranyi (Moreau, 1888)

Family Polymixiidae
Polymixia lowei Gunther, 1859

Family Bregmacerotidae
Bregmaceros atlanticus Goode & Bean, 1886

Family Gadidae
Urophycis brasiliensis (Kaup, 1858)
U. mystaceus Ribeiro, 1903

Family Merlucciidae
Macruronus magellanicus Lönnberg, 1907
Merluccius hubbsi Marini, 1933

Family Ophidiidae
Benthocometes robustus (Goode & Bean, 1886)
Genypterus brasiliensis Regan, 1903
Raneya fluminensis (Ribeiro, 1903)

Family Macrouridae
Coelorhynchus coelorhynchus (Risso, 1810)
C. marinii Hubbs, 1934
Malacocephalus laevis (Lowe, 1843)
M. occidentalis (Goode & Bean, 1885)
Ventrifossa ori (Smith, 1968)

Family Batrachoididae
Porichthys porosissimus (Valenciennes, 1837)
Thalassophryne montevidensis Berg, 1893

Family Lophiidae
Lophius gastrophysus Ribeiro, 1915

Family Ogcocephalidae
Ogcocephalus vespertilio (Linnaeus, 1758)

Family Gobiesocidae
Gobiesox strumosus Cope, 1870

Family Antennariidae
Phrynelox scaber Cuvier, 1817

Family Exocoetidae
Hiporhamphus unifasciatus (Ranzani, 1842)
Cypselurus sp.

Family Anablepidae
Jenynsia lineata (Jenyns, 1842)

Family Poeciliidae
Cnesterodon decemmaculatus (Jenyns, 1842)
Phalloceros caudimaculatus (Hensel, 1868)
Phalloptychus januaris (Hensel, 1868)
Poecilia vivipara Schneider, 1841

Family Belonidae
Strongylura marina (Walbaum, 1792)
S. timucu (Walbaum, 1792)

Family Scomberesocidae
Scomberesox saurus (Walbaum, 1792)

Family Atherinidae
Austroatherina incisa (Jenyns, 1842)
Odontesthes bonariensis (Valenciennes, 1835)
O. argentinensis (Valenciennes, 1835)
Atherinella brasiliensis (Chernoff, 1986)

Family Lamprididae
Lampris guttatus (Brunnich, 1788)

Family Trachichthyidae
Hoplostethus occidentalis Woods, 1973
Paratrachichthys atlanticus Menezes, 1971

Family Berycidae
Beryx splendens Lowe, 1834

Family Anoplogasteridae
Anoplogaster cornuta (Valenciennes, 1833)

Family Zeidae
Zenopsis conchifer (Lowe, 1850)

Family Grammicolepididae
Xenolepidichthys dalgleishi Gilchrist, 1922

Family Caproidae
Antigonia capros Lowe, 1843

Family Fistulariidae
Fistularia petimba Lacépède, 1803

Family Macrorhamphosidae
Macrorhamphosus scolopax (Linnaeus, 1758)
Notopogon fernadezianus (Delfin, 1899)

Family Syngnathidae
Hippocampus erectus Perry, 1810
H. reidi Ginsburg, 1933
Sygnathus elucens Poey, 1867
S. folletti Herald, 1942
S. pelagicus Linnaeus, 1758

Family Scorpaenidae
Helicolenus lahillei Norman, 1937
Pontinus rathbuni Goode & Bean, 1896
Setarches guentheri Johnson, 1862

Family Peristediidae
Peristedion altipinne Regan, 1903
Peristedion sp.

Family Triglidae
Bellator brachychir (Regan, 1914)
Prionotus nudigula Ginsburg, 1950
P. punctatus (Bloch, 1797)

Family Congiopodidae
Congiopodus peruvianus (Cuvier, 1829)

Family Dactylopteridae
Dactylopterus volitans (Linnaeus, 1758)

Family Centropomidae
Centropomus ensiferus Poey, 1860
C. parallelus Poey, 1860
C. undecimalis (Bloch, 1792)

Family Serranidae
Acanthistius brasilianus (Cuvier, 1828)
A. patachonicus (Jenyns, 1840)
Anthias menezesi Anderson & Heemstra, 1980
Diplectrum formosum (Linnaeus, 1766)
D. radiale (Quoy & Gaimard, 1824)
Dules auriga Cuvier, 1829
Epinephelus flavolimbatus Poey, 1865
E. itajara (Lichtenstein, 1822)
E. marginatus (Lowe, 1834)
E. morio (Valenciennes, 1828)

E. nigritus (Holbrook, 1855)
E. niveatus (Valenciennes, 1828)
Hemanthias vivanus (Jordan & Swain, 1884)
Holanthias martinicensis (Guichenot, 1868)
Pikea rosea (Gunther, 1880)
Serranus atrobranchus (Cuvier, 1829)
S. flaviventris (Cuvier, 1829)

Family Polyprionidae
Polyprion americanus (Schneider, 1801)

Family Priacanthidae
Cookeolus japonicus (Bloch & Schneider, 1801)
Priacanthus arenatus Cuvier, 1829

Family Apogonidae
Apogon pseudomaculatus Longley, 1932
Synagrops bella (Goode & Bean, 1895)
S. spinosus Schultz, 1940

Family Branchiostegidae
Caulolatilus chrysops (Valenciennes, 1833)
Lopholatilus villarii Ribeiro, 1915

Family Pomatomidae
Pomatomus saltatrix (Linnaeus, 1766)

Family Rachycentridae
Rachycentron canadus (Linnaeus, 1766)

Family Echeneididae
Echeneis naucrates Linnaeus, 1758
Remora brachyptera (Lowe, 1839)
R. osteochir (Cuvier, 1829)

Family Carangidae
Caranx chrysos (Mitchill, 1815)
C. hippos (Linnaeus, 1766)
C. latus Agassiz, 1831
Chloroscombrus chrysurus (Linnaeus, 1766)
Decapterus tabl Berry, 1968
Naucrates ductor (Linnaeus, 1758)
Oligoplites saliens (Bloch, 1793)
O. saurus (Schneider, 1801)
Parona signata (Jenyns, 1842)
Pseudocaranx dentex (Bloch & Schneider, 1801)
Selar crumenophthalmus (Bloch, 1793)
Selene setapinnis (Mitchill, 1815)
S. vomer (Linnaeus, 1758)
Seriola lalandi Valenciennes, 1833
S. rivoliana Valenciennes, 1833
Trachinotus carolinus (Linnaeus, 1766)
T. falcatus (Linnaeus, 1758)
T. goodei Jordan & Evermann, 1896
T. marginatus Cuvier, 1832
Trachurus lathami Nichols, 1920

Family Caristiidae
Caristius sp.

Family Pomadasyidae
Boridia grossidens (Cuvier, 1830)
Orthopristis ruber (Cuvier, 1830)
Pomadasys corvinaeformis (Steindachner)

Family Coryphaenidae
Coryphaena equiselis Linnaeus, 1758
C. hippurus Linnaeus, 1758)

Family Bramidae
Brama brama (Bonnaterre, 1788)
Pteraclis aesticola (Jordan & Snyder, 1901)

Family Lutjanidae
Pristipomoides freemani Anderson, 1966

Family Lobotidae
Lobotes surinamensis (Bloch, 1790)

Family Gerreidae
Diapterus rhombeus (Cuvier, 1829)
Eucinostomus gula (Cuvier, 1830)
E. melanopterus (Blecker, 1863)
Eugerres brasilianus (Cuvier, 1830)
Ulaema lefroyi (Goode, 1874)

Family Sparidae
Diplodus argenteus (Valenciennes, 1830)
Pagrus pagrus (Linnaeus, 1758)

Family Sciaenidae
Ctenosciaena gracilicirrhus (Metzelaar, 1919)
Cynoscion guatucupa (Cuvier, 1829)
C. jamaicensis (Vaillant & Bocourt, 1883)
Macrodon ancylodon (Schneider, 1801)
Menticirrhus americanus (Linnaeus, 1758)
M. littoralis (Holbrook, 1860)

Micropogonias furnieri (Desmarest, 1823)
Paralonchurus brasiliensis (Steindachner, 1875)
Pogonias cromis (Linnaeus, 1766)
Stellifer brasiliensis (Schultz, 1945)
S. rastrifer (Jordan, 1889)
Umbrina canosai Berg, 1895

Family Ephippidae
Chaetodipterus faber (Broussonet, 1782)

Family Cichlidae
Cichlaurus facetus (Jenyns)
Geophagus brasiliensis (Quoy & Gaimard)

Family Mullidae
Mullus argentinae Hubbs & Marini, 1935

Family Kyphosidae
Kyphosus incisor (Cuvier, 1831)

Family Chaetodontidae
Chaetodon guyanensis Durand, 1960

Family Cheilodactylidae
Sciaenoides bergi (Norman, 1937)

Family Mugilidae
Mugil curema Valenciennes, 1836
M. gaimardianus Desmarest, 1831
M. platanus Gunther, 1880

Family Sphyraenidae
Sphyraena barracuda (Walbaum, 1792)
S. guachancho Cuvier, 1829
S. tome Fowler, 1903

Family Polynemidae
Polydactylus oligodon (Gunther, 1860)
P. virginicus (Linnaeus, 1758)

Family Labridae
Decodon puellaris (Poey, 1860)

Family Opistognathidae
Lonchopisthus meadi Menezes & Figueiredo, 1971

Family Uranoscopidae
Astroscopus sexspinosus (Steindachner, 1877)

Family Percophidae
Bembrops heterurus (Ribeiro, 1903)
Percophis brasiliensis Quoy & Gaimard, 1824

Family Pomacentridae
Abudefduf saxatilis (Linnaeus, 1758)

Family Pinguipedidae
Pseudopercis numida Ribeiro, 1903
P. semifasciata (Cuvier, 1829)

Family Blenniidae
Hypleurochilus fissicornis (Quoy & Gaimard, 1824)

Family Eleotrididae
Guavina guavina (Valenciennes, 1837)

Family Gobiidae
Gobioides broussonnetii Lacépède, 1800
Gobionellus oceanicus (Pallas, 1770)
G. shufeldti (Jordan & Eigenmann, 1886)
G. smaragdus (Valenciennes)

Family Gempylidae
Promethichthys prometeus (Cuvier, 1832)
Thyrsitops lepidopoides (Cuvier, 1831)

Family Trichiuridae
Benthodesmus elongatus (Clarke, 1879)
Evoxymetopon taeniatus Poey, 1863
Lepidoupus sp.
Trichiurus lepturus Linnaeus, 1758

Family Scombridae
Auxis thazard (Lacépède, 1803)
Euthynnus alletteratus (Rafinesque, 1810)
Katsuwonus pelamis (Linnaeus, 1758)
Sarda sarda (Bloch, 1793)
Scomber japonicus Houttuyn, 1780
Thunnus alalunga (Bonaterre, 1788)
T. albacares (Bonnaterre, 1788)
T. atlanticus (Lesson, 1830)
T. obesus (Lowe, 1839)
T. thynnus (Linnaeus, 1758)

Family Xiphiidae
Xiphias gladius Linnaeus, 1758

Family Istiophoridae
Istiophorus albicans (Latreille, 1804)
Makaira indica (Cuvier, 1831)
M. nigricans Lacépède, 1803
Tetrapturus albidus Poey, 1860

Family Centrolophidae
Centrolophus niger (Gmelin, 1788)

Family Nomeidae
Cubiceps athenae Haedrich, 1965

Family Ariommidae
Ariomma bondi Fowler, 1930

Family Stromateidae
Peprilus paru (Linnaeus, 1758)
Stromateus brasiliensis Fowler, 1906

Family Gobiesocidae
Gobiesox strumosus Cope, 1870

Family Bothidae
Citharichthys cornutus (Günther, 1880)
C. spilopterus (Valenciennes)
Etropus longimanus Norman, 1933
Mossolene antillarum Norman, 1933
Paralichthys brasiliensis (Ranzani, 1840)
P. isoceles Jordan, 1890
P. orbignyanus (Valenciennes, 1839)
P. patagonicus Jordan, 1886
Xistreuris rasile (Jordan, 1890)

Family Pleuronectidae
Oncopterus darwini Steindachmer, 1875

Family Soleidae
Achirus garmani (Jordan & Goss, 1889)
A. lineatus (Linnaeus, 1758)

Family Cynoglossidae
Symphurus ginsburgi Menezes & Benvegnú, 1976
S. jenynsi Evermann & Kendall, 1907
S. trewavasae Chabanaud, 1948

Family Balistidae
Balistes capriscus Gmelin, 1788
B. vetula Linnaeus, 1758

Family Monacanthidae
Monacanthus ciliatus (Mitchill, 1818)
Stephanolepis setifer (Bennett)

Family Tetraodontidae
Lagocephalus laevigatus (Linnaeus, 1766)

Family Diodontidae
Chilomycterus spinosus (Linnaeus, 1758)

Family Molidae
Mola mola (Linnaeus, 1758)

SUPERCLASS TETRAPODA
CLASS AMPHIBIA
SUBCLASS ANURA
Family Bufonidae
Bufo arenarum arenarum Hensel, 1867

Family Leptodactylidae
Odontophrynus sp.

CLASS REPTILIA
Order Squamata
Family Iguanidae
Liolaemus occipitalis Boulenger, 1885

Family Colubridae
Lystrophis sp.

Order Testudines (Chelonia)
Suborder Cryptodira
Family Chelonidae
Caretta carreta (Linnaeus, 1758)
Chelonia mydas (Linnaeus, 1758)
Erectmochelys imbricata (Linnaeus, 1766)

Family Dermochelyidae
Dermochelys coriacea (Vandelli, 1761)

CLASS AVES
Order Sphenisciformes
Family Spheniscidae
Eudyptes chrysolophus (Brandt, 1837)
E. crestatus (Forster, 1781)
Spheniscus magellanicus (Forster, 1781)

Order Podicipediformes
Family Podicipedidae
Podiceps major (Boddaert, 1783)

Order Procellariiformes
Family Diomedeidae
Diomedea chlororhynchos Gmelin, 1789
D. chrysostoma Forster, 1758
D. epomophora (Lesson, 1825)

D. exulans (Linnaeus, 1758)
D. melanophrys Temminck, 1828
Phoebetria palpebrata (Forster, 1758)

Family Procellariidae
Calonectris diomedea (Scopoli, 1769)
Daption capense (Linnaeus, 1758)
Fulmarus glacialoides (Smith, 1840)
Macronectes sp.
Pachyptila belcheri Mathews, 1912
P. desolata (Gmelin, 1789)
P. vittata (Forster, 1777)
Procellaria aequinioctialis aequinioctialis Linnaeus, 1758
P. aequinioctialis conspicillata Gould, 1844
P. cinerea Gmelin, 1789
Pterodroma brevirostris (Lesson, 1831)
P. incerta (Schlegel, 1863)
P. lessoni (Garnot, 1826)
P. mollis (Gould, 1844)
Puffinus gravis (O'Reilly, 1818)
P. griseus (Gmelin, 1789)
P. puffinus (Brunnich, 1764)

Family Hydrobatidae
Oceanites oceanicus (Kuhl, 1820)

Family Pelecanoididae
Pelecanoides magellani (Mathews, 1912)

Order Pelecaniformes
Family Phalacrocoracidae
Phalacrocorax olivaceus (Humboldt, 1805)

Family Fregatidae
Fregata magnificens Mathews, 1914

Order Ciconiiformes
Family Ardeidae
Ardea cocoi Linnaeus, 1766
Butorides striatus (Linnaeus, 1785)
Egretta thula (Molina, 1782)

Order Phoenicopteriformes
Family Phoenicopteridae
Phoenicopterus chilensis Molina, 1782

Order Anseriformes
Family Anatidae
Cygnus melancorhyphus (Molina, 1782)

Order Falconiformes
Family Falconidae
Milvago chimango (Vieillot, 1816)
Polyborus plancus (Miller, 1777)

Order Gruiformes
Family Rallidae
Fulica armillata Vieillot, 1817

Order Charadriiformes
Family Haematopodidae
Haematopus palliatus Temminck, 1820

Family Charadriidae
Charadrius collaris Vieillot, 1818
C. falklandicus Latham, 1790
C. modestus Lichtenstein, 1823
C. semipalmatus Bonaparte, 1825
Oreopholus ruficollis (Wagler, 1929)
Pluvialis dominica (Müller, 1766)
P. squatarola (Linnaeus, 1758)

Family Scolopacidae
Actitis macularia (Linnaeus, 1766)
Arenaria interpres (Linnaeus, 1758)
Bartramia longicauda (Bechstein, 1812)
Calidris alba (Pallas, 1764)
C. bairdii (Coues, 1861)
C. canutus rufa (Wilson, 1813)
C. fuscicollis (Vieillot, 1819)
C. melanotos (Vieillot, 1819)
C. pusilla (Linnaeus, 1766)
Catatrophorus semipalmatus (Gmelin, 1789)
Limosa haemastica (Linaeus, 1758)
Micropalama himantopus (Bonaparte, 1826)
Numenius phaeopus (Linnaeus, 1758)
Tringa flavipes (Gmelin, 1789)
T. melanoleuca (Gmelin, 1789)
T. solitaria Wilson, 1813
Tryngites subruficollis (Vieillot, 1819)

Family Recurvirostridae
Himantopus himantopus melanurus Vieillot, 1817

Family Phalaropidae
Steganopus tricolor Vieillot, 1819

Family Chionidae
Chionis alba (Gmelin, 1789)

Family Stercorariidae
Catharacta skua antarctica (Lesson, 1831)
C. skua chilensis (Bonaparte, 1857)
Stercorarius longicaudus Vieillot, 1819
S. parasiticus (Linnaeus, 1758)
S. pomarinus (Temminck, 1815)

Family Laridae
Gelochiledon nilotica Mathews, 1912
Larus belcheri atlanticus Olrog, 1958
L. cirrocephalus Vieillot, 1818
L. dominicanus Lichtenstein, 1823
L. maculipennis Lichenstein, 1823
Phaetusa simplex (Gmelin, 1789)
Sterna eurygnatha Saunders, 1876
S. *hirundinacea* Lesson, 1831
S. hirundo Linnaeus, 1758
S. maxima Boddaert, 1783
S. paradisea Pontoppidan, 1763
S. superciliaris Vieillot, 1819
S. trudeaui Audubon, 1838
S. vittata Gmelin, 1789

Family Rynchopidae
Rynchops nigra Linnaeus, 1758

Order Passeriformes
Family Motacillidae
Anthus correndera Vieillot, 1818

CLASS MAMMALIA
Order Rodentia
Family Cricetidae
Calomys laucha (Olfers, 1818)

Family Ctenomyidae
Ctenomys flamarioni Travi, 1981

Order Cetacea
Suborder Mysticeti
Family Balaenopteridae
Balaenoptera acutorostrata Lacépède, 1804
B. edeni Anderson, 1878
B. musculus (Linnaeus, 1758)
Megaptera novaeangliae (Borowski, 1781)

Family Balaenidae
Eubalaena australis Desmoulins, 1822

Suborder Odontoceti
Family Physeteridae
Physeter macrocephalus Linnaeus, 1758

Family Kogiidae
Kogia breviceps (Blainville, 1838)
K. simus Owen, 1866

Family Ziphiidae
Hyperoodon planifrons (Flower, 1882)
Mesoplodon densirostris (Blainville, 1817)
Ziphius cavirostris (G. Cuvier, 1823)

Family Delphinidae
Delphinus delphis Linnaeus, 1758
Globicephala melas (Traill, 1809)
Grampus griseus (G. Cuvier, 1812)
Orcinus orca (Linnaeus, 1758)
Pseudorca crassidens (Owen, 1846)
Sotalia fluviatilis (Gervais, 1853)
Stenella attenuata
S. clymene (Gray 1846)
S. *coeruleoalba* (Meyen, 1833)
S. frontalis (G. Cuvier, 1829)
S. longirostris (Gray, 1828)
Steno bredanensis (Lesson, 1828)
Tursiops truncatus (Montagu, 1821)

Family Pontoporiidae
Pontoporia blainvillei (Gervais & Orbigny, 1844)

Family Phocoenidae
Australophocaena dioptrica (Lahille, 1912)
Phocoena spinipinnis Burmeister, 1865

Order Carnivora
Family Otariidae
Arctocephalus australis (Zimmerman, 1783)
A. gazella (Peters, 1875)
A. tropicalis (Gray, 1872)
Otaria flavescens (Shaw, 1800)

Family Phocidae
Lobodon carcinophagus (Hombron & Jacquinot, 1842)
Hydrurga leptonyx (Blainville, 1820)
Mirounga leonina Linnaeus, 1728

References

Aasen O, Castello JP (1968a) Informe sobre un reconocimiento en la plataforma continental entre Cabo San Antonio (Argentina) y Rio Grande (Brasil) con especial referencia al bonito (*Sarda sarda*). Ser Inf Tec FAO/PDP Mar del Plata (Argent) 9:1-15

Aasen O, Castello JP (1968b) Informe sobre estudios y experimentos de marcación del bonito en aguas de la plataforma entre Mar del Plata y Uruguay en abril-mayo 1968. Ser Inf Tec FAO/PDP Mar del Plata (Argent) 12:1-12

Aboussouan A (1969) Sur une petite collection de larves de téléostéens recoltées au large du Brésil (Campagne Calypso, 1962). Vie Milieu Ser A XX(3):595-610

Abreu PC (1987) Variações temporais de biomassa fitoplanctônica (clorofila *a*) e relação com fatores abióticos no canal de acesso ao estuário da Lagoa dos Patos. MSc Thesis, Univ Rio Grande, Brazil

Abreu PC (1992) Phytoplankton production and the microbial food web of the Patos Lagoon Estuary, southern Brazil. PhD Thesis, Univ Bremen, Germany

Abreu PC, Biddanda B, Odebrecht C (1992) Bacterial dynamics of the Patos Lagoon estuary, Southern Brazil (32° S, 52° W): relationship with phytoplankton production and suspended material. Estuar Coast Shelf Sci 35:621-635

Abreu PC, Odebrecht C, González A (1994a) Particulate and dissolved phytoplankton production of the Patos Lagoon estuary, southern Brazil: comparison of methods and influencing factors. J Plankt Res 16:737-753

Abreu PC, Granéli E, Odebrecht C, Kitzmann D, Proença LA, Resgalla C Jr (1994b) Effect of fish and mesozooplankton manipulation on phytoplankton community in the Patos Lagoon estuary, southern Brazil. Estuaries 17:575-584

Abreu PC, Hartmann C, Odebrecht C (1995a) Nutrient rich salt-water and its influence on the phytoplankton of the Patos Lagoon estuary. Estuar Coast Shelf Sci 40:219-229

Abreu PC, Granéli HW, Odebrecht C (1995b) Produção fitoplanctônica e bacteriana na região da pluma estuarina da Lagoa dos Patos, RS, Brasil. Atlantica (Rio Grande) 16:(in press)

Absalão RS (1986) Discriminação ambiental entre associações de moluscos macrobentônicos ao largo de Rio Grande, RS, Brasil. Situação inverno-primavera. MSc Thesis, Univ Rio Grande, Brazil

Absalão RS (1991) Environmental discrimination among soft-bottom mollusc associations off Lagoa dos Patos, South Brazil. Cont Shelf Sci 32:71-85

Acuña A, Castello JP (1986) Estructura de la población, crecimiento y reproducción de *Engraulis anchoita* (Hubbs, Marini 1935) en el sur de Brasil. Revta Biol Mar (Valparaíso) 22(1):23-60

Adam P (1990) Saltmarsh ecology. Cambridge University Press, Cambridge

Almeida LESB, Casalas AB, Rosauro NML, Schettini EBC (1990) Simulação da circulação induzida por vento na Lagoa dos Patos. Congresso Latino-americano de Hidráulica, Montevideo, Memórias 3:1251-1263

Almeida MTA, Baumgarten MGZ, Rodrigues RMS (1993) Identificação das possíveis fontes de contaminação das águas que margeiam a cidade do Rio Grande. Documentos Técnicos, Oceanografia, Univ Rio Grande, Rio Grande 6:1-63

Alverson DL, Freeberg MH, Murawski SA, Pope JG (1994) A global assessment of the fisheries bycatch and discards. FAO Tech Pap 339

Ambrose WG (1984a) Role of predatory infauna in structuring soft-bottom communities. Mar Ecol Prog Ser 17:109-115

Ambrose WG (1984b) Influences of predatory polychaetes and epibenthic predators on the structure of a soft-bottom community in a Maine estuary. J Exp Mar Biol Ecol 81:115-145

Andriguetto JM, Haimovici M (1991) Abundance and distribution of *Loligo sanpaulensis* (Brackoniecki 1984) (Cephalopoda, Myoposida) in southern Brazil. Sci Mar (Barc) 55(4):611-618

Andriguetto JM, Haimovici M (in press) Reproductive cycle of *Loligo sanpaulensis* (Brakoniecki 1984) (Cephalopoda, Loliginidae) in southern Brazil. Sci Mar (Barc)

Antero da Silva JN (1994) Tuna fisheries in Brazil by leased Japanese longliner fleet from 1977 to 1991. Report of the second ICCAT billfish workhop. Coll Vol Sc Pap SCRS 92(33):180-188

Araújo FJ (1984) Hábitos alimentares de três espécies de bagres marinhos (Ariidae) no estuário da Lagoa dos Patos, RS, Brasil. Atlantica (Rio Grande) 7:47-63

Arena G, Barea L, Barreiro D, Beathyate G, Marín Y (1992) Utilización de redes de baja apertura en la pesca de lenguado (*Paralichthys* spp.) Inst Nac Pesca Inf Tec (Montevideo) 37:1-22

Arkhipkin AI (1990) Edad y crecimiento del calamar (*Illex argentinus*). Frente Mar (Montevideo) 6:25-36

Arntz WE, Fahrbach E (1991) El Niño, Klimaexperiment und Natur. Physikalische Ursachen und biologische Folgen. Birkhäuser Verlag, Basel

Ashmole NP, Ashmole MJ (1967) Comparative feeding ecology of sea birds of a tropical oceanic island. Bull Peabody Mus Nat Hist (Yale) 24:1-131

Asmus HE, Asmus ML, Drews CR (1984) Levantamento e modificações dos ambientes do ecossistema estuarial da Lagoa dos Patos. Súmula I Seminário sobre pesquisa da Lagoa dos Patos, novembro de 1994, Porto Alegre, pp 170-181

Asmus HE, Asmus ML, Tagliani PR (1989) O estuário da Lagoa dos Patos: um problema de planejamento costeiro. Anais III Encontro Brasil Gerenciamento Costeiro, Universidade Federal do Ceará, Fortaleza, pp 71-95

Asmus HE, Asmus ML, Matarezi J (1991) Uma visão crítica da metodologia para o levantamento ambiental costeiro do Brasil. Anais III Encontro Nac Estudos Meio Ambiente, Londrina, pp 207-237

Asmus ML (1984) Estrutura da comunidade associada a *Ruppia maritima* no estuário da Lagoa dos Patos, RS, Brasil. MSc Thesis, Univ Rio Grande, Brazil

Asmus ML (1989) Pradarias de gramíneas marinhas (*Ruppia maritima*) como áreas vitais na região estuarial da Lagoa dos Patos. Anais III Encontro Brasil Gerenciamento Costeiro, Fortaleza, pp 291-299

Assis LF (1994) Avaliação de impactos e prospectiva ambiental. BIO Eng Sanit Amb ABES (Rio de Janeiro) 2(5):85-92

Azam F, Fenchel T, Field JG, Gray JS, Meyer-Reil LA, Thingstad F (1983) The ecological role of water-column microbes in the sea. Mar Ecol Prog Ser 10:257-263

Baisch PR, Niencheski LF, Lacerda LD (1988) Trace metal distribution in sediments of the Patos Lagoon estuary, Brazil. In: Seeliger U, Lacerda LD, Patchineelam SR (eds) Metals in coastal environments of Latin America. Springer Verlag, Berlin Heidelberg New York, pp 59-64

Baisch P, Jouanneau JM, Asmus HE (1989) Chemical composition of sediments from the Patos Lagoon. XIII Int Geoch Expl Symp II Braz Geoch Cong 1:11-20

Bamber RN, Henderson PA (1988) Pre-adaptative plasticity in atherinids and the estuarine seat of teleost evolution. J Fish Biol 33:17-23

Baptista JR (1984) Flutuações diárias e horárias dos elementos dissolvidos, material em suspensão e características físicas da água na parte sul do estuário da Lagoa dos Patos e praia do Cassino. MSc Thesis, Univ Rio Grande, Brazil

Barattini LD, Escalante R (1971) Catalogo de las aves uruguayas, 2a parte: Anseriformes. Intendencia Municipal de Montevideo, Museo "Damaso Antonio Larrañaga", Montevideo

Barbieri LRR (1986) Distribuição espacial e temporal de Sciaenidae juvenis na região estuarina da Lagoa dos Patos, RS, Brasil. MSc Thesis, Univ Rio Grande, Brazil

Barbour MG (1992) Life at the leading edge: the beach plant syndrome. In: Seeliger U (ed) Coastal Plant Communities of Latin America. Academic Press, New York, pp 291-307

Barbour MG, de Jonge TM, Johnson AF (1976) Synecology of beach vegetation along the Pacific coast of the United States of America: a first approximation. J Biogeogr 3:55-69

Barbour MG, Rejmanek M, Johnson AF, Pavlik BM (1987) Beach vegetation and plant distribution patterns along the northern Gulf of Mexico. Phytocoenologia 15:201-233

Barcellos BN (1957) Ictiofauna do Rio Grande do Sul. I. Sobre *Heptranchias perlo* (Bonnaterre 1788). Rev Brasil Biol (Rio de Janeiro) 17(2):173-178

Barcellos BN (1963) Ictiofauna marinha do Rio Grande do Sul. VI. Cação galha preta *Carcharhinus maculipinnis* (Poey 1865). Bolm Inst Ciênc Nat Univ Fed Rio Grande Sul (Porto Alegre) 20(13):3-13

Barcellos LJP, Peres MB, Wahlrich R, Barison MB (1991) Relatório sobre a otimização bioeconômica dos recursos pesqueiros marinhos do Rio Grande do Sul. Ed Univ Rio Grande, Rio Grande

Barcellos LP, Pinedo MC (1980) Sobre a ocorrência do tubarão espinhoso *Echinorhinus brucus* (Bonnaterre 1788) para a costa sul brasileira (Squaliformes: Squalidae). Iheringia Ser Zool (Porto Alegre) 56:71-74

Barth R (1966) Estudos planctológicos em material da Operação Tridente III. Publ Inst Pesq Marin Nota Técnica (Rio de Janeiro) 29:1-43

Barth R (1968) Estudos planctonológicos entre Cabo Frio e Mar del Plata. Publ Inst Pesq Marin (Rio de Janeiro) 26:1-14

Bastos EO, Perazzolo M, Baptista JMR (1993a) Variação do número de estômatos e micropêlos em *Paspalum vaginatum* Sw. em relação às condições abióticas numa marisma do estuário da Lagoa dos Patos, RS - Brasil. Acta Bot Brasil (S Paulo) 6(2):105-117

Bastos EO, Perazzolo M, Gorgen AUG (1993b) Ocorrência e estrutura de glândulas de sal em espécies halófitas no município do Rio Grande, RS. Iheringia Ser Bot (Porto Alegre) 43:3-14

Batista VS (1988) Idade e crescimento do cação *Mustelus schmitti*. MSc Thesis, Univ Rio Grande, Brazil

Baumgarten MGZ (1987) Avaliação do *Balanus improvisus* como bioindicador dos níveis metálicos do estuário da Lagoa dos Patos. MSc Thesis, Univ Rio Grande, Brazil

Baumgarten MGZ, Niencheski LF (1990a) O estuário da laguna dos Patos: variações de alguns parâmetros fisico-químicos da água e metais associados ao material em suspensão. Ciencia Cult (S Paulo) 42:390-396

Baumgarten MG, Niencheski LF (1990b) Avaliação da capacidade bioindicadora de *Balanus improvisus* para os metais chumbo, cobre e manganês presentes no estuário da Lagoa dos Patos (RS, Brasil). Atlantica (Rio Grande) 12(2):5-19

Baumgarten MGZ, Niencheski LF, Kuroshima KN (1995) Qualidade das águas que margeiam a cidade do Rio Grande: nutrientes e detergentes dissolvidos. Atlantica (Rio Grande) 17(in press)

Bé AWH, Tolderlund DS (1971) Distribution and ecology of living planktonic foraminifera in surface waters of the Atlantic and Indian Oceans. In: Funncle BM, Reedel WR (eds) The micropaleontology of the oceans. Cambridge Univ Press, London

Belton W (1984) Birds of Rio Grande do Sul, Brazil. 1. Rheidae through Furnariidae. Bull Am Mus Nat Hist 178(4):371-631

Belton W (1985) Birds of Rio Grande do Sul, Brazil. 2. Formicariidae through Corvidae. Bull Am Mus Nat Hist 180(1):1-242

Bemvenuti CE (1983) Efeitos da predação sobre as características estruturais de uma comunidade macrozoobentônica numa enseada estuarina da Lagoa dos Patos, RS, Brasil. MSc Thesis, Univ Rio Grande, Brazil

Bemvenuti CE (1987a) Macrofauna bentônica da região estuarial da Lagoa dos Patos, RS, Brasil. Publ Acad Ciências Est (S Paulo) 54(l):428-459

Bemvenuti CE (1987b) Predation effects on a benthic community in estuarine soft sediments. Atlantica (Rio Grande) 9(1):5-32

Bemvenuti CE (1988) Impacto da predação sobre *Heteromastus similis* Southern, 1921 e *Nephtys fluviatilis* Monro, 1937 (Annelida, Polychaeta), em fundos moles estuarinos. Atlantica (Rio Grande) 10(1):85-102

Bemvenuti CE (1992) Interações biológicas da macrofauna bentônica numa enseada estuarina da Lagoa dos Patos, RS, Brasil. PhD Thesis, Univ S Paulo, Brazil

Bemvenuti CE (1994) O poliqueta *Nephtys fluviatilis* (Monro 1937) como predador da infauna na comunidade de fundos moles. Atlantica (Rio Grande) 16:87-98

Bemvenuti CE, Capítoli RR, Gianuca NM (1978) Estudos de ecologia bentônica na região estuarial da Lagoa dos Patos. II. Distribuição quantitativa do macrobentos infralitoral. Atlantica (Rio Grande) 3:23-32

Bemvenuti CE, Cattaneo SA, Netto SA (1992) Características estruturais da macrofauna bentônica em dois pontos da região estuarial da Lagoa dos Patos, RS, Brasil. Atlantica (Rio Grande) 14:5-28

Bemvenuti MA (1984) Ciclo de vida de peixe-rei (Atherinidae) no estuário da Lagoa dos Patos, RS, Brasil. MSc Thesis, Univ Rio Grande, Brazil

Bemvenuti MA (1987) Abundância, distribuição e reprodução de peixes-rei (Atherinidae) na região estuarina da Lagoa dos Patos, RS, Brasil. Atlantica (Rio Grande) 9(1):5-32

Bemvenuti MA (1990) Hábitos alimentares de peixes-rei (Atherinidae) na região estuarina da Lagoa dos Patos, RS, Brasil. Atlantica (Rio Grande) 12(1):79-102

Bemvenuti MA (1993) Redescrição do peixe-rei *Odontesthes argentinesis* (Valenciennes) Pisces: Atherinidae, na costa do Rio Grande do Sul. Atlantica (Rio Grande) 15:17-35

Benvegnu-Lé GQ (1973) Relação dos peixes teleósteos. Relatório sobre a segunda pesquisa oceanográfica e pesqueira do Atlântico Sul entre Torres e Maldonado (Lat. 29^0-34^0 S). Publ Esp Inst Oceanogr (S Paulo) (3)l:489-500

Benvegnu-Lé GQ (1978) Distribuição dos peixes teleósteos marinhos demersais na plataforma continental do Rio Grande do Sul. PhD Thesis, Univ S Paulo, Brazil

Bergesch M (1990) Variações de biomassa e composição do fitoplâncton na área estuarina rasa da Lagoa dos Patos e suas relações com fatores de influência. MSc Thesis, Univ Rio Grande, Brazil

Bergesch M, Odebrecht C (1987) Variações sazonais da biomassa do fitoplâncton total e fracionado, em categorias de tamanho, em área estuarina rasa da Lagoa dos Patos, RS. Publ Acad Ciências Est (S Paulo) 54(2):288-298

Bergesch M, Odebrecht C, Abreu PC (1995) Microalgas do estuário da Lagoa dos Patos: interação entre o sedimento e a coluna de água. Oecol Brasil (Rio de Janeiro) 1:273-289

Bernardi H, Seeliger U (1989) Population biology of *Blutaparon portulacoides* (St Hill.) Mears on southern Brazilian backshores. Ciencia Cult (S Paulo) 41(11):1110-1113

Bernardi H, Cordazzo CV, Costa CSB (1987) Efeito de ressacas sobre *Blutaparon portulacoides* (St Hill.) Mears, nas dunas costeiras do sul do Brasil. Ciencia Cult (S Paulo) 39(5/6):545-547

Bersano JG (1994) Zooplâncton da zona de arrebentação de praias arenosas, situadas ao sul de Rio Grande, RS. Primavera de 1990, verão de 1991. MSc Thesis, Univ Rio Grande, Brazil

Best PB, Payne R, Rowntree V, Palazzo JT, Both MC (1993) Long-range movements of South Atlantic right whales *Eubalaena australis*. Mar Mamm Science 9(3):227-234

Betito R (1984) Dinâmica da população de *Jenynsia lineata* (Cyprinodontiformes, Anablepidae) na restinga de Rio Grande, estuáro da Lagoa dos Patos (RS, Brasil). MSc Thesis, Univ Rio Grande, Brazil

Bianchini A, Wasielesky WFB, Miranda KC Fo (1996) Toxicity of nitrogenous compounds to juveniles of the flatfish *Paralichthys orbignyanus*. Bull Environ Contamin Toxicol 56(3):453-459

Bicalho H (1883) Descrição geral e observações preliminares. Estudos e projetos de melhoramentos. Obras do Porto e Barra do Rio Grande, Oficinas Gráficas da Federação (Porto Alegre) 3:1-455

Björnberg TKS (1963) On the marine free living copepods of Brazil. Bolm Inst Oceanogr (S Paulo) 13:3-142

Blaber SJM, Brewer DT, Salini JP (1989) Species composition and biomasses of fishes in different habitats of a tropical northern Australia Estuary: their occurrence in the adjoining sea and estuarine dependence. Estuar Coast Shelf Sci (29):509-531

Blaxter JHS, Hempel G (1963) The influence of egg size on herring larvae (*Clupea harengus*). J Cons Int Explor Mer 28(2):211-240

Boff MH, Marchiori MA (1984) The effect of temperature on larval development of the pink shrimp *Penaeus paulensis*. Atlantica (Rio Grande) 7:7-13

Boltovskoy D, Jankilevich S (1981) Chaetognatha. In: Boltovskoy D (ed) Atlas del Zooplancton del Atlántico Sudoccidental y métodos de trabajo con el zooplancton marino. Publ Esp INIDEP, Mar del Plata, pp 759-793

Boltovskoy E (1959) Foraminíferos recientes del sur de Brasil y sus relaciones con los de Argentina e India del oeste. Serv Hidrogr Nav (Argent) H 1005:1-120

Boltovskoy E (1970) Masas de água (característica, distribución, movimentos) en la superficie del Atlántico Sudoeste según indicadores biológicos - foraminíferos. Serv Hidrogr Nav (Argent) H 643:1-99

Boltovskoy E (1981) Foraminifera. In: Boltovskoy D (ed) Atlas del zooplancton del Atlántico Sudoccidental y métodos de trabajo con el zooplancton marino. Publ Esp INIDEP, Mar del Plata, pp 317-352

Bonecker ACT, Hubold G (1990) Distribution and abundance of larval gonostomatid fishes in the southwest Atlantic. Meeresforschung 33:38-51

Bonilha F (1974) Circulação em lagos: um modelo matemático simplificado aplicado ao Rio Guaíba e Lagoa dos Patos. MSc Thesis, Univ Fed Rio Grande Sul, Brazil

Bonilha LE (1996) Modelo ecológico da coluna d'água do estuário da Lagoa dos Patos (RS-Brasil) MELP: uma abordagem sistêmica e integrada. MSc Thesis, Univ Rio Grande, Brazil

Bonilha LE, Asmus ML (1994) Modelo ecológico do fitoplâncton e zooplâncton do estuário da Lagoa dos Patos, RS. Publ Acad Ciências Est (S Paulo) 87(1):347-362

Borzone CA (1988) Estudo da macrofauna bentônica infralitoral da região costeira adjacente à Barra de Rio Grande. MSc Thesis, Univ Rio Grande, Brazil

Borzone CA, Gianuca NM (1990) A zonação infralitoral em praias arenosas expostas. Publ Acad Ciências Est (S Paulo) 71(3):280-295

Borzone CA, Griep GH (1991) Características do sedimento superficial infralitoral da região costeira adjacente à desembocadura da Lagoa dos Patos, RS, Brasil. Pesquisas (Porto Alegre) 18(1):71-78

Boschi EE (1969) Estudio biológico-pesquero del camarón *Artemesia longinaris* Bate, de Mar del Plata. Bol Inst Biol Mar (Mar del Plata) 3:1-23

Boyd JF, Simmons RC (1974) Continuous laboratory production of fertile *Fundulus heroclitus* (Walbaum) eggs lacking chorionic fibrils. J Fish Biol 6:389-394

Brackoniecki TF (1984) A full description of *Loligo sanpaulensis* (Brackoniecki 1984) and a redescription of *Loligo gahi* (d'Orbigny 1835), two species from the southwest Atlantic. Bull Mar Sci 34(3):435-448

Brandhorst W, Castello JP (1971) Evaluación de los recursos de anchoíta (*Engraulis anchoita*) frente a la Argentina y Uruguay-I-Las condiciones oceanográficas, sinopsis del conocimiento actual sobre la anchoíta y el plan para su evaluación. Proy Des Pesq Ser Inf Tec Mar del Plata (Argent) 29:1-63

Brinton E (1975) Euphausiids of southeast Asian waters. Naga Rept Scripps Inst Oceanogr 4(5):1-287

Bromberg E (1992) Dinâmica osmo e ionorregulatória de *Chasmagnathus granulata* Dana, 1851 (Crustacea, Decapoda, Grapsidae) submetido ao estresse hipo e hiperosmótico no inverno e verão. MSc Thesis, Univ S Paulo, Brazil

Brown AC, McLachlan A (1990) Ecology of sandy shores. Elsevier, Amsterdam

Brownell RL Jr (1975) Progress report on the biology of the franciscana dolphin, *Pontoporia blainvillei*, in Uruguayan waters. J Fish Res Bd Can 32:1073-1078

Brownell RL Jr (1989) Franciscana *Pontoporia blainvillei* (Gervais and d'Orbigny 1844). In: Ridgway SH, Harrison R (eds) Handbook of marine mammals, vol 4. Academic Press, London, pp 45-47

Brummet AR, Dumont JN (1981) A comparison of chorions from eggs of northern and southern populations of *Fundulus heroclitus*. Copeia 3:607-614

Buckup PA (1984) Distribuição e abundância de engraulídeos (Osteichthyes, Clupeiformes) na região estuarial da Lagoa dos Patos, RS, Brasil. MSc Thesis, Univ Rio Grande, Brazil

Bursa AS (1968) Starch in the oceans. J Fish Res Bd Can 25 (6):1269-1284

Burton PJK (1974) Feeding and the feeding apparatus of waders: a study of anatomy and adaptations in the Charadrii. Br Mus (Nat Hist), vol 719

Cabrera A, Yepes J (1940) Los cetáceos. In: Cabrera A, Yepes J (1940) (ed) Mamíferos sul-americanos. Companía Argentina Editores, Buenos Aires, pp 107-119

Cacciari PL (1986) Estudo descritivo sobre vórtices de núcleos quentes e frios incorporados à circulação oceânica, no limite oeste do Atântico Sul. MSc Thesis, Univ S Paulo, Brazil

Cafruni AMS (1983) Estudo autoecológico de *Ruppia maritima* no estuário da Lagoa dos Patos. MSc Thesis, Univ Rio Grande, Brazil

Cafruni AMS, Krieger J, Seeliger U (1978) Observação sobre *Ruppia maritima* L. no sul do Brasil. Atlantica (Rio Grande) 3:85-90

Calderón JJD (1994) Morfologia, distribuição, abundância e reprodução de duas espécies do gênero *Squalus* no sul do Brasil. MSc Thesis, Univ Rio Grande, Brazil

Callan EMcC (1964) Ecology of sand dunes with special reference to the insect communities. In: Davis JH (ed) Ecological studies in southern Africa. Junk, The Hague, pp 174-185

Calliari LJ (1980) Aspectos sedimentológicos e ambientais da região sul da Lagoa dos Patos. MSc Thesis, Univ Fed Rio Grande Sul, Brazil

Calliari LJ, Abreu JGN (1984) Litologia da plataforma continental interna adjacente à cidade do Rio Grande (RS), através da interpretação de registros de sonar de varredura lateral e amostragem superficial. Anais XXXIII Cong Brasil Geol, Rio de Janeiro, pp 1553-1564

Calliari LJ, Fachin S (1993a) Laguna dos Patos. Influência nos depósitos lamíticos Costeiros. Pesquisas (Porto Alegre) 20(1)57-69

Calliari LJ, Klein AHF (1993b) Características morfodinâmicas e sedimentológicas das praias oceânicas entre Rio Grande e Chuí, RS. Pesquisas (Porto Alegre) 20(1):48-56

Calliari LJ, Esteves LS, Oliveira CPL, Tozzi HAM, Silva R, Cardoso JN (1994) Padrões sonográficos e sedimentológicos de um afloramento de beachrock na plataforma interna do Rio Grande do Sul. COMEMIR/OSNLR. Not Tec CECO Univ Fed Rio Grande Sul (Porto Alegre) 7:27-32

Capítoli RR (1983) Seqüência temporal de colonização e desenvolvimento da comunidade incrustante na região mixohalina da Lagoa dos Patos, RS, Brasil. MSc Thesis, Univ Rio Grande, Brazil

Capítoli RR, Haimovici M (1993) Alimentación del besugo *Pagrus pagrus* en el extremo sur del Brasil. Frente Mar (Montevideo) 14:81-86

Capítoli RR, Ortega EL (1993) Contribuição ao conhecimento do espectro trófico de *Cyrtograpsus angulatus*, Dana, nas enseadas vegetadas de águas rasas da Lagoa dos Patos, RS, Brasil. Nauplius (Rio Grande) 1:81-87

Capítoli RR, Bemvenuti CE, Gianuca NM (1977) Occurrence and bioecologic observations on *Metasesarma rubripes* crab in the estuarine region of Lagoa dos Patos. Atlantica (Rio Grande) 2(1):50-62

Capítoli RR, Bemvenuti CE, Gianuca NM (1978) Estudos de ecologia bentônica na região estuarial da Lagoa dos Patos. I. As comunidades bentônicas. Atlantica (Rio Grande) 3:5-22

Cappetta H (1987) Extinctions et renouvellements fauniques chez les sélaciens post-jurassiques. Mém Soc Geol Fr Nouv Ser 150:113-131

Carnavale NJ, Torres P, Boccanelli SI, Lewis JP (1987) Halophilous communities and species distributions along environmental gradients in southeastern Santa Fé province, Argentina. Coenoses (Trieste) 2:49-60

Carreto JR, Negri RM, Benavides HR (1986) Algunas características del florecimiento del fitoplancton en el frente del Río de la Plata. I: Los sistemas nutritivos. Rev Inv Des Pesq (Mar del Plata) 5:7-29

Carvalho CT (1961) *Stenodelphis blainvillei* na costa meridional do Brasil, com notas osteológicas (Cetacea, Platanistidae). Revta Bras Biol (Rio de Janeiro) 21(4):443-454

Carvalho CT (1975) Ocorrência de mamíferos marinhos no Brasil -ordem Cetacea. Bolm Tec Inst Flor Secr Agr (S Paulo) 16:13-32

Cassano V (1987) Relações filogenéticas de dez espécies de Sciaenidae (Teleostei:Perciformes): estudo eletroforético. MSc Thesis, Univ Rio Grande, Brazil

Cassano V, Levy JA (1990) Sistemática bioquímica de *Menticirrhus americanus* e *M. littorales* (Teleostei:Perciformes:Scianidae). Bolm Inst Oceanogr (S Paulo) 38(1):87-91

Castel J (1992) The meiofauna of coastal lagoon ecosystems and their importance in the food web. Vie Milieu 42(2):125-135

Castello HP, Pinedo MC (1977a) Botos na Lagoa dos Patos. Nat Rev Publ Fund Zoobot (Porto Alegre) 2:46-49

Castello HP, Pinedo MC (1977b) Primeiro registro de *Arctocephalus tropicalis* para a costa do Rio Grande do Sul (Pinnipedia, Otariidae). Atlantica (Rio Grande) 2(2):111-119

Castello HP, Pinedo MC (1977c) Os visitantes ocasionais de nosso litoral. Nat Rev Publ Fund Zoobot (Porto Alegre) 3:40-46

Castello HP, Pinedo MC (1986) Sobre algunos avistajes en el mar de distintas especies de cetáceos en el sur de Brasil. Actas 1a Reunión de Trabajo de Expertos en Mamíferos Acuáticos de América del Sur, 25-29 junho 1984, B Aires, pp 61-68

Castello JP (1985) La ecología de los consumidores del estuario de la Lagoa dos Patos, Brasil. In: Yáñez-Arancibia (ed) Fish community ecology in estuaries and coastal lagoons: towards and ecosystem integration. Univ Nac Aut Mexico Press, Mexico, pp 383-406

Castello JP (1986) Distribución, crecimiento y maduración sexual de la corvina juvenil *Micropogonias furnieri* en el estuario de la Lagoa dos Patos, Brasil. Physis (B Aires) 44(106):21-36

Castello JP, Cousseau MB (1976) Observaciones sobre la caballa en las temporadas de pesca del período 1969 a 1975 (Pisces, *Scomber japonicus marplatensis*). Physis (B Aires) 35(91):195-203

Castello JP, Habiaga RP (1982) Resultados preliminares de avaliação de pequenos peixes pelágicos usando técnicas hidro-acústicas. Anais I Simp Naval Sonar (Rio de Janeiro) 2:1-19

Castello JP, Habiaga RP (1988) The skipjack tuna fishery in southern Brazil. ICCAT, Madrid, Coll Vol Sc Pap 30(1):6-19

Castello JP, Krug LC (1978) Distribution, growth and spawning groups of the manjuba (*Lycengraulis grossidens*) in the estuary of the Lagoa dos Patos, Rio Grande do Sul, Brasil. Atlantica (Rio Grande) 3:33-46

Castello JP, Möller OO Jr (1977) On the oceanographic conditions in the Rio Grande do Sul state. Atlantica (Rio Grande) 2(2):25-110

Castello JP, Möller OO Jr (1978) On the relationship between rainfall and shrimp production in the estuary of the Patos Lagoon (Rio Grande do Sul, Brazil). Atlantica (Rio Grande) 3:67-74

Castello JP, Vasconcellos MC (1995) Growth rate of anchovy (*Engraulis anchoita*) larvae caught off Santa Marta Grande Cape, Brazil. Arch Fish Mar Res 42(3):263-281

Castello JP, Duarte AK, Möller OO Jr, Niencheski LF, Odebrecht C, Weiss G, Habiaga RP, Bellotto VR, Kitzmann D, Souto C, Souza RB, Ciotti AM, Fillmann G, Schwingel PR, Bersano JC, Cirano M, Freire K, Lima I, Mello R, Monteiro A, Resgalla C Jr, Soares I, Suzuki M (1990) On the importance of coastal and subantarctic waters for the shelf ecosystem off Rio Grande do Sul. Publ Acad Ciências Est (S Paulo) 71(1):112-129

Caussanel C (1970) Contribution à l'étude du peuplement d'une plage et d'une dune landaise. Vie Milieu 21:59-104

Cendrero A, Diaz de Teran JR (1987) The environment map system of the University of Cantabria, Spain. In: Arndt P, Luttig G (eds) Mineral resources extraction, environmental protection and land use planning in the industrial and developing countries. Schweizerbart Verlag, Stuttgart, pp 149-181

Cerqueira VR, Haimovici M (1990) Dinâmica populacional do gordinho *Peprilus paru* (Pisces, Stromateidae), no litoral sul do Brasil. Rev Brasil Biol (Rio de Janeiro) 50(3):599-613

Chao LN, Pereira LE, Vieira JP, Bemvenuti MA, Cunha LPR (1982) Relação preliminar dos peixes estuarinos e marinhos da Lagoa dos Patos e região costeira adjacente, Rio Grande do Sul, Brasil. Atlantica (Rio Grande) 5(1):67-75

Chao LN, Pereira LE, Vieira JP (1985) Estuarine fish community of the dos Patos Lagoon, Brazil: a baseline study. In: Yáñez-Arancibia A (ed) Fish community ecology in estuaries and coastal lagoons: towards an ecosystem integration. Univ Nac Aut Mexico Press, Mexico, pp 429-450

Chao NL, Vieira JP, Barbieri LRR (1986) Lagoa dos Patos as a nursery ground for shore fishes off southern Brazil. In: IOC/FAO Workshop on recruitment in tropical coastal demersal communities. Ciudad del Carmen, Campeche, México, 21-25 April 1986. Intergov Oceanogr Comm Workshop Rep 44 [Suppl]: 144-150

Chebatoroff J (1952) Vegetacíon de los suelos salinos. Revta Urug Geogr (Montevideo) 6:71-100

Chiaradia A (1991) Interação entre aves marinhas e cardumes de bonito listrado (*Katsuwonus pelamis*) na costa sul do Brasil. Atlantica (Rio Grande) 13(1):115-118

Chomenko LA, Schafer A (1984) Interpretação biogeográfica da distribuição do gênero *Littoridina* (Hydrobiidae) nas lagoas costeiras do Rio Grande do Sul, Brasil. Amazoniana (Kiel) 9(1):127-146

Christiansen HE, Cousseau MB (1985) Aportes a la determinación de la frecuencia reproductiva de la anchoíta, *Engraulis anchoita*. Physis (B Aires) 43(104):7-17

Ciechomski J, Cassia MC (1980) Reproducción y fecundidad del surel (*Trachurus picturatus australis*). Rev Inv Des Pesq (Mar del Plata) 2(2):39-45

Ciotti AM (1990) Fitoplâncton da plataforma continental do Sul do Brasil: clorofila *a*, feopigmentos e análise preliminar da produção primária (outubro de 1987 e setembro de 1988). MSc Thesis, Univ Rio Grande, Brazil

Ciotti AM, Odebrecht C, Fillmann G, Möller OO Jr (1995) Freshwater outflow and Subtropical Convergence influence on phytoplankton biomass on the southern Brazilian continental shelf. Cont Shelf Res 15(14):1737-1756

Clarke MR, Macleod N Castello HP, Pinedo MC (1980) Cephalopod remains from the stomach of a sperm whale stranded at Rio Grande do Sul in Brazil. Mar Biol 59:235-239

Closs D, Barberena MC (1960) Foraminíferos recentes da Praia do Cassino (Rio Grande, RS). Bolm Escol Geol (Porto Alegre) 5:1-29

Closs D, Barbarena MC (1962) Foraminíferos recentes das praias do litoral sul-brasileiro. 1. Arroio Chuí (RS)-Araranguá (SC). Bolm Inst Ciênc Nat (Porto Alegre) 16:7-55

Commito JA, Schrader PB (1985) Benthic community response to experimental additions of the polychaete *Nereis virens*. Mar Biol 86:101-107

Compagno LJV (1984) Sharks of the world. FAO Fish Synop 125(4):1-655

Corcuera J, Monzón F, Crespo EA, Aguilar A, Raga JA (1994) Interactions between marine mammals and coastal fisheries of Necochea and Claromecó (Buenos Aires Province, Argentina). In: Perrin WP, Donovan GP, Barlow J (eds) Gillnets and cetaceans, IWC special issue, vol 15. Red House, Cambridge, pp 283-290

Cordazzo CV (1985) Taxonomia e ecologia da vegetação de dunas costeiras ao sul do Cassino (RS). MSc Thesis, Univ Rio Grande, Rio Grande, Brazil

Cordazzo CV (1994) Comparative population studies of four dominant plants of southern Brazilian coastal dunes. PhD Thesis, Univ East Anglia, England

Cordazzo CV, Seeliger U (1987) Composição e distribuição da vegetação nas dunas costeiras ao sul de Rio Grande (RS). Ciencia Cult (S Paulo) 39(3):121-124

Cordazzo CV, Seeliger U (1988a) Guia ilustrado da vegetação costeira no extremo sul do Brasil. Ed Univ Rio Grande, Rio Grande

Cordazzo CV, Seeliger U (1988b) Phenological and biogeographical aspects of dune plant communities in southern Brazil. Vegetatio 75:169-173

Cordazzo CV, Seeliger U (1993) Zoned habitats of southern Brazilian coastal foredunes. J Coast Res 9(2):317-323

Cordazzo CV, Souza ZD (1993) Germinação de *Senecio crassiflorus* (Compositae). Revta Brasil Biol (Rio de Janeiro) 53(1):81-86

Corrêa ICS (1983) Depósitos de material carbonático da plataforma continental interna do Rio Grande do Sul, Brasil. In: Arana PM (ed) Proceedings of the international conference on marine resources of the Pacific. Viña del Mar, Chile, pp 155-163

Corrêa ICS (1987) Sedimentos superficiais da plataforma e talude continental do Rio Grande do Sul. Pesquisas (Porto Alegre) 19:95-104

Corrêa ICS (1990) Analyse morphostructurale et evolution paléogeographique de la plateforme continentale Atlantique Sud-Brésilienne (Rio Grande do Sul-Brésil). PhD Thesis, Univ Bordeaux, France

Correia KV (1983) As medusas da plataforma dos estados do Paraná e Santa Catarina (Operação Conversut III). Sistemática e distribuição. MSc Thesis, Univ Fed Paraná, Brazil

Cory RL (1967) Epifauna of the Patuxent River Estuary, Maryland, for 1963 and 1964. Chesapeake Sci 8:71-89

Costa CSB (1987) Aspectos da ecologia populacional das plantas dominantes das dunas costeiras do Rio Grande do Sul, Brasil. MSc Thesis, Univ Rio Grande, Brazil

Costa CSB (in press) Production ecology of *Scirpus maritimus* in southern Brazil. Ciencia Cult (S Paulo)

Costa CSB, Davy AJ (1992) Coastal salt marsh communities of Latin America. In: Seeliger U (ed) Coastal plant communities of Latin America. Academic Press, New York, pp 179-199

Costa CSB, Seeliger U (1988a) Demografia de folhas de *Spartina ciliata* Brong. em dunas e brejos costeiros. Revta Brasil Bot (S Paulo) 11:85-94

Costa CSB, Seeliger U (1988b) Demografia de folhas de *Hydrocotyle bonariensis* Lam, uma planta herbácea rizomatosa perene, nas dunas costeiras. Rev Brasil Biol (Rio de Janeiro) 48(3):443-451

Costa CSB, Seeliger U (1989) Vertical distribution and biomass allocation of *Ruppia maritima* L. in a southern Brazilian estuary. Aquat Bot 33:123-129

Costa CSB, Seeliger U (1990) Quantitative phenology and horizontal distribution of the rhizomatous perennial herb *Hydrocotyle bonariensis* Lam in coastal sand dunes. Vida Silv Neotrop (Washington DC) 2(2):36-42

Costa CSB, Seeliger U, Cordazzo CV (1984) Aspectos da ecologia populacional de *Panicum racemosum* (Spreng) nas dunas costeiras do Rio Grande do Sul, Brasil. In: Lacerda LD, Araujo DSD, Turq B, Cerqueira R (eds) Restingas: origem, estruturas e processos. Ed Univ Fed Fluminense, Niterói, pp 395-411

Costa CSB, Seeliger U, Kinas PG (1988a) The effect of wind velocity and direction on the salinity regime in the Patos Lagoon estuary. Ciencia Cult (S Paulo) 40(9):909-912

Costa CSB, Seeliger U, Cordazzo CV (1988b) Distribution and phenology of *Andropogon arenarius* Haeckel in coastal dunes of Rio Grande do Sul, Brazil. Revta Brasil Biol (Rio de Janeiro) 8(3):527-536

Costa CSB, Seeliger U, Cordazzo CV (1988c) Dinâmica populacional e distribuição do *Androtrichum trigynum* (Spreng) (Cyperaceae) nos brejos e dunas costeiras do Rio Grande do Sul. Acta Limnol Brasil 2:813-842

Costa CSB, Seeliger U, Cordazzo CV (1991) Leaf demography and decline of a *Panicum racemosum* population in coastal sand dunes of southern Brazil. Can J Bot 69:1593-1599

Costa CSB, Cordazzo CV, Seeliger U (1995) Shore pertubation and dune plant distribution. J Coast Res 11(4):

Costa PA, Fernandes CF (1993) Reproductive cycle of *Loligo sanpaulensis* (Cephalopoda: Loliginidae) in the Cabo Frio region, Brazil. Mar Ecol Prog Ser 101:91-97

Coull BC, Bell SS (1979) Perspectives of marine meiofaunal ecology. In: Livingston RR (ed) Ecological processes in coastal marine systems. Marine Science Series, Plenum Press, New York

Cousin JCB, Balouet G, Baudin-Laurencin F (1986) Alterations histologiques observées chez des larvaes de turbot (*Scophthalmus maximus*) en élevage intensif. Aquaculture 52:173-189

Cousseau MB (1967) Contribución al conocimiento de la biología del surel (*Trachurus picturatus australis*) del área de Mar del Plata (Pisces, Fam. Carangidae). Bol Inst Biol Mar (Mar del Plata) (15):1-38

Coutinho R (1982) Taxonomia, distribuição, crescimento sazonal, reprodução e biomassa das algas bentônicas no estuário da Lagoa dos Patos (RS). MSc Thesis, Univ Rio Grande, Brazil

Coutinho R, Seeliger U (1984) The horizontal distribution of the benthic algal flora in the Patos Lagoon estuary, Brazil. J Exp Mar Biol Ecol 80:247-257

Coutinho R, Seeliger U (1986) Seasonal occurrence and growth of benthic algae in the Patos Lagoon estuary, Brazil. Estuar Coast Shelf Sci 23:889-900

Cowardin LM, Carter V, Golet FC, LaRoe ET (1979) Classification of wetlands and deepwater habitats of the United States. US Fish Wildlife Ser FWS/OBS-79/31:1-131

Croxall JP, Prince PA (1990) Recoveries of wandering albatrosses *Diomedea exulans* ringed at South Georgia 1985-1986. Ringing Migr 11:43-51

Csirke J (1987) Los recursos pesqueros patagónicos y las pesquerias de altura en el Atlántico Sudoccidental. FAO Doc Tec Pesca (Rome) 286:1-78

Cunha AM, Fonseca O (1918) O microplâncton das costas meridionaes do Brazil. Mem Inst Oswaldo Cruz 10:99-103

Cunha LPR (1981) Variação sazonal da distribuição, abundância e diversidade dos peixes na zona de arrebentação da praia do Cassino, RS, Brasil. MSc Thesis, Univ Rio Grande, Brazil

Cunha LPR (1987) Importância da zona de arrebentação de praias para o desenvolvimento de juvenis de *Trachinotus* (Pisces, Carangidae): aspectos da bioecologia e distribuição geográfica do gênero, com ênfase nas espécies que ocorrem no litoral sul/sudeste do Brasil e no Atlântico Ocidental. PhD Thesis, Univ S Paulo, Brazil

Cushing DH (1988) The flow of energy in marine ecosystems, with special reference to the continental shelf. In: Postma H, Zijlstra JJ (ed) Continental shelves. Elsevier, Amsterdam, pp 203-230. (Ecosystems of the world, vol 27)

Cushing DH (1989) A difference in structure between ecosystems in strongly stratified waters and in those that are only weakly stratified. J Plankton Res 11(1):1-13

Davy AJ, Costa CSB (1992) Development and organization of salt marsh communities. In: Seeliger U (ed) Coastal plant communities of Latin America. Academic Press, New York, pp 157-178

Day JW Jr, Hall CAS, Kemp WM, Yáñez-Arancibia A (1989) Estuarine ecology. Wiley, New York

Deacon GER (1933) A general account of the hydrology of the South Atlantic Ocean. Discov Rep 7:171-238

Dean TA, Bellis VJ (1975) Seasonal and spatial distribution of epifauna in the Pamlico River Estuary, North Carolina. J Elishe Mitchell Scient Soc (Chapel Hill) 91(1):1-12

Defant A (1961) Physical oceanography, vol 1. Pergamon Press, New York

Delaney PJV (1965) Fisiografia e geologia da superfície da planície costeira do Rio Grande do Sul. Publ Esp Esc Geol Univ Fed Rio Grande Sul (Porto Alegre) 6:1-63

de Buen F (1952) El tiburón vitamínico. Revta Fac Human Cienc (Montevideo) 7:87-116

de Menezes MF (1972) As tartarugas marinhas do Brasil. Arq Cienc Mar (Fortaleza) 12(1):17-20

DHN (1974) Atlas de cartas piloto do Oceano Atlântico: Trindade ao Rio da Prata. Diretoria de Hidrografia e Navegação, Rio de Janeiro

D'Incao F (1991) Pesca e biologia de *Penaeus paulensis* na Lagoa dos Patos, RS. Atlantica (Rio Grande) 13(1):159-169

D'Incao F, Silva KG, Ruffino ML, Braga AC (1990) Hábito alimentar do caranguejo *Chasmagnathus granulata* Dana, 1851, na Barra do Rio Grande, RS (Decapoda, Grapsidae). Atlantica (Rio Grande) 12(2):85-93

D'Incao F, Ruffino ML, Silva KG, Braga AC (1992) Responses of *Chasmagnathus granulata* Dana (Decapoda:Grapsidae) to salt marsh environmental variations. J Exp Mar Biol Ecol 161:179-188

Dillenburg LR, Rosa LMG, Oliveira PL (1986) Anatomia foliar de *Blutaparon portulacoides* (St Hil) Mears (Amaranthaceae) sob condições salinas e não salinas. Iheringia Ser Bot (Porto Alegre) 35:151-164

Dixon SW (1980) Between coastal marshes and coastal waters-a review of twenty years of speculation and research on the role of salt marshes in estuarine productivity and water chemistry. In: Hamilton P, Macdonald KB (eds) Estuarine and wetland processes with emphasis on modelling. Plenum Press, New York, pp 437-525

DNPVN (1941) Enchentes de maio de 1941. Diretoria Nacional de Portos e Vias de Navegação. Relatório Técnico, Porto Alegre

Dohms V (1983) Zusammensetzung und Verteilung des Phytoplanktons in Beziehung zu den Wassermassen der subtropischen Konvergenzzone im Südwestatlantik (Januar-März 1981). PhD Thesis, Univ Christian-Albrechts, Germany

Doing H (1985) Coastal foredune zonation and succession in various parts of the world. Vegetatio 61:65-75

Duarte AK (1986) Distribuição, abundância e flutuações sazonais do copepodo calanoide *Acartia tonsa* na enseada estuarina Saco do Justino e Canal do Estuário da Lagoa dos Patos (RS-Brasil). MSc Thesis, Univ Rio Grande, Brazil

Duarte AK, Gloeden I, Resgalla C Jr (1991) Flutuações espaciais e temporais da comunidade zooplanctônica da Lagoa dos Patos (RS, Brasil). Biol Acuat Not Cient (La Plata) 15(2):126-127

Dyer MI, Vinogradov BV (1990) Le rôle des reserves de la biosphere dans l'étude des paysages et des écosystèmes. Nat Res 26(1):19-28

Ehrlich KF, Blaxter JHS, Pembertorn R (1976) Morphlogical and histological changes during the growth and starvation of herring and plaice larvae. Mar Biol 35:105-118

Emilsson I (1961) The shelf and coastal waters off southern Brazil. Bolm Inst Oceanogr (S Paulo) 11(2):101-112

EPAGRI (1992) Arroz irrigado: recomendações técnicas de pesquisa para o sul do Brasil. Empresa Pesq Agropec Difusão Tecnol Santa Catarina, Florianópolis

Escalante R (1970) Aves marinhas del Rio de la Plata y aguas vecinas del Oceano Atlântico. Barreiro y Ramos SA, Montevideo

Escalante R (1983) Catálogo de las aves uruguayas, 3a parte: Galliformes y Gruiformes. Intendencia Municipal de Montevideo, Museo "Damaso Antonio Larrañaga", Montevideo

Escofet A, Gianuca NM, Maitia S, Scarabino V (1979) Playas arenosas del Atlántico Sudoccidental entre los 29° y 43° S: consideraciones generales y esquema biocenológico. Memorias del Seminario sobre Ecología Bentónica e Sedimentación de la Plataforma Continental del Atlántico Sur. UNESCO, Montevideo, pp 245-258

Eskuche U (1973) Pflanzengesellschaften der Küstendünen von Argentinien, Uruguay und Südbrasilien. Vegetatio 28(3/4):201-250

Evans DL, Signorini SR, Miranda LB (1983) A note on the transport of the Brazil Current. J Phys Oceanogr 13:1732-1738

Fabre-Domérgue P, Biétrix E (1897) La période critique post-larvaire des poissons marins. Bull Mus Hist (Paris) 3:57-58

Faccini UF, Asmus ML, Asmus HE, Fenstenseifer HC, Konrad H, Froes JC, Nowatski CH (1995) Caracterização, diagnóstico e planejamento da bacia hidrográfica do rio Camaquã (RS): metodologia e processo de integração de dados em um estudo multidisciplinar. Anais VI Simp Sul Brasil Geol, Porto Alegre, 327-329

Fagundes Netto E, Oliveira ES (1991) Aspectos da biologia de *Merluccius hubbsi* (Marini 1933) (Pisces, Merlucciidae) na região de Arraial do Cabo, Rio de Janeiro, Brasil. Neritica (Curitiba) 6(1/2)139-156

Fahrig L, Hayden B, Dolan R (1993) Distribution of barrier island plants in relation to overwash disturbance: a test of life history theory. J Coast Res 9(2):403-412

FAO/UNEP (1981) Conservation of the genetic resources of fish: problems and recommendations. Report of the expert consultation on the genetic resouces of fish, Rome, 9-13 june 1980. FAO Fish Tech Pap 217

FEPAM (1995) Enquadramento dos recursos hídricos da parte sul do estuário da Laguna dos Patos. Fundação Estadual de Proteção Ambiental Henrique Roessler, RS. Diario Ofic Est Rio Grande do Sul (Porto Alegre) 25/05/95:12-13

Ferreira S, Seeliger U (1985) The colonization process of algal epiphytes on *Ruppia maritima* L. Bot Mar 28:245-249

Ferreira BP, Vooren CM (1991) Age, growth and structure of vertebra of *Galeorhinus galeus* in southern Brasil. US Fish Bull 89:19-31

Figueiredo AG (1975) Geologia dos depósitos calcários biodetríticos da plataforma do Rio Grande do Sul. MSc Thesis, Univ Fed Rio Grande Sul, Brazil

Figueiredo JL (1977) Manual de peixes marinhos do sudeste do Brasil. I. Introdução, cações, raias e quimeras. Museu de Zoologia, S Paulo

Figueiredo JL (1981) Estudo das distribuições endêmicas de peixes da Província Zoogeográfica Argentina, PhD Thesis, Univ S Paulo, Brazil

Figueiredo JL, Menezes NA (1978) Manual de peixes marinhos do sudeste do Brasil. II. Teleostei (1). Museu de Zoologia, S Paulo

Figueiredo JL, Menezes NA (1980) Manual de peixes marinhos do sudeste do Brasil. III. Teleostei (2). Museu de Zoologia, S Paulo

Figueroa ME, Martos MJ, Fernandez-Palacios JM, Garcia JCR (1985) Papel geomorfológico de *Spartina maritima* (Curtis) Fernald y *Spartina densiflora* Brong. en las marismas del Río Odiel. Anais Ia Reunião Quartenário Ibérico, Lisboa, pp 367-378

Fillmann G (1990) Caracterização química das massas de água da palataforma continental do sul do Brasil. MSc Thesis, Univ Rio Grande, Brazil

Fischer W (1978) FAO species identification sheets for fisheries purposes. Western Central Atlantic (fishing area 31). FAO, Rome, vol V

Fortune MA, Kousky VE (1983) Two severe reezes in Brazil: precursors and synoptic evolution. Mon Weath Rev 111:181-196

Frazier J (1984) Tortugas marinas en el oceano Atlántico Sur Occidental. Assoc Herp Arg Ser Divulg (B Aires) 2:2-21

Freire KMF (1995) Alimentação de larvas de *Engraulis anchoita* (Teleostei: Engraulididae) na plataforma continental do Rio Grande do Sul, Brasil. MSc Thesis, Univ Rio Grande, Brazil

Furtado VV (1973) Sedimentos de superfície de fundo. Relatório sobre a segunda pesquisa oceanográfica e pesqueira do Atlântico Sul entre Torres e Maldonado (Lat 29°-35° S). Publ Esp Inst Oceanogr (S Paulo) 3(2):83-115

Gafrée CL (1927) Relatório técnico da Barra de Rio Grande entre 6 de dezembro de 1926 e 16 de janeiro de 1927, Porto Alegre

Gagliardi R, Cousseau MB (1970) Estudios biológico-pesqueros sobre la caballa (Scombridae: *Scomber japonicus marplatensis*). Ser Ing Tec FAO/PDP (Argent) 28:1-36

Gan MA (1992) Ciclogênese e ciclones sobre a América do Sul. PhD Thesis, Inst Nac Pesq Espaciais, São José dos Campos

Garcia VMT, Odebrecht C, Rörig LR (1994) Florações nocivas de fitoplâncton na costa brasileira. Intergov Oceanogr Comm Workshop Rep 101 [Annex III]:9-14

Garcia-Baptista M (1993) Contribuição ao conhecimento das algas psâmicas de Praia Azul: considerações taxonômicas e ecológicas. MSc Thesis, Univ Fed Rio Grande do Sul, Brazil

Garfield N (1990) The Brazil Current at subtropical latitudes. PhD Thesis, Univ Rhode Island, USA

Gemeinhardt K (1932) Die Silicoflagellaten des Südatlantischen Ozeans. Wiss Ergn Dt Atl Exped Meteor 1925-1927 (Berl) 12:274-311

Gianuca NM (1971) Relatório preliminar sobre a 1a prospecção do marisco branco (*Mesodesma mactroides* Deshayes 1854) na costa do Rio Grande do Sul. Bolm Inf Ind Pesca (Rio Grande) 10:38-43

Gianuca NM (1983) A preliminary account of the ecology of sandy beaches in southern Brazil. In: McLachlan A, Erasmus T (eds) Sandy beaches as ecosystems. Junk, The Hague, pp 413-419

Gianuca NM (1985) The ecology of a sandy beach in southern Brazil. PhD Thesis, Univ Southampton, England

Gianuca NM (1987) Zonação e produção nas praias arenosas do litoral sul e sudeste do Brasil: síntese dos conhecimentos. Publ Acad Ciências Est (S Paulo) 54(1):313-332

Gianuca NM (1988) Recursos naturais das praias arenosas do sul do Brasil. Inf UNESCO Cienc Mar (Montevideo) 47:89-94

Giller PS (1984) Community structure and the niche. Chapman and Hall, London

Godfrey PJ (1977) Climate, plant response and development of dunes on barrier beaches along the U.S. east coast. Int J Biometeorol 21(3):203-215

Godoi SS (1983) Estudo das variações sazonais da frente oceânica subtropical entre a Corrente do Brasil e a Corrente das Malvinas, utilizando dados oceanográficos e do satélite SMS-S. MSc Thesis, Inst Nac Pesq Espaciais, Brazil

Godolphin MF (1976) Geologia do Holoceno costeiro do município de Rio Grande-RS. MSc Thesis, Uni Fed Rio Grande Sul, Brazil

Gomes A, Tricart JLF, Trautmann J (1987) Estudo ecodinâmico da Estação Ecológica do Taim e seus arredores. Ed Univ Fed Rio Grande do Sul, Porto Alegre

González AGS (1994) Modelos semi-analíticos para determinar a produção primária fitoplanctônica através de sensoriamento remoto: uma aplicação à nivel regional. MSc Thesis, Univ Rio Grande, Brazil

Gonzalez Alberdi P, Nani A (1967) Contribuición al conocimiento de la biologia del pargo blanco *Umbrina canosai*, de la region de Mar del Plata. Carpas Doc Tec (Rio de Janeiro) 10:1-36

Gordon AL (1989) Brazil-Malvinas Confluence (1984). Deep Sea Res 36:359-384

Gorri C (1995) Distribuição espaço-temporal e estrutura populacional de Euphausiacea (Crustacea) na região do extremo sul do Brasil (31°40' S-33°45' S). MSc Thesis, Univ Rio Grande, Brazil

Groot RS (1986) A functional ecosystem evaluation method as a tool in environmental planning and decision making. Nature Conservation Department, Agricultural University, Wageningen, The Netherlands

Haagensen DA (1976) Caribbean zooplankton, Part II. Official Naval Res, Washington DC, pp 551-712

Habiaga RP, Castello JP (1986) Experiências de pesca com rede de arrasto de meia água na plataforma continental do Rio Grande do Sul. Anais IV Cong Brasil Eng Pesca, Curitiba, pp 235-247

Haeder DP, Worrest RC (1991) Effects of enhanced solar ultraviolet radiation on aquatic ecosystems. Photochem Photobiol 53(5):717-725

Haimovici M (1977) Age growth and aspects of the biology of the croacker *Micropogon opercularis* (Quoy, Gaimard 1824). Atlantica (Rio Grande) 2(1):21-49

Haimovici M (1982) Estructura y dinámica poblacional del pargo blanco *Umbrina canosai* (Pisces, Sciaenidae) del litoral de Rio Grande do Sul, Brasil. PhD Thesis, Univ B Aires, Argentina

Haimovici M (1985) Class Cephalopoda. In: Rios EC (ed) Seashells of Brazil. Ed Univ Rio Grande, Rio Grande, pp 283-288

Haimovici M (1987) Estratégia de amostragens de comprimentos de teleósteos demersais nos desembarques da pesca de arrasto no litoral sul do Brasil. Atlantica (Rio Grande) 9(1):65-82

Haimovici M (1988a) Crecimiento de la pescadilla real *Macrodon ancylodon* (Sciaenidae) en el sur de Brasil. Publ Com Tec Mix Fr Mar (Montevideo) 4:99-106

Haimovici M (1988b) Analisis de cohortes del stock de pargo blanco *Umbrina canosai* explotado en el sur de Brasil, Uruguay y Argentina. Publ Com Tec Mix Fr Mar (Montevideo) 4:33-40

Haimovici M (1988c) *Eledone gaucha*, a new species of eledonid octopod (Cephalopoda: Octopodidae) from southern Brazil. Nautilus 102(2):82-87

Haimovici M, Andriguetto JM (1986) Cefalópodes costeiros capturados na pesca de arrasto do litoral sul do Brasil. Arq Biol Tecnol (Curitiba) 29(3):473-495

Haimovici M, Cousin JCB (1989) Reproductive biology of the castanha *Umbrina canosai* (Pisces, Sciaenidae) in southern Brazil. Rev Brasil Biol (Rio de Janeiro) 49(2):523-537

Haimovici M, Habiaga RG (1982) Rejeição a bordo na pesca de arrasto de fundo no litoral de Rio Grande do Sul num cruzeiro de primavera. Pub Cient Com Tec Mix Fr Mar (Montevideo) 4:33-40

Haimovici M, Krug LC (1992) Alimentação e reprodução da enchova *Pomatomus saltatrix* no litoral sul do Brasil. Rev Brasil Biol (Rio de Janeiro) 52(3):503-513

Haimovici M, Krug LC (1996) Fishery and biology of the enchova *Pomatomus saltatrix* in southern Brazil. Australian Jour Mar Freshwater Res 47(2):(in press)

Haimovici M, Mendonça JT (1996) Descartes na pesca de arrasto e tangones dirigida à linguados e camarões no Rio Grande do Sul. Atlantica (Rio Grande) (in press)

Haimovici M, Palacios RP (1981) Observações sobre a seleção a bordo e rejeição na pesca de arrasto de fundo no Rio Grande do Sul. Anais II Cong Brasil Eng Pesca, julho 1981, Recife, pp 401-412

Haimovici M, Perez JAA (1990) Observaciones sobre la distribuición y maduración sexual del calamar argentino *Illex argentinus* en el sur de Brasil. Sci Mar (Barc) 54(2):179-185

Haimovici M, Perez JAA (1991a) Abundância e distribuição de cefalópodes em cruzeiros de prospecção pesqueira demersal na plataforma externa e talude continental do Sul do Brasil. Atlantica (Rio Grande) 13(1):189-200

Haimovici M, Perez JAA (1991b) The coastal cephalopod fauna of southern Brazil. Bull Mar Sci 49(1-2):221-230

Haimovici M, Reis EG (1984) Determinação de idade e crescimento da castanha *Umbrina canosai* (Pisces, Sciaenidae) do sul do Brasil. Atlantica (Rio Grande) 7:25-46

Haimovici M, Umpierre RG (1996) Variaciones estacionales en la estructura poblacional y cambios de crecimiento de la corvina *Micropogonias furnieri* (Desmarest 1823) del sur de Brasil. Atlantica (Rio Grande) (in press)

Haimovici M, Vieira PC (1986) Captura e esforço na pesca de arrasto de fundo no litoral sul do Brasil, no período 1975-1984. Anais IV Cong Brasil Eng Pesca, Curitiba, pp 215-234

Haimovici M, Pereira SD, Vieira PC (1989a) La pesca demersal en el sur de Brasil en el período 1975-1985. Frente Mar (Montevideo) 5:151-163

Haimovici M, Arruda MC, Teixeira RL (1989b) Alimentação da castanha *Umbrina canosai* no litoral sul de Brasil. Rev Brasil Biol (Rio de Janeiro) 49(2):511-522

Haimovici M, Perez JAA, Costa PAS (1989c) A review of cephalopods ocurring in the waters of Rio de Janeiro state, Brazil with first record of four species. Rev Brasil Biol (Rio de Janeiro) 49(2):503-510

Haimovici M, Martins AS, Teixeira RL (1993) Distribuición, alimentación y observaciones sobre la reproducción de la merluza *Merluccius hubbsi* en el sur de Brasil. Frente Mar (Montevideo) 14:81-86

Haimovici M, Martins AS, Figueiredo JL, Vieira PC (1994a) Demersal bony fish of the outer shelf and upper slope of the southern Brazil Subtropical Convergence Ecosystem. Mar Ecol Prog Ser 108:59-77

Haimovici M, Perez JAA, Santos RA (1994b) Class Cephalopoda Cuvier, 1798. In: Rios EC (ed) Seashells of Brazil, 2nd ed. Ed Univ Rio Grande, Rio Grande, pp 311-320, plates 108-113

Haimovici M, Vidal EAG, Perez JAA (1995) Larvae of *Illex argentinus* (Castellanos 1960) in five surveys on the continental shelf of southern Brazil. Shellfish life histories and shellfisheries models ICES. Mar Sci Symp Can 199:414-424

Haimovici M, Martins AS, Vieira PC (1996) Distribuição e abundância de teleósteos demersais sobre a plataforma continental do sul do Brasil. Rev Brasil Biol (Rio de Janeiro) 56(1):27-50

Haimovici M, Brunetti NE, Rodhouse P, Csirke J, Leta R (in press) *Illex argentinus* In: Rodhouse PG, Dawe EG, O'Dor RK (eds) Squid recruitment dynamics. FAO Fish Tech Pap

Haines EB (1979) Interactions between Georgia salt marshes and coastal waters: a changing paradigm. In: Livingstone RJ (ed) Ecological processes in coastal and marine systems. Plenum Press, New York, pp 35-46

Harán NS, Fenucci JL, Díaz AC (1992) Efectos de la temperatura y la salinidad sobre el crecimiento y la supervivencia del camarón (*Artemesia longinaris*) y del langostino (*Pleoticus muelleri*). Frente Mar (Montevideo) 11(A):79-83

Harrington BA, Antas PTZ, Silva F (1986) Northward shorebird migration on the Atlantic coast of southern Brazil. Vida Silv Neotrop (Whashington DC) 1:45-54

Hartmann C, Calliari L, Charpy-Roubaud CJ, Baumgarten MGZ, Kantin R (1980) Estudo do material em suspensão e do material dissolvido das águas de superfície da plataforma do Rio Grande do Sul, entre Torres e Rio Grande (Operação GEOMAR XIII, 19 a 27 de novembro de 1979). Anais XXXI Cong Brasil Geol Balneário Camburiú 2:956-967

Hartmann C, Sano EE, Paz RS, Möller OO Jr (1986) Avaliação de um período de cheia (junho de 1984) na região sul da Laguna dos Patos, através de dados de sensoriamento remoto, meteorológicas e oceanográficas. Anais IV Simp Bras Sens Remoto, 10-15 agosto Gramado, pp 685-694

Hentschel E (1932) Die biologischen Methoden und das biologische Beobachtungsmaterial der "Meteor" Expedition. Wiss Ergn Dt Atl Exped Meteor 1925-1927 (Berl) 10:1-344

Herz R (1977) Circulação de águas de superfície na Lagoa dos Patos. PhD Thesis, Univ S Paulo, Brazil

Hesp PA (1991) Ecological processes and plant adaptations on coastal dunes. J Arid Environ 21:165-191

Hillman K, Walker DI, Larkum AWD, McComb AJ (1989) Productivity and nutrient limitations. In: Larkum AWD, McComb AJ, Shepherd SA (eds) Biology of seagrasses. Elsevier, Amsterdam, pp 635-685

Holland AF, Mountford NK, Hiegel MH, Kaumeyer KR, Mihursky JA (1980) Influence of predation on infaunal abundance in Upper Chesapeake Bay, USA. Mar Biol 57:221-235

Hubold G (1980a) Hydrography and plankton off southern Brazil and Rio de la Plata, August-November 1977, Atlantica (Rio Grande) 4: 1-21

Hubold G (1980b) Second report on hydrography and plankton off southern Brazil and Rio de la Plata, autumn cruise: April-June 1978, Atlantica (Rio Grande) 4:23-42

Hubold G, Ehrlich MD (1981) Distribution of egg and larvae of five clupeoid fish species in the southwest Atlantic between 25° S and 40° S. Meeresforschung 29:17-29

Hueck K (1955) Plantas e formações organogênicas das dunas do litoral paulista. Inst Bot Secr Agricult (S Paulo)

IAI (Inter-American Institute for Global Change Research) (1992) Report of the meeting of scientific experts. Emily Lind Baker, Washington DC

Ibagy AS, Sinque C (1995) Distribution of Sciaenidae eggs and larvae (Perciformes-Teleostei) in the coastal area of Rio Grande do Sul, Brazil. Arq Biol Tecnol (Curitiba) 38(1):248-270

IBAMA (1993) Peixes Demersais. Coleção Meio Ambiente, Série Estudos Pesca, Brasília. Minist Meio Ambiente 8:1-93

IBAMA (1995) Relatório IV Reunião grupo permanente de estudos sobre peixes demersais. Coleção Meio Ambiente Ser Est Pesca, Brasília. Minist Meio Ambiente 16:1-127

IBGE (1986) Levantamento dos recursos naturais. Fund Inst Brasil Geogr Estat, vol 33, Rio de Janeiro

IBGE (1993) Anuário Estatístico Brasileiro. Fund Inst Brasil Geogr Estat, vol 27 to 53, Rio de Janeiro

IPAGRO (1977) Observações meterológicas no Estado do Rio Grande do Sul. Bolm Tecn IPAGRO (Porto Alegre) 3:1-272

Isaac VJ (1988) Synopsis of biological data on the whitemouth croaker *Micropogonias furnieri* (Desmarest, 1823). FAO Fish Synop 150:1-35

IUCN (1991) Dolphins, porpoises and whales of the world. The IUCN red data book. Compiled by Klinowska M. International Union for Conservation of Nature and Natural Resources, Gland, Switzerland

Ivankov VN, Kurdyayena VP (1973) Systematic differences and the ecological importance of the membranes in fish eggs. J Ichtyol 13:864-873

Jablonski S, Matsuura Y (1985) Estimate of exploitation rates and population size of skipjack tuna off the southern coast of Brazil. Bolm Inst Oceanogr (S Paulo) 33(1):29-38

Juanicó M (1979) Contribuição ao estudo da biologia dos Cephalopoda Lolliginidae do Atlántico Sul Ocidental, entre Rio de Janeiro e Mar del Plata. PhD Thesis, Univ S Paulo, Brazil

Juanicó M (1980) Developments in South American squid fisheries. Mar Fish Rev 42(7/8):10-14

Juras AA, Yamaguti N (1985) Food and feeding habits of the king weakfish (*Macrodon ancylodon*) (Bloch and Schneider 1801) caught in the southern coast of Brazil (Lat 29° S-32° S). Bolm Inst Oceanogr (S Paulo) 33(2):149-157

Juras AA, Yamaguti N (1989) Sexual maturity and fecundity of the king weakfish *Macrodon ancylodon* (Bloch and Schneider 1801) caught in the southern coast of Brazil (Lat 29° S-32° S). Bolm Inst Oceanogr (S Paulo) 33(2):149-157

Kantin R (1982) Estudo de algumas características hidrogeológicas da margem continental sul-brasileira: estado do conhecimento atual. Anais XXXII Cong Brasil Geol (Salvador) 4:1710-1719

Kantin R (1983) Hydrologie et qualité des eaux de la region sud de la Lagune dos Patos (Brasil) et de la plataforme continental adjacente. PhD Thesis, Univ Bordeaux, France

Kantin R, Baumgarten MGZ (1982) Observações hidrográficas no estuário da Lagoa dos Patos: distribuição e flutuações dos sais nutrientes. Atlantica (Rio Grande) 5:76-92

Kantin R, Baumgarten MGZ, Niencheski LF (1982) Distribuição do material particulado em suspensão na costa sul do Estado do Rio Grande do Sul, entre Rio Grande e Chui. Anais XXXII Cong Brasil Geol (Salvador) 4:1720-1726

Karentz D (1991) Ecological considerations of Antartic ozone depletion. Antarct Sci 3(1):3-11

Käsler R (1932) Die Verbreitung der Dinophysiales im Südatlantischen Ozean. Wiss Ergn Dt Atl Exped Meteor 1925-1927 (Berl) 12:165-237

Kasuya T, Brownell Jr RL (1979) Age determination, reproduction and growth of franciscana dolphin, *Pontoporia blainvillei*. Sci Rep Whales Res Inst (Tokyo) 31:45-67

Kawakami E (1975) Alimentação de Pleuronectiformes (análise comparativa e bionomia). MSc Thesis, Univ S Paulo, Brazil

Kennish MJ (1986) Ecology of estuaries, vol 1, Physical and chemical aspects. CRC Press, Boca Raton

Kennish MJ (1990) Ecology of estuaries, vol 2, biological aspects. CRC Press, Boca Raton

Kirchhoff VWJH, Schuch NJ, Hilsenrath E (1994) Buraco de ozônio: novidades no sul. Ciência Hoje (S Paulo) 17(99):6-7

Kjerfve B (1986) Comparative oceanography of coastal lagoons. In: Wolfe DA (ed) Estuarine variability. Academic Press, New York, pp 63-81

Kleerekoper H (1945) O peixe-rei. Minist Agr Serv Inf Agricola (Rio de Janeiro) 167:1-98

Knak RB (1986) Ecologia experimental de *Ruppia maritima* L. em condições controladas de cultivo. MSc Thesis, Univ Rio Grande, Brazil

Knox GA (1986) Estuarine ecosystems: a system approach, vol 1 and 2. CRC Press, Boca Raton

Koch EW, Seeliger U (1988) Germination ecology of two *Ruppia maritima* L. populations in southern Brazil. Aquat Bot 31:321-327

Koswmann RO, Costa MPA, Vicalvi MA, Coutinho PMN, Gamboa LAP (1977) Modelo da sedimentação holocênica na plataforma continental sul-brasileira. Projeto REMAC. Evolução sedimentar holocênica da plataforma continental e do talude do sul do Brasil. Rio de Janeiro, Petrobrás, CENPES, DINTEP, pp 7-26

Kousky VE, Ferreira NJ (1981) Interdiurnal surface pressure variations in Brazil: their spatial distribution, origins and effects. Mont Weath Rev 109:1999-2008

Kremer LM, Rosa ZM (1983) Dinoflagelados do microfitoplâncton de Tramandaí, RS, Brasil. Iheringia Ser Bot (Porto Alegre) 30:3-35

Krug LC (1984) Idade, crescimento, alimentação e pescaria da enchova, *Pomatomus saltatrix* da costa do Rio Grande do Sul. MSc Thesis, Univ Rio Grande, Brazil

Krug LC, Haimovici M (1989) Crescimento da enchova *Pomatomus saltatrix* do sul do brasil Atlantica (Rio Grande) 11(1):47-61

Krug LC, Haimovici M (1991) Análise da pesca da enchova, *Pomatomus saltatrix* no sul do Brasil. Atlantica (Rio Grande) 13(1):119-130

Lacerda LD, Cunha CT, Seeliger U (1986) Distribuição de nutrientes em perfis de sedimentos em brejos costeiros tropicais e temperados. Acta Limnol Brasil 1:387-399

Lana PC (1986) Macrofauna bêntica de fundos sublitorais não consolidados da Baía de Paranaguá (Paraná). Neritica (Curitiba) 1(3):79-89

Lange CB, Mostajo EL (1985) Phytoplankton (diatoms and silicoflagellates) from the Southwestern Atlantic Ocean. Bot Mar 28:469-476

Legeckis R, Gordon AL (1982) Satellite observations of the Brazil and Falkland Currents: 1975 to 1976 and 1978. Deep Sea Res 29(3A):375-401

Legendre L, Le Fèvre J (1995) Microbial food webs and the export of biogenic carbon in oceans. Aquat Microb Ecol 9:69-77

Legrand CD (1959) Comunidades psamófilas de la región de Carrasco (Uruguay). An Mus Hist Nat (Montevideo) 6(7):1-53

Lenanton RC, Potter IC (1987) Contribution of estuaries to commercial fisheries in temperate western Australia and the concept of estuarine dependence. Estuaries 10(1):28-35

Lessa R, Vooren CM, Lahaye J (1986) Desenvolvimento e ciclo sexual das fêmeas, migrações e fecundidade da viola *Rhinobatos horkelii* (Mueller and Henle 1841) do Sul do Brasil. Atlantica (Rio Grande) 8:5-34

Levy JA, Cassano VPF (1994) Biochemical-genetic comparison of *Cynoscion jamaicensis* and *Cynoscion striatus* (Teleostei: Perciformes: Scianidae) on South Brazil. Comp Biochem Physiol 107B(4):515-517

Levy JA, Conceição M (1989) Biochemical evidence for two sibling species of genus *Myliobatis* (Chodrichthyes: Myliobatidae) in South Brazil. Comp Biochem Physiol 94B:687-690

Levy JA, Haimovici M, Conceição MB (1988) Genetic evidence of two species for the genus *Eledone* (Cephalopoda: Octopidae) in South Brazil. Comp Biochem Physiol 90(2B):275-277

Lima ID (1992) Distribuição e abundância de anchoita (*Engraulis anchoita*) em relação aos processos oceanográficos na plataforma continental do sul do Brasil. MSc Thesis, Univ Rio Grande, Brazil Lima ID, Castello JP (1994) Distribución y abundancia de *Engraulis anchoita* en la costa sur de Brasil. Frente Mar (Montevideo) 15:87-100

Lima ID, Castello JP (1995) Distribution and abundance of southwest Atlantic anchovy spawners (*Engraulis anchoita*) in relation to oceanographic processes in the southern Brazilian shelf. Fish Oceanogr 4(1):1-16

Lima ID, Garcia AE, Möller OO Jr (1996) Ocean surface processes on the southern Brazilian shelf: characterization and seasonal variability. Cont Shelf Res (in press)

Lima JEM, Branco RM (1991) Análise das operações de pesca do caranguejo de profundidade (*Geryon quinquedens*, Smith 1879) por barcos japoneses arrendados na região sul do Brasil - 1958/85. Atlantica (Rio Grande) 13:178-188

Lindholm T (1985) *Mesodinium rubrum*-a unique photosynthetic ciliate. Adv Aquat Microbiol 3:1-48

Liss PS (1976) Conservative and non-conservative behavior of dissolved constituents during estuarine mixing. In: Burton JD, Liss PS (ed) Estuarine chemistry. Academic Press, New York

Loneragan NR, Potter IC, Lenanton RC (1989) Influence of site, season and year on contributions made by marine, estuarine, diadromous and freshwater species to the fish fauna of a temperate australian estuary. Mar Biol 103:461-479

Long SP, Mason CF (1983) Saltmarsh ecology. Blackie, Glasgow

Long T, Paim PSG (1987) Modelo de evolução histórica e holocênica do estuário da Lagoa dos Patos, RS. Anais I Cong Assoc Brasil Estudos Quaternário (ABEQUA), Porto Alegre, pp 133-139

Longhurst A, Sathyendranath S, Platt T, Caverhill C (1995) An estimate of global primary production in the ocean from satellite radiometer data. J Plankt Res 17(6):1245-1271

Lonning S (1981) Comparative electron microscope studies of the chorion of the fish egg. Rapp P-v Réun Cons Perm Int Expl Mer 178:560-564

Luedemann EF (1971) Relatório sobre resultados obtidos com lançamento de garrafas de deriva, realizados durante o Programa Rio Grande do Sul. Contr Inst Oceanogr (S Paulo) 14:1-183

Luedemann EF (1978) Contribuição ao estudo da variação da circulação em superfície no setor sul da plataforma continental brasileira na primavera. MSc Thesis, Univ S Paulo, Brazil

Lüning K (1990) Seaweeds: their environment, biogeography, and ecophysiology. Wiley, Inc, New York

Machado PA (1979) Dinoflagellate bloom on the Brazilian South Atlantic coast. In: Taylor DL, Seeliger HH (eds) Toxic dinoflagellate blooms. Elsevier, New York, pp 29-32

Mafalda PO Jr (1989) Identificação, ocorrência e abundância de larvas da subordem Scombroidei (Pisces-Perciformes) no Atlântico Sul Ocidental (25°-40° S). MSc Thesis, Univ Rio Grande, Brazil

Maggione R (1992) Estudo genético-populacional da corvina *Micropogonias furnieri*, entre Macaé e Chuí. MSc Thesis, Univ Rio Grande, Brazil

Maggione R, Pereira A, Jerez B, Marins L, Conceição M, Levy JA (1994) Estudio preliminar de la estructura genética de la corvina (*Micropogonias furnieri*) entre Rio Grande (Brasil) y El Rincon (Argentina). Frente Mar (Montevideo) 15:127-131

Magliocca A (1971) Relatório sobre a variação temporal e espacial do oxigênio dissolvido, na costa do Estado do Rio Grande do Sul. Contr Esp Inst Oceanogr Ser Ocean Fis (S Paulo) 15:1-171

Magliocca A (1973) Distribuição vertical de oxigênio e alguns nutrientes. Relatório sobre a segunda pesquisa oceanográfica e pesqueira do Atlântico Sul entre Torres e Maldonado (Lat 29°-35° S). Publ Esp Inst Oceanogr (S Paulo) 3(3):1-40

Magliocca A, Miranda LB, Pinheiro EA (1982) Variação sazonal de oxigênio dissolvido, temperatura e salinidade na costa sul brasileira (28°-35° S; 48°-54° W). Bolm Inst Oceanogr (S Paulo) 31(1):1-9

Malaval MB (1922) Travaux du port et la Barre de Rio Grande do Sul (Brésil). Libraire de 1 Enseignement Technique. Ed Eyrolles, Paris

Maluf S, Matsuura Y, Stech JL (1984) First attempt to identify areas favorable for the surface occurrence of skipjack (*Katsuwonus pelamis*) off the southeastern coast of Brazil. SCRS/83/50, ICCAT Coll Vol Sci Pap 20(1):201-211

Mann KH, Lazier JRN (1991) Dynamics of marine ecosystems. Biological-physical interactions in the oceans. Blackwell, Oxford

Manooch CS, Hassler WW (1978) Synopsis of the biological data on the red porgy *Pagrus pagrus* (Linneaus). NOAA Tec Rep NMFS Circ 412. FAO Fish Synop 116:1-19

Marchant S, Higgins PJ (1990) Handbook of Australian, New Zealand and Antarctic Birds, vol 1(A). Oxford University Press, Melbourne

Marchiori MA, Boff MH (1983) Induced maturation, spawning and larvae culture of the pink shrimp *Penaeus paulensis* Perez Farfante. Mem Asoc Latinoam Acuicult 5:331-337

Marchiori MA, Magalhães Filho CV, Yunes JS, Levy JA (1982) Estudos sobre a alimentação artificial do camarão rosa *Penaeus paulensis*. Atlantica (Rio Grande) 5:43-48

Marcovaldi MA (1991) Sea turtle conservation program in Brazil expands activities. Mar Turtle Newslett (San Diego) 52:2-3

Marins L, Delevedove G, Levy JA (1995) Estudo genético-bioquímico do marisco branco (*Mesodesma mactroides* Desh. 1854) na costa do Rio Grande do sul. I. Determinação dos padrões isoenzimáticos e análises ontogenética. Bolm Inst Oceanogr (S Paulo) 11:145-152

Márquez MR (1990) FAO species catalogue, vol 11, Sea turtles of the world. An annotated and illustrated catalogue of sea turtle species known to date. FAO Fish Synop (Rome) 125(11):1-81

Martins AS (1992) Bioecologia do peixe-espada *Trichiurus lepturus* Linnaeus, 1758, no sul do Brasil. MSc Thesis, Univ Rio Grande, Brazil

Martins ILR (1963) Contribuição à sedimentologia da Lagoa dos Patos, 1. Saco do Rincão e Mendanha. Bolm Esc Geol Univ Fed Rio Grande Sul (Porto Alegre) 13:1-43

Martins ILR (1966) Contribuição à sedimentologia da Lagoa dos Patos, 2. Sacos do Umbú, Arraial e Mangueira. Notas e Estudos Esc Geol Univ Fed Rio Grande Sul (Porto Alegre) 1:27-44

Martins ILR (1971) Sedimentologia do canal do Rio Grande. MSc Thesis, Univ Fed Rio Grande Sul, Brazil

Martins LRS, Melo U, França AMC, Santana CI, Martins ILR (1972) Distribuição faciológica da margem continental sul-riograndense. XXVI Congr Brasil Geol, Belém. Anais Soc Brasil Geol, S Paulo, pp 115-132

Martins Juras IAG (l980) Estudo sobre o crescimento de *Macrodon ancylodon* (Bloch and Schneider, l801) capturada nas costas de Rio Grande do Sul (Lat 29°-32° S). MSc Thesis, Univ S Paulo, Brazil

Matsuura Y (1982a) Distribution and abundance of skipjack (*Katsuwonus pelamis*) larvae in eastern Brazilian waters. Bolm Inst Oceanogr (S Paulo) 31(2):5-7

Matsuura Y (1982b) Perspectiva de pesca de bonitos e atuns no Brasil. Ciência Cult (S Paulo) 34(3):333-339

Matsuura Y (1986) Distribution and abundance of skipjack larvae off the coast of Brazil. In: Symons PK, Miyake PM, Sakagawa GT (eds) Proc ICCAT Conf Intern Skipjack year Program, Madrid, pp 285-289

Matsuura Y, Garcia ACS, Katsuragawa M, Suzuki K (1993) Distribution and abundance of two species of codlet (Teleostei, Bregmacerotidae) larvae from south-eastern Brazilian Bight. Fish Oceanogr 2(2):82-90

Matthiensen A, Tellechea E, Levy JA (1993) Biochemical chracterization for the genetic interpretation of esterase isozymes in *Micropogonias furnieri* (Pisces, Sciaenidae) in south Brazil. Comp Biochem Physiol 104B(2):349-352

Mauchline J (1980) The biology of mysids and euphausiids, vol 18. Academic Press, London

Mauchline J, Fisher LR (1969) The biology of euphausiids, vol 7. Academic Press, New York

McCall PL (1977) Community patterns and adaptative strategies of the infaunal benthos of Long Island Sound. J Mar Res 35(2):221-226

McClusky DS (1981) The estuarine ecosystem. Wiley, Halsted Press, New York

McHugh JL (1966) Management of estuarine fishes. In: Smith RF, Swarta AH, Massmann WH (ed) A symposium on estuarine fishes. Am Fish Soc Wash Spec Publ 3:133-154

McHugh JL (1967) Estuarine nekton. In: Lauff GH (ed) Estuaries. Am Assoc Adv Sci, Washington DC, pp 581-620

McHugh JL (1985) The estuarine ecosystem integrated, foreword. In: Yáñes-Arancibia A (ed) Fish community ecology in estuaries and coastal lagoons: towards an ecosystem integration. Univ Nac Auton Mex Press, Mexico, pp 9-16

McLachlan A (1980) Exposed sandy beaches as semi-close ecosystems. Mar Environ Res 4:59-63

McLachlan A (1983) Sandy beach ecology-a review. In: McLachlan A, Erasmus T (eds) Sandy beaches as ecosystems. Junk, The Hague, pp 321-380

McLusky DS (1981) The estuarine ecosystem. Wiley, New York

Meffe GK (1986) Conservation genetics and the management of endangered fishes. Fisheries 11:14-23

Mello RM (1992) Análise de conteúdos estomacais, intensidade de alimentação, idade e crescimento do espadarte, *Xiphias gladius* (Xiphoidei: Xiphiidae) no sul do Brasil. MSc Thesis, Univ Rio Grande, Brazil

Mello RM, Castello JP, Freire KFM (1992) Asociación de especies pelágicas marinas en el sur de Brasil durante invierno y primavera. Frente Mar (Montevideo) 11:63-69

Mello RM, Vaske T Jr, Castello JP (1993) Pesquería de atunes y pez espada por la flota arrendada de Rio Grande, Brasil. Frente Mar (Montevideo) 14:125-133

Meneghetti JO (1973) Zooplankton from southern Brazil. 2. Its composition, distribution and density in space and time. Iheringia Ser Zool (Porto Alegre) 43:100-125

Menezes NA (1971) Relação dos peixes ósseos coletados durante os cruzeiros do navio oceanográfico "Professor W. Besnard" às costas do Rio Grande do Sul. Contrib Inst Oceanogr (S Paulo) Ser Oceanogr Biol 25:44-61

Menezes NA, Figueiredo JL (1980) Manual de peixes marinhos do sudeste do Brasil. IV. Teleostei (3). Museu de Zoologia, S Paulo

Menezes NA, Figueiredo JL (1985) Manual de peixes marinhos do sudeste do Brasil. V. Teleostei (4). Museu de Zoologia, S Paulo

Miller J, Dunn ML (1980) Feeding strategies and patterns of movements in juvenile estuarine species. In: Kennedy VS (ed) Estuarine perspectives. Academic Press, New York, pp. 437-448

Miranda LB (1972) Propriedades e variáveis físicas das águas da plataforma continental do Rio Grande do Sul. PhD Thesis, Univ S, Brazil

Miranda LB, Castro Filho BM (1979) Aplicação do diagrama T-S estatístico-volumétrico à análise de massas de água da plataforma continental do Rio Grande do Sul. Bolm Inst Oceanogr (S Paulo) 28(1):185-200

Miranda LB, Luedemann EF, Miyao SW (1973) Distribuição da temperatura, salinidade e circulação geral em superfície. Relatório sobre a segunda pesquisa oceanográfica e pesqueira do Atlântico Sul entre Torres e Maldonado (Lat 29°-35° S). Publ Esp Inst Oceanogr (S Paulo) 3(2):1-82

Miranda RB (1994) Efeitos da temperatura e da salinidade sobre a tolerância e a ionorregulação de *Chasmagnathus granulata* Dana, 1851. MSc Thesis, Univ Rio Grande, Brazil

Miyake M, Hayasi S (1978) Manual de operaciones, para las estadísticas y el muestreo de túnidos y especies afines en el Océano Atlántico. ICAAT 2nd ed, Madrid

Mobberley (1956) Taxonomy and distribution of the genus *Spartina*. Iowa St Coll J Sci 30(4):471-574

Möller LM, Simões-Lopes PC, Secchi ER, Zerbini AN (1994) Uso de fotoidentificação no estudo do deslocamento de botos *Tursiops truncatus* (Cetacea, Delphinidae) na costa sul do Brasil. Anais VI Reunião de Trabalho de Especialistas em Mamíferos Aquáticos da América do Sul, 24-28 outubro, Florianópolis, pp.5-8

Möller OO Jr, Paim PSG, Soares ID (1991) Facteurs et mechanismes de la circulation des eaux dans l'estuarie de la Lagune dos Patos (RS, Bresil). Bull Inst Geol Basin Aquitaine (Bordeaux) 49:15-21

Möller OO Jr, Lorenzzetti JA, Stech JL, Mata MM (1996) Patos Lagoon summertime circulation and dynamics. Cont Shelf Res 16(3):335-351

Monserrat J, Bianchini A (1995) Effects of temperature and salinity on the toxicity of a commercial formulation of methyl parathion to *Chasmagnathus granulata* (Decapoda, Grapsidae). Braz J Med Biol Res 28:74-78

Monteiro CAF (1968) Clima. Biblioteca Geográfica Brasileira, IBGE CNG (Rio de Janeiro) 4(1):114-166

Monteiro C Neto, Cunha LPR (1990) Seasonal and ontogenetic variation in food habits of juvenile *Trachinotus marginatus* Cuvier, 1832 (Teleostei, Carangidae) in the surf zone of Cassino Beach, RS, Brazil. Atlantica (Rio Grande) 12:45-54

Montú M (1980) Zooplâncton do estuário da Lagoa dos Patos. I. Estrutura e variações temporais e espaciais da comunidade. Atlantica (Rio Grande) 4:53-72

Montú M, Gloeden I (1982) Morphological alterations in *Acartia tonsa* (Saco da Mangueira, Lagoa dos Patos, Brasil). Arq Biol Tecnol (Curitiba) 25(3/4):361-369

Montú M, Gloeden I (1986) Atlas dos Cladocera e Copepoda (Crustacea) do estuário da Lagoa dos Patos (Rio Grande, Brasil). Neritica (Curitiba) 1(2):1-134

Moreno AMM (1994) Distribuição e biomassa da fanerógama submersa *Ruppia maritima* L. no estuário da Lagoa dos Patos, Rio Grande, RS, Brasil. MSc Thesis, Univ Rio Grande, Brazil

Moreno-Casasola P, Espejel I (1986) Classification and ordination of coastal sand dune vegetation along the Gulf and Caribbean Sea of Mexico. Vegetatio 66:147-182

Motta VF (1969) Relatório-diagnóstico sobre a melhoria e o aprofundamento do acesso pela Barra de Rio Grande. Relat Tec Instituto Pesquisas Hidráulicas, Univ Fed Rio Grande Sul, Porto Alegre

Muelbert JH (1986) Estrutura e dinâmica do ictioplâncton da área de canal no Estuário da Lagoa dos Patos, RS, no período de outono/81 à verão/83. MSc Thesis, Univ Rio Grande, Brazil

Muelbert JH, Weiss G (1991) Abundance and distribution of fish larvae in the channel area of the Patos Lagoon Estuary, Brazil. In: Hoyt RD (ed) Larval fish recruitment and research in the Americas. NOAA Tech Rep NMFS 95:43-54

Nakamura H (1963) Survey report on the Brazilian fishing grounds by the Tôkó Maru translated from Japanese; 1958 edn. Tokyo Fishery University, Tokyo

Nakatani K (1982) Estudos sobre ovos e larvas de *Engraulis anchoita* (Hubbs, Marini 1935) (Teleostei: Engraulidae) coletados na região entre Cabo Frio (23° S) e Cabo de Santa Marta Grande (29° S). MSc Thesis, Univ S Paulo, Brazil

Navas-Pereira D (1973) Análise qualitativa e quantitativa do zooplâncton. Publ Esp Inst Oceanogr (S Paulo) 3(1):29-92

Nelson WG (1979a) An analysis of structural pattern in an eelgrass (*Zostera marina* L.) amphipod community. J Exp Mar Biol Ecol 39: 231-264

Nelson WG (1979b) Experimental studies of decapod and fish predation on seagrass macrobenthos. Mar Ecol Prog Ser 5:141-149

Nery LEM, Santos EA (1993) Carbohydrate metabolism during osmoregulation in *Chasmagnathus granulata* Dana, 1851 (Crustacea-Decapoda). Comp Biochem Physiol 106B:747-753

Nery LEM, Santos EA, Bianchini A, Gonçalves A (1993) Effects of crustacean hyperglycemic hormone from *Carcinus maenas* and *Orconectes limosus* on blood and muscle glucose and glycogen concentrations of *Chasmagnathus granulata*. Braz J Med Biol Res (Ribeirão Preto) 26:1291-1296

Nichols MM, Allen G (1981) Sedimentary processes in coastal lagoons. UNESCO Techn Pap Mar Sci 33:27-80

Nichols MM, Howard-Strobel M (1991) Evolution of an urban estuarine harbor: Norfolk, Virginia. J Coast Res 7(3):745-757

Niencheski LF, Möller OO Jr, Odebrecht C, Fillmann G (1988) Distribuição espacial de alguns parâmetros físicos e químicos na Lagoa dos Patos - Porto Alegre a Rio Grande (verão 1986). Acta Limnol Brasil 2:79-97

Niencheski LFH, Windom HL (1994) Nutrient flux and budget in Patos Lagoon Estuary. Sci Total Environ 149(1-2):53-60

Niencheski LFH, Baptista JR, Hartmann C, Fillmann G (1986) Caracterização hidrológica de 3 regiões distintas no estuário da Lagoa dos Patos (RS). Acta Limnol Brasil 1:47-64

Niencheski LFH, Windom HL, Smith R (1994) Distribution of particulate trace metal in Patos Lagoon estuary (Brazil). Mar Pollut Bull 28(2):96-102

Nobre CA, Cavalcanti MAG, Nobre P, Kayano MT, Rao VB, Bonatti JP, Satyamurti P, Uvo CB, Cohen JC (1986) Aspectos da climatologia dinâmica do Brasil. Climanálise número especial

Odebrecht C, Djurfeldt L (1996) The role of nearshore mixing on the phytoplankton size structure off Cape Santa Marta Grande, southern Brazil (spring 1989). Arch Fish Mar Res 43(3):13-26

Odebrecht C, Seeliger U, Coutinho R, Torgan L (1987) Florações de *Microcystis* (cianobactérias) na Lagoa dos Patos, RS. Publ Acad Ciências Est (S Paulo) 54(2):280-287

Odebrecht C, Möller OO Jr, Niencheski LF (1988) Biomassa e categorias de tamanho do fitoplâncton total na Lagoa dos Patos, RS, Brasil (verão de 1986). Acta Limnol Brasil 2:367-386

Odebrecht C, Segatto AZ, Freitas CA (1995a) Surf zone chlorophyll *a* variability at the southern Brazilian coast. Estuar Coast Shelf Sci 41:81-90

Odebrecht C, Rörig L, Garcia VM, Abreu PC (1995b) Shellfish mortality and a red tide event in southern Brazil. In: Lassus P, Arzul G, Erard-Le Denn E, Gentien P, Marcaillou-Le Baut C (eds) Harmful marine algal blooms. Lavoisier Publishing, Paris, pp 213-218

Odum EP (1971) Fundamentals of ecology. 3rd edn. Saunders Company, Philadelphia

Ogren L, Berry F, Bjorndal K, Kumpf H, Mast R, Medina G, Reichart H, Witham R (1989) Proceedings of the second western Atlantic turtle symposium. NOAA Techn Mem NMFS-SEFC-226

Ogura N (1972) Rate and extent of decomposition of dissolved organic matter in surface waters. Mar Biol 13:89-93

Oliveira MEA, Nhunch G (1986) Avaliação sazonal do conteúdo de PO_4, biomassa e densidade em *Scirpus californicus*, no Saco de Tapes (RS). Acta Limnol Brasil 1:299-316

Olson DB, Podestá GP, Evans RH, Brown OB (1988) Temporal variations in the separation of Brazil and Malvinas currents. Deep Sea Res 35:1971-1990

Orensanz JM, Gianuca NM (1974) Contribuição ao conhecimento dos anelídeos poliquetas do Rio Grande do Sul, Brasil. Com Mus Ciên Pontif Univ Católica Rio Grande Sul (Porto Alegre) 4:1-37

Ormond WT (1960) Ecologia das restingas do sudeste do Brasil: comunidades vegetais das praias arenosas. Arq Mus Nacional (Rio de Janeiro) 50:185-236

Orth RJ, Heck KL Jr, Montfrans JV (1984) Faunal communities in seagrass beds: a review of the influence of plant structure and prey characteristics on predator-prey relationships. Estuaries 7(4A):339-350

Ostrensky A Neto (1991) Toxicidade da amônia e nitrito no processo produtivo de pós-larvas do camarão-rosa, *Penaeus paulensis* Pérez-Farfante, 1967. MSc Thesis, Univ Fed Paraná, Brazil

Paes ET (1989) Biologia e ecologia de *Donax gemmula* Morrison, 1971 (Bivalvia, Donacidae) na zona de arrebentação da praia do Cassino, Rio Grande, RS e observações sobre a fauna acompanhante. MSc Thesis, Univ Rio Grande, Brazil

Paim PS, Long T, Asmus HE (1987) Aspectos geológicos e geomorfológicos da região estuarina da Lagoa dos patos. Anais I Congr Assoc Brasil Estudos Quaternário (ABEQUA), Porto Alegre, pp 249-261

Palacio JF (1977) A study of the coastal cephalopods from Brazil with reference to brazilian zoogeography. PhD Thesis, Univ Miami, USA

Palacio JF (l978) *Vosseledone charrua*: a new patagonian cephalopod (Octopodidae) with notes on related genera. Bull Mar Sci 28(2):282-296

Palacio JF (1982) Revisión zoogeográfica marina del sur del Brasil. Bolm Inst Oceanogr (S Paulo) 31(1):69-92

Palma JJC (1984) Fisiografia da área oceânica. In: Schobbenhau C (ed) Geologia do Brasil. Texto explicativo do Mapa Geológico do Brasil e da área oceânica adjacente incluindo depósitos minerais. Departamento de produção Mineral, Rio de Janeiro, pp 429-440

Parodi LR (1940) La distribución geográfica de los talares en la Provincia de Buenos Aires. Darwiniana (San Isidro) 4:33-56

Parsons TR, Takahashi M, Hargrave B (1984) Biological oceanographic processes. Pergamon Press, Oxford

Paz RS (1985) Fatores meteorológicos e sua influência ecológica: um exemplo no sistema estuarial da Lagoa dos Patos. Anais III Encontro Brasil Gerenciamento Costeiro. Univ Fed Ceará, Fortaleza, pp 301-311

Perazzolo M, Pinheiro F (1991) Aspectos anatômicos e adaptativos das partes vegetativas de *Spartina densiflora* Brong. (gramínea) da marisma do estuário da Lagoa dos Patos, RS. Acta Bot Brasil (S Paulo) 5(2): 3-16

Pereira CS (1977) Correntes costeiras: estudo de um modelo estacionário aplicado ao litoral do Rio Grande do Sul (Lat 29°-35° S) e análise de um modelo transiente. MSc Thesis, Univ São Paulo, Brazil

Pereira IGT (1983) Morfometria, caracteres merísticos, osteologia e alimentação dos estágios larvais e distribuição de ovos e larvas de Merluza, *Merluccius hubbsi* (Gadiformes: Merlucciidae), na plataforma continental do Rio Grande do Sul. MSc Thesis, Univ Rio Grande, Brazil

Pereira LE (1986) Variação diurna e sazonal da comunidade de peixes e crustáceos decápodas na entrada do estuário da Lagoa dos Patos, RS, Brasil. MSc Thesis, Univ Rio Grande, Brazil

Pereira LE (1994) Variação diurna e sazonal da comunidade de peixes e crustáceos decápodos na entrada do estuário da Lagoa dos Patos, RS, Brasil. Atlantica (Rio Grande) 14:5-21

Pereira OA (1957) Contribuição ao conhecimento da ecologia e evolução de *Ecpantheria indecisa* Walker, 1855 (Lepidoptera, Arctiidae). Rev Brasil Biol (Rio de Janeiro) 17(2):211-218

Peres MB, Vooren CM (1991) Sexual development, reproductive cycle, and fecundity of the school shark *Galeorhinus galeus* off southern Brazil. US Fish Bull 89:655-667

Perez JAA (1990) Distribuição, reprodução e alimentação de *Eledone massyae*, Voss 1964, e *Eledone gaucha*, Haimovici 1988 (Cephalopoda: Octopodidae) no Sul do Brasil. MSc Thesis, Univ Rio Grande, Brazil

Perez JAA, Haimovici M (1991a) Cephalopod collection of the Museu de Zoologia of Universidade de São Paulo, São Paulo, Brazil. Papéis Avulsos Museu Zool (S Paulo) 37(16):251-258

Perez JAA, Haimovici M (1991b) Sexual maturation and reproductive cycle of *Eledone massyae*, Voss l964 (Cephalopoda: Octopodidade) in Southern Brazil. Bull Mar Sci 49 (1/2):270-289

Perez JAA, Haimovici M (1993) Cefalópodes coletados em quatro cruzeiros de prospecção pesqueira demersal no talude continental de Sul do Brasil entre Chui (34^0 30' S) e o Cabo de Santa Marta Grande (28^{0}40' S). Atlantica (Rio Grande) 15:49-72

Perez JAA, Haimovici M (1995) The descriptive ecology of two South American eledonids (Cephalopoda; Octopodinae). Bull Mar Sci 56(3):752-766

Perez JAA, Haimovici M, Cousin JB (1990) Some observations on sperm storing mechanisms and fertilization in females of eledonids from Brazil. Malacologia 32(1):147-152

Pérez-Macri G, Crespo EA (1989) Survey of the franciscana *Pontoporia blainvillei*, along the Argentine coast, with a preliminary evaluation of mortality in coastal fisheries. In: Perrin WF, Brownell RL Jr, Kaiya Z, Jiankang L (eds) Biology and conservation of the river dolphins. IUCN Spec Surv Comm Occas Pap 3:57-63

Perrin WF (1988) Dolphins, porpoises, and whales. An action plan for the conservation of biological diversity: 1988-1992. IUCN/SSC Cetacean Specialist Group 1-30

Perrotta RG (1987) Estudio biológico-pesquero de la caballa (*Scomber japonicus marplatensis*) en el área marplatense y zona adyacente. PhD Thesis, Univ Nac La Plata, Argentina

Perrotta RG, Aubone A, Sanchez F (1990) Estudio comparado de los caracteres morfométricos y merísticos de la caballa (*Scomber japonicus* Houttuyn, 1782) (Teleostei: Scombridae) del sur de Bràsil y del área marplatense (Mar Argentino). Sci Mar (Barc) 54(1):47-53

Persich GR (1993) Ciclo anual do fitoplâncton e alguns parâmetros abióticos no Saco da Mangueira, estuário da Lagoa dos Patos. MSc Thesis, Univ Rio Grande, Brazil

Peters N (1932) Die Bevölkerung des Südatlantischen Ozeans mit Ceratien. Wiss Ergn Dt Atl Exped Meteor 1925-1927 (Berl) 12:1-69

Pfadenhauer J (1978) Contribuição ao conhecimento da vegetação e de suas condições de crescimento nas dunas do Rio Grande do Sul, Brasil. Revta Brasil Biol (Rio de Janeiro) 38(4):827-836

Pfadenhauer J (1980) Die Vegetation der Küstendünen von Rio Grande do Sul, Südbrasilien. Phytocoenologia 8(3/4):321-364

Pfadenhauer J, Ramos RF (1979) Um complexo de vegetação entre dunas e pântanos próximos a Tramandaí, Rio Grande do Sul, Brasil. Iheringia Ser Bot (Porto Alegre) 25:17-26

Phonlor G (1984a) Morfologia e biologia dos ovos de Engraulidae do sul do Brasil (Teleostei, Clupeiformes). Rev Brasil Biol (Rio de Janeiro) 44(4):467-487

Phonlor G (1984b) Morfologia e biologia de larvas de anchoita, *Engraulis anchoita* (Hubbs, Marini 1935) (Osteichthyes, Engraulidae). Atlantica (Rio Grande) 7:85-98

Phonlor G, Sampaio LA (1992) Effect of salinity on growth and survival of *Odontesthes argentinensis* larvae. Arq Biol Tecnol (Curitiba) 35(1):153-158

Phonlor G, Sampaio LA (in press) Growth of the estuarine silverside *Odontesthes* sp. (Pisces: Atherinidae) larvae. Bolm Inst Oceanogr (S Paulo)

Phonlor G, Vinagre LEC (1989) Efeito do retardo da primeira alimentação sobre o crescimento e a sobrevivência da larva de *Odontesthes argentinensis* (Val 1835). Atlantica (Rio Grande) 11(1):63-75

Phonlor G, Vinagre LEC (1990) Fertilização artificial, incubação, crescimento e sobrevivência das larvas de *Odontesthes humensis* criadas em laboratório. Rev Brasil Biol (Rio de Janeiro) 50(2):335-343

Pinedo MC (1982) Análise dos conteúdos estomacais de *Pontoporia blainvillei* (Gervais & d'Obrbigny 1944) e *Tursiops gephyreus* (Lahille, 1908) (Cetacea, Platanistidae e Delphinidae) na zona estuarial e costeira de Rio Grande, RS, Brasil. MSc Thesis, Univ Rio Grande, Brazil

Pinedo MC (1985) A note on a stranding of the humpback whale on the southern coast of Brazil. Sci Rep Whales Res Inst (Tokyo) 36:165-168

Pinedo MC (1986) Mortalidade de *Pontoporia blainvillei*, *Tursiops gephyreus*, *Otaria flavescens* e *Arctocephalus australis* na costa do Rio Grande do Sul, Brasil, 1976-1983. Actas, 1ª Reunión de Trabajo de Expertos en Mamíferos Acuáticos de América del Sur. Buenos Aires, 25-29 junho 1984, pp 187-199

Pinedo MC (1987) First record of a dwarf sperm whale, *Kogia simus*, from southwest Atlantic, with references to osteological characters, food habits and reproductive period. Sci Rep Whales Res Inst (Tokyo) 38:171-186

Pinedo MC (1989) Primeiro registro de *Phocoena spinipinnis* (Cetacea, Phocoenidae) para o litoral do Rio Grande do Sul, Brasil, com medidas osteológicas e análise do conteúdo estomacal. Atlantica (Rio Grande) 11(1):85-99

Pinedo MC (1990) Ocorrência de pinípedes na costa brasileira. Garcia Orta Ser Zool (Lisboa) 15(2):37-48

Pinedo MC (1991) Development and variation of the franciscana, *Pontoporia blainvillei*. PhD Thesis, Univ California, USA

Pinedo MC (1994a) Review of small cetacean fishery interactions in southern Brazil with special reference to the franciscana *Pontoporia blainvillei*. In: Perrin WP, Donovan GP, Barlow J (eds) Gillnets and cetaceans. Red House, Cambridge, pp 251-259, (IWC special issue, vol 15)

Pinedo MC (1994b) Impacts of incidental fishery mortality on age structure of *Pontoporia blainvillei* in southern Brazil and Uruguay. In: Perrin WP, Donovan GP, Barlow J (eds) Gillnets and cetaceans. Red House, Cambridge, pp 261-264, (IWC special issue, vol 15)

Pinedo MC (1995) Development and external variation in external morphology of the franciscana, *Pontoporia blainvillei*. Rev Brasil Biol (Rio de Janeiro) 55(1):85-96

Pinedo MC, Castello HP (1980) Primeiros registros dos golfinhos *Stenella coeruleoalba*, *Stenella* cf *plagiodon* e *Steno bredanensis* para o sul do Brasil, com notas osteológicas. Bolm Inst Oceanogr (S Paulo) 29(2):313-317

Pinedo MC, Rosas FCW (1989) Novas ocorrências de *Pseudorcacrassidens* (Cetacea, Delphinidae) para o Atlântico Sul Ocidental, com observaçães sobre medidas cranianas e alimentação. Atlantica (Rio Grande) 11(1):77-83

Pinedo MC, Praderi R, Brownell RL Jr (1989) Review of the biology and status of the franciscana *Pontoporia blainvillei.* In: Perrin WF, Brownell RL Jr, Kaiya Z, Jiankang L (eds) Biology and conservation of the river dolphins. IUCN Spec Surv Comm Occas Pap 3:46-51

Pinedo MC, Rosas FCW, Marmontel M (eds) (1992) Cetáceos e pinípedes do Brasil. Uma revisão dos registros e guia para a identificação das espécies. UNEP/Fundação Univ Amazonas, Manaus

Piola AR, Gordon AL (1989) Intermediate waters in the southwest South Atlantic. Deep Sea Res 36:1-16

Pla C, Garcia-Marin JL, Vila A (1990) Utilisation des methodes genetiques comme instrument de gestion piscicole. Bull Soc Zool Fr 114(3):31-45

Platt T, Li WKW (eds) (1986) Photosynthetic picoplankton. Can Bull Fish Aquat Sci 214:1-583

Podestá GP (1990) Migratory pattern of Argentine hake *Merluccius hubbsi* and oceanic processes in the Southwestern Atlantic Ocean. Fish Bull 88:167-177

Pomeroy LR (1974) The ocean's food web, a changing paradigm. Bioscience 24:499-504

Potter IC, Loneragan NR, Lenanton RCJ, Chrystal PJ, Grant CJ (1983) Abundance, distribution and age struture of fish populations in a western Australian estuary. J Zool Proc Zool Soc (Lond) 200:21-50

Praderi R, Pinedo MC, Crespo EA (1989) Conservation and management of *Pontoporia blainvillei* in Uruguay, Brazil and Argentina. In: Perrin WF, Brownell RL Jr, Kaiya Z, Jiankang L (eds) Biology and conservation of the river dolphins. IUCN Spec Surv Comm Occas Pap 3:52-56

Prince PA, Wood AG, Barton T, Croxall JP (1992) Satellite tracking of wandering albatrosses (*Diomedea exulans*) in the South Atlantic. Antarct Sci 4(1):31-36

Prodohl P, Levy JA (1989) Genetic study of Atherinidae fishes of Mangueira Lagoon (RS-Brazil). Comp Biochem Physiol 94:423-426

Proença LA (1990) Ciclo anual da produção primária, biomassa do fitoplâncton e carbono orgânico particulado em área rasa da porção sul da Lagoa dos Patos. MSc Thesis, Univ Rio Grande, Brazil

Proença LA, Hama LL, Odebrecht C (1994) Contribution of microalgae to particulate organic carbon in the shallow area of Lagoa dos Patos Estuary, southern Brazil. Atlantica (Rio Grande) 16:191-199

Pryor K, Lindbergh J, Lindbergh S, Milano R (1990) A dolphin-human fishing cooperative in Brazil. Mar Mammal Sci 6(1):77-81

Queiroz EL (1986) Estudo comparativo da alimentação de *Sympterigia acuta* Garman 1877 e *S. bonapartei* Mueller e Henle 1841 (Pisces: Rajiformes) com relação a distribuição, abundância, morfologia e reprodução nas águas litorâneas do Rio Grande do Sul, Brasil. MSc Thesis, Univ Rio Grande, Brazil

Queiroz EL (1995) Coocorrência entre *Sympterigia acuta* (Garman 1877) e *Sympterigia bonapartei* (Muller e Henly 1841) (Elasmobranchii, Rajiformes) espécies endêmicas e simpátricas na plataforma continental do Rio Grande do Sul, Brasil. Rev Brasil Biol (Rio de Janeiro) 55(2):23-303

Rabelo da Cunha S (1994) Modelo ecológico das marismas de *Spartina alterniflora* Loisel (Poaceae) do estuário da Lagoa dos Patos, RS. MSc Thesis, Univ Rio Grande, Brazil

Ramirez F (1971) Eufáusidos de algunos sectores del Atlántico Sudoccidental. Physis (B Aires) 30(81):385-405

Ramirez F, Zamponi M (1981) Hydromedusae. In: Boltovskoy D (ed) Atlas del zooplancton del Atlántico Sudoccidental y métodos de trabajo con el zooplancton marino. Publ Esp INIDEP, Mar del Plata, pp 443-469

Reid JL (1989) On the geostrophic circulation of the South Atlantic Ocean: flow patterns, tracers, and transport. Prog Oceanog 23:149-244

Reid JL, Nowlin WD Jr, Patzert WC (1977) On the characteristics and circulation of the southern Atlantic Ocean. J Phys Oceanogr 7:62-91

Reis EG (1986a) Age and growth of the marine catfish *Netuma barba* (Silurformes, Ariidae), in the estuary of the Patos Lagoon (Brasil). Fish Bull 84(3):679-686

Reis EG (1986b) Reproduction and feeding habits of the marine catfish *Netuma barba* (Silurformes, Ariidae), in the estuary of Lagoa dos Patos, Brazil. Atlantica (Rio Grande) 8:35-55

Reis EG (1992) An assessment of the exploitation of the white croaker *Micropogonias furnieri* (Pisces, Sciaenidae) by the artisanal and industrial fisheries in coastal waters of southern Brazil. PhD Thesis, Univ East Anglia, England

Reis EG, Vieira PC, Duarte VS (1994) Pesca artesanal de teleósteos no estuário da Lagoa dos Patos e costa do Rio Grande do Sul. Atlantica (Rio Grande) 16:55-68

Reise K (1985) Tidal flat ecology. An experimental approach to species interactions. Springer Verlag, Berlin Heidelberg New York

Reitz PR (1961) Vegetação da zona marítima de Santa Catarina. Sellowia (Itajaí) 13:17-115

Resende SML (1988) Nonbreeding strategies of migratory birds at Lagoa do Peixe. MSc Thesis, Univ Cornell, USA

Resgalla C Jr (1993) Influência das massas de água na distribuição espacial e temporal de Pteropoda, Cladocera e Chaetognata na plataforma sul do Brasil (31°40'S-33°45'S). MSc Thesis, Univ Rio Grande, Brazil

Resgalla C Jr, Montú M (1993) Cladoceros marinhos da plataforma continental do Rio Grande do Sul, Brasil. Nauplius (Rio Grande) 1:63-79

Resgalla C Jr, Montú M (1994) Distribuição espacial e temporal de Pteropoda Euthecosomata (Mollusca, Gasteropoda) na plataforma do Sul do Brasil. Atlantica (Rio Grande) 16:99-123

Rieger PJ, D'Incao F (1991) Distribuição das larvas de *Loxopagurus loxochelis* (Decapoda, Diogenida) na região adjacente à Barra de Rio Grande, RS. Neritica (Curitiba) 6:93-106

Riley GA (1967) The plankton of estuaries. In: Lauff GH (ed) Estuaries. Publ Am Assoc Adv Sci (Washington) 83:316-326

Rios EC (1994) Seashells of Brazil. Ed Univ Rio Grande, Rio Grande

Roberts MJ, Long SP, Tieszen LL, Beadle CL (1993) Measurement of plant biomass and net primary production of herbaceous vegetation. In: Hall DO, Scurlock JMO, Bolhàr-Nordenkampf RO, Leegood RC, Long SP (eds) Photosynthesis and production in a changing environment: a field and laboratory manual. Chapman and Hall, London, pp 1-21

Rocha CT, Costa CSB (1988) Ordenação e distribuição das macrófitas vasculares de um pequeno lago de águas doces e rasas. Ciência Cult (S Paulo) 40(2):164-173

Roettger EV (1970) Recent foraminifera from the continental shelf of Rio Grande do Sul collected by the hydrographic vessel "Canopus". Iheringia Ser Zool (Porto Alegre) 38:3-71

Rojas-Galaviz JL, Yáñez-Arancibia A, Day JW Jr, Vera-Herrera FR (1992) Estuarine primary producers, Laguna de Terminos-a case study. In: Seeliger U (ed) Coastal plant communities of Latin America. Academic Press, New York, pp 141-154

Roper CFE, Sweeney M, Nauen CE (1984) FAO species catalogue vol 3. Cephalopods of the World. FAO Fish Synop 125(3):1-227

Rosa ZM (1979) Diatomáceas marinhas e estuarinas de Tramandaí, RS, Brasil. MSc Thesis, Univ Fed Rio Grande do Sul, Brazil

Rosa ZM, Buselato TC (1981) Sobre a ocorrência de floração de *Gyrodinium aureolum* Hulburt (Dinophyceae) no litoral sul do Estado do Rio Grande do Sul, Brasil. Iheringia Ser Bot (Porto Alegre) 28:169-179

Rosas FCW (1989) Aspectos da dinâmica populacional e interações com a pesca, do leão marinho do sul, *Otaria flavescens* (Shaw 1800) (Pinnipedia, Otariidae), no litoral sul do Rio Grande do Sul, Brasil. MSc Thesis, Univ Rio Grande, Brazil

Rosas FCW, Pinedo MC (1989) Nota sobre a ocorrência do cachalote pigmeu, *Kogia* breviceps, no litoral do Rio Grande do Sul, Brasil. Atlantica (Rio Grande) 11(1):109-113

Rosas FCW, Haimovici M, Pinedo MC (1993) Age and growth of the South American sea lion, *Otaria flavescens* (Shaw 1800), in southern Brazil. J Mammal 74(1):141-147

Rosas FCW, Pinedo MC, Marmmontel M, Haimovici M (1994) Seasonal movements and distribution pattern of the southern sea lion, *Otaria flavescens*, (Shaw, 1890) off the Rio Grande do Sul coast, Brazil. Mammalia 58(1):51-59

Rosauro NML (1986) Seiches e suas ocorrências na Lagoa dos Patos e Rio Guaíba (RS. Brasil). Revta Brasil Engenharia (Rio de Janeiro) 4(2):73-131

Rosauro NML, Endres LAM (1991) Análise estatística dos dados de nível e vento da Lagoa dos Patos. IX Simp Brasil Recursos Hídricos, Rio de Janeiro, pp 477-486

Rosen PS (1975) Origin and processes of cuspate spits shorelines. In: Cronin ED (ed) Estuarine research, vol 2. Academic Press, New York, pp 77-92

Roux C (1973) Poissons téléostéens du plateau continental brésilien. In: Roux C (ed) Résultats scientifiques des campagnes de la "Calypso". Fascicule X. Campagne de la "Calypso" au large des cotes atlantiques de l'Amerique du Sud (1961-1962). Première partie, vol 26. Masson and Cie, Paris, pp 23-207

Rozema J, Dorel F, Janissen R, Lenssen G, Broekman R, Arp W, Drake BG (1991) Effect of elevated atmospheric CO_2 on growth, photosynthesis and water relations of salt marsh grass species. Aquat Bot 39:45-55

Ruffino M, Castello JP (1992a) Alterações na fauna acompanhante da pesca do camarão barba-ruça (*Artemesia longinaris*) nas imediações da Barra de Rio Grande, RS. Neritica (Curitiba) 7(1):43-55

Ruffino M, Castello JP (1992b) Dinâmica poblacional del camarón *Artemesia longinaris* del sur de Brasil. Frente Mar, (Montevideo) 12(A):71-81

Ryther JH (1969) Photosynthesis and fish production in the sea. Science 166:72-76

Saccardo SA (1980) Biologia e bionomia de *Trachurus lathami* (Nichols 1920) (Teleostei: Carangidae) na plataforma continental brasileira entre 23^0 S (RJ) e 30^0 S (RS). PhD Thesis, Univ S Paulo, Brazil

Saccardo SA (1987) Morfologia, distribuição e abundância de *Trachurus lathami* (Nichols 1920) (Teleostei: Carangidade) na região sudeste-sul do Brasil. Bolm Inst Oceanogr (S Paulo) 35(1):65-95

Sadowski V (1973) Relação dos peixes cartilaginosos. Publ Esp Inst Oceanogr (S Paulo) 3(1):483-488

Sampaio LA, Phonlor G (1992) Freshwater tolerance of marine and estuarine larvae of *Odontesthes argentinensis* (Pisces: Atherinidae). Arq Biol Tecnol (Curitiba) 35(1):85-90

Sanchez RP, Ciechomski JD, Lasta C, Guerrero RA (1991) A drift study of vertical distribution and mortality of *Engraulis anchoita* eggs and larvae. ICES CM 1991/L:22. Biol Ocean Comm Sess V:1-24

Santee ML, Read WG, Waters JW, Froidevaux GL, Manney GL, Flower DA, Jarnot RF, Harwood RS, Peckham GE (1995) Interhemispheric differences in polar stratospheric HNO_3, H_2O, ClO, and O_3. Science 267:849-852

Santos EA (1982) The inducer of the feeding response in *Penaeus paulensis* (Crustacea-Decapoda). Physiol Behav 31:733-734

Santos EA, Colares EP (1986) Blood glucose regulation in an intertidal crab, *Chasmagnathus granulata* Dana, 1851. Comp Biochem Physiol 83A:673-675

Santos EA, Colares EP (1990) Blood glucose changes in the blue crab *Callinectes sapidus* Rathbun on transfer from sea water to air. Brazil J Med Biol Res (Ribeirão Preto) 23:333-335

Santos EA, Nery LEM (1987) Blood glucose regulation in an estuarine crab, *Chasmagnathus granulata* Dana, 1851 exposed to different salinities. Comp Biochem Physiol 87A:1033-1035

Santos EA, Santos EP (1989) Preliminary observations on the effects of thyroxine on yolk-sac larvae of *Odontesthes argentinensis* (Pisces:Atherinidae). J Fish Biol 35:457-458

Santos EA, Stefanello TM (1991) The hemolymph of *Chasmagnathus granulata* Dana, 1851 (Decapoda, Grapsidae) as a target tissue of the crustacean hyperglycemic hormone. Brazil J Med Biol Res (Ribeirão Preto) 24: 267-270

Santos EA, Baldisseroto B, Bianchini A, Colares EP, Nery LEM, Manzoni G (1987) Respiratory mechanisms and metabolic adaptations of an intertidal crab, *Chasmagnathus granulata* Dana, 1851. Comp Biochem Physiol 88A:21-25

Santos EA, Nery LEM, Manzoni GC (1988) Action of the crustacean hyperglycemic hormone of *Chasmagnathus granulata* Dana, 1851 (Decapoda:Grapsidae). Comp Biochem Physiol 89A:329-333

Santos EP, Yamaguti N (1965) Migração da pescada foguete *Macrodon ancylodon*. Bolm Inst Oceanogr (S Paulo) 14(l):129-138

Santos PP (1990) Ecologia de *Scolelepis gaucha* (Polychaeta, Spionidae) nas praias do Rio Grande do Sul: uma abordagem dinâmica da interação população-ambiente. MSc Thesis, Rio Grande, Brazil

Santos RA (1992) Relações tróficas do calamar argentino *Illex argentinus* (Castellanos, 1960) (Teuthoidea: Ommastrephidae) no sul do Brasil. MSc Thesis, Univ Rio Grande, Brazil

Schaefer CT, Lewin J (1984) Persistent blooms of surf diatoms along the Pacific coast, USA. IV. Diatom productivity and its relation to standing stock. Mar Biol 83:205-217

Schafer A (1985) Fundamentos de ecologia e biogeografia das águas continentais. Ed Univ Fed Rio Grande Sul, Porto Alegre

Schmack EJ (1993) The vulnerability of the east coast of South America to sea level rise and possible adjustment strategies. In: Warrick RA, Barzoiv EM, Wigley TML (eds) Climate and sea level change: observation, projections and implications. Cambridge University Press, Cambridge, pp 336-347

Schmitt ASC, Santos EA (1993a) Behaviour and haemolymphatic ionic composition of the intertidal crab *Chasmagnathus granulata* Dana, 1851 (Crustacea-Decapoda) during emersion. Comp Biochem Physiol 106A:337-342

Schmitt ASC, Santos EA (1993b) Lipid and carbohydrate metabolism of the intertidal crab *Chasmagnathus granulata* Dana, 1851 (Crustacea-Decapoda) during emersion. Comp Biochem Physiol 106A:329-336

Schneider SH (1989) The changing climate. Sci Am 261(3):38-47

Schubauer JP, Hopkinson CS (1984) Above- and belowground emergent macrophyte production and turnover in a coastal marsh ecosystem, Georgia. Limnol Oceanogr 29(5):1052-1065

Schwarzbold A, Schafer A (1984) Gênese e morfologia das lagoas costeiras do Rio Grande do Sul, Brasil. Amazoniana (Kiel) 9(1):87-104

Schwingel PR (1991) Alimentação de *Engraulis anchoita* (Clupeiformes, Engraulididae) na plataforma continental do Rio Grande do Sul, Brasil. MSC Thesis, Univ Rio Grande, Brazil

Schwingel PR, Castello JP (1990) Validación de la edad y el crecimiento de la corvina (*Micropogonias furnieri*) en el sur de Brasil. Frente Mar (Montevideo) 7:19-24

Schwingel PR, Castello JP (1995) Alimentación de la anchoíta, *Engraulis anchoita* en el sur de Brasil. Frente Mar (Montevideo) 15:67-86

Seckendorff RW, Zavala-Camin LA (1985) Reprodução, crescimento e distribuição da cavalinha (*Scomber japonicus*) no sudeste-sul do Brasil. Bolm Inst Pesca (S Paulo) 12(2):1-13

Seeliger U (1992a) Coastal foredunes of southern Brazil: physiography, habitats, and vegetation. In: Seeliger U (ed) Coastal plant communities of Latin America. Academic Press, New York, pp 367-381

Seeliger U (ed) (1992b) Coastal plant communities of Latin America. Academic Pres, New York

Seeliger U, Cordazzo CV (1982) Field and experimental evaluation of *Enteromorpha* as qualitative and quantitative monitoring organism for Cu and Hg in estuaries. Environ Pollut 29(3):197-206

Seeliger U, Knak RB (1982a) Estuarine metal monitoring in southern Brazil. Mar Pollut Bull 13(7):253-254

Seeliger U, Knak RB (1982b) Origin and concentration of copper and mercury in water and biota of the Patos Lagoon estuary, Brazil. Atlantica (Rio Grande) 5(1):35-42

Seeliger U, Cordazzo CV, Koch EW (1984) Germination and algal free laboratory culture of widgeon grass *Ruppia maritima* L. Estuaries 7(1):176-178

Seeliger U, Lacerda LD, Patchineelam SR (eds) (1988) Metals in coastal environments of Latin America. Springer Verlag, Berlin Heidelberg New York

Seguin G (1965) Contribution à la connaissance du plancton des eaux côtieres du Brésil (Copepodes et Amphipodes exceptés) et comparaison avec celui du Sénegal (campagne de la Calypso: Janvier-Fevier 1962). Pelagos 2(3):1-44

Semenov VN (1978) Geographical distribution of benthos on the South American shelf as a function of the distribution of coastal waters. Oceanology 18(1):77-87

Semenov VN, Berman IS (1977) Biogeographic aspects of the distribution and dynamics of the water masses off the South American coast. Oceanology 17(6):710-718

Sharp GD (1988) Fish populations and fisheries. In: Postma H, Zijlstra JJ (eds) Continental shelves. Ecosystems of the world, vol 27. Elsevier, Amsterdam, pp 155-202

Sherman K, Alexander LK, Golda BD (1993) Large marine ecosystems: stress, mitigation and sustainability. AAAS Press, Washington DC

Short AD, Hesp PA (1982) Wave, beach and dune interactions in southeastern Australia. Mar Geol 48:259-284

Sick H (1984) Ornitologia brasileira. Ed Univ Brasília, Brasília

Sieburth JMcN, Smetacek V, Lenz J (1978) Pelagic ecosystem structure: heterotrophic compartments of the plankton and their relationship to plankton size fractions. Limnol Oceanogr 23:1256-1263

Silva CP, Pereira CMP, Dorneles LPP (1993) Espécies de gramíneas e crescimento de *Spartina densiflora* Brong. em uma marisma da Laguna dos Patos, RS, Brasil. Cad Pesq Ser Bot (Santa Cruz do Sul) 5(1):95-108

Silva KG (1996) Estudo comparativo dos parâmetros populacionais da reprodução dos cações-anjo *Squatina argentina* (Marini 1930), *Squatina guggenheim* (Marini 1936) e *Squatina occulta* (Vooren e Silva 1991) no sul do Brasil. MSc Thesis, Univ Rio Grande, Brasil

Silva LCF (1981) Condições oceanográficas do extremo oeste da Convergência Subtropical do Oceano Atlântico Sul baseado nos dados obtidos nas Operações Conversut I e II. MSc Thesis, Univ São Paulo, Brazil

Smith LB, Wasshausen DC, Klein RM (1982) *Panicum racemosum* Spreng. In: Reitz R (ed) Flora Ilustrada de Santa Catarina. Herbário Barbosa Rodrigues, Itajaí, pp 687-690

Soares AG (1992) Estrutura espaço-temporal das associações macrobênticas da zona de arrebentação interna da praia do Cassino, Rio Grande do Sul, Brasil. MSc Thesis, Univ Fed Paraná, Brazil

Solé-Cava AM, Levy JA (1987) Biochemical evidence for a third species of angel shark off the east coast of South America. Biochem Syst Ecol 15(1):139-144

Solé-Cava AM, Vooren CM, Levy JA (1983) Isozymic differentiation of two sibling species of *Squatina* (Chondrichthyes) in south Brazil. Comp Biochem Physiol 75B(2):355-358

Soriano-Sierra EJ (1990) Ecossistemas de marismas, II. A fitocenosis. Publ Acad Ciencias Est (S Paulo) 71(2):142-149

Souto CFM (1986) Estudo comparativo da reprodução nos cações *Mustelus schmitti* (Springer 1939) e *Mustelus canis* (Mitchill 1815) na plataforma continental do Rio Grande do Sul. MSc Thesis, Univ Rio Grande, Brazil

Souza-Santos LP (1991) Influência da salinidade no metabolismo de *Donax hanleyanus* Philippi, 1847 (Bivalvia:Donacidae). MSc Thesis, Univ Rio Grande, Brazil

Stech JL (1990) Um estudo comparativo da dinâmica de circulação de inverno entre as plataformas continentais das costas sudeste do Brasil e dos Estados Unidos utilizando um modelo numérico. PhD Thesis, Univ S Paulo, Brazil

Stech JL, Lorenzetti JA (1992) The response of the south Brazil Bight to the passage of wintertime cold fronts. J Geophys Res 97(C6):9507-9520

Stramma L (1989) The Brazil Current transport south of 23° S. Deep Sea Res 36:639-646

SUDEPE (1974) Relatório da Primeira Reunião do Grupo de Trabalho e Treinamento (GTT) sobre avaliação de estoques. PDP/SUDEPE Bras Ser Doc Tec 7:1-149

Summerson HC, Peterson CH (1984) Role of predation in organizing benthic communities of a temperate-zone seagrass bed. Mar Ecol Prog Ser 15:63-77

Sunyé PS (1993) Gestação e desenvolvimento embrionário dos cações-anjo *Squatina guggenheim* (Marini 1936) e *Squatina occulta* (Vooren and Silva 1991) (Pisces, Squatinidae). MSc Thesis, Univ Rio Grande, Brazil

Sverdrup HU, Johnson MW, Fleming RH (1942) The oceans: their physics, chemistry and general biology. Prentice-Hall, New York

Tagliani PR (1995) Estratégia de planificação ambiental para o sistema ecológico da restinga da Lagoa dos Patos-Planície Costeira do Rio Grande do Sul. PhD Thesis, Univ Fed São Carlos, Brazil

Talbot MMB, Bate GC (1988) The relative quantities of live and detrital organic matter in a beach-surf ecosystem. J Exp Mar Biol Ecol 121:255-264

Talbot MMB, Bate GC (1989) Beach morphodynamics and surf-zone diatom populations. J Exp Mar Biol Ecol 129:231-241

Talbot MMB, Bate GC, Campbell EE (1990) A review of the ecology of surf-zone diatoms, with special reference to *Anaulus australis*. Oceanogr Mar Biol An Rev 28:155-175

Tandler A (1985) Overview: food for the larval stages of marine fish. Live or inert? Isr J Zool 33:161-166

Tanji S (1974) Estudo do conteúdo estomacal da pescada foguete, *Macrodon ancylodon* e da corvina *Micropogon furnieri*. Bolm Inst Pesca (S Paulo) 3:(2):21-36

Teixeira C, Aidar E, Fernandez RM (1973) Estudo preliminar sobre a distribuição de clorofila e o potencial de produção primária. Relatório sobre a segunda pesquisa oceanográfica e pesqueira do Atlântico Sul entre Torres e Maldonado (29°-35° S). GEDIP e Inst Oceanográfico, S Paulo

Teixeira da Silva E (1995) Modelo ecológico de fundos vegetados dominados por *Ruppia maritima* (Pomatomogetonaceae) do estuário da Lagoa dos Patos, RS. MSc Thesis, Univ Rio Grande, Brazil

Teixeira RL, Haimovici M (1989) Distribuição, reprodução e hábitos alimentares de *Prionotus punctatus* e *P. nudigula* (Pisces, Triglidae) entre Solidão (30°43' S) e Chuí (33°45' S), Rio Grande do Sul, Brasil. Atlantica (Rio Grande) 11(1):13-45

Theilacker GH (1978) Effect of starvation on the histological and morphological characteristics of jack mackerel *Trachurus symmetricus* larvae. Fish Bull 76(2):403-414

Thomsen H (1962) Masas de água características del Oceano Atlántico (Parte Sudoeste). Secr Mar (B Aires) H632:1-27

Toldo EE Jr (1991) Morfodinâmica da Laguna dos Patos, Rio Grande do Sul. Pesquisas (Porto Alegre) 18(1):58-63

Tommasi LR (1969) Nota sobre os fundos detríticos do circalitoral inferior da plataforma continental brasileira ao sul do Cabo Frio (RJ). Bolm Inst Oceanogr (S Paulo) 18(1):55-62

Tommasi LR (1983) Observações sobre a irritação respiratória humana ocorrida em 1978 no litoral sul do Rio Grande do Sul. Ciencia Cult 35(2):225-232

Tommasi LR, Adaime RR, Rodrigues S, Bordin G (1973) Bentos da plataforma continental do Rio Grande do Sul. Contr Inst Oceanogr Ser Ocean Biol (S Paulo) 25:95-237

Tommazelli LJ (1993) Regime dos ventos e a taxa de migração das dunas eólicas costeiras do Rio Grande do Sul, Brasil. Pesquisas (Porto Alegre) 20(1):18-26

Torgan LC, Garcia MI (1990) Ocorrência de *Skeletonema subsalsum* (A.Cleve) Bethge (Baciallariophyceae) no sul do Brasil e suas implicações taxonômicas e ecológicas. Acta Limnol Brasil 3(1):439-457

Torres D, Guerra CG, Cardenas JC (1984) Primeiros registros de *Arctocephalus gazella* y nuevos hallagos de *Arctocephalus tropicalis* y *Leptonychotes weddelli* en el archipelago de Juan Fernandez. Inst Antart Chil Ser Cient 31:115-148

Tseng YC (1976) Estudo do extremo oeste da Convergência Subtropical do Oceano Atlântico Sul, usando imagens do satélite Nimbus V e dados oceanográficos de 1972 a 1973, INPE-940-TPT/038, São José dos Campos

Turner RE (1976) Geographic variation in salt marsh macrophyte production: a review. Contrib Mar Sci Univ Texas 20:47-68

Urien CM (1972) Rio de la Plata Estuary environments. The Geological Society of America. Memoir 133:213-234

Utter FM (1991) Biochemical genetics and fishery management: an historical perspective. J Fish Biol 39[Suppl A]:1-20

Valentini H, Castro PMG, Servo GJM, Castro LAB (l991a) Evolução da pesca das principais espécies demersais da costa sudeste do Brasil, pela frota de arrasteiros de parelha baseada em São Paulo, de l968 a l987. Atlantica (Rio Grande) 13(1):87-96

Valentini H, D'Incao F, Rodrigues LF, Rebelo JE Neto, Rahan E (1991b) Análise da pesca do camarão-rosa (*Penaeus paulensis* e *Penaeus brasiliensis*) nas regiões sudeste e sul do Brasil. Atlantica (Rio Grande) 13:171-178

van Heerdt PF, Morzer Bruyns MF (1960) A biocenological investigation of the yellow dune region of Terschelling. Tijdschr Entomol (Amsterdam) 103:225-275

van der Meulen F, Wanders EA, van Huis JC (1985) A landscape map for coastal dune management. Meijendel, The Netherlands. ITC J 2:85-92

Vannucci M (1957) On Brazilian hydromedusae and their distribution in relation to different water masses. Bolm Inst Oceanogr (S Paulo) 8(1/2):23-209

Vannucci M (1963) Preliminary results in the study of the zooplankton standing stock of south Brazilian coast at 25^0 Lat S. Contrções avulsas Inst Oceanogr (S Paulo) 3:1-28

Vargas-Boldrin C (1980) Estrutura, ciclo de vida e bionomia de *Cynoscion striatus* (Cuvier 1829) (Teleostei:Sciaenidae) ao sul de Cabo Frio, Brasil. PhD Thesis, Univ S Paulo, Brazil

Vaske T Jr (1991) Seabirds mortality in longline fishing for tuna in southern Brazil. Ciencia Cult (S Paulo) 45(3):388-390

Vaske T Jr (1992) Idade, crescimento e alimentação de albacora de lage, *Thunnus albacares* (Bonnaterre, 1788) (Perciformes: Scombridae), explorada no sul do Brasil. MSc Thesis, Univ Rio Grande, Brazil

Vaz-Ferreira R (1982a) *Otaria flavescens* (Shaw), South American seal lion. FAO Fish Ser Mamm Seas 4(5):477-495

Vaz-Ferreira R (1982b) *Arctocephalus australis*, Zimmermann, South American fur seal. FAO Fish Ser Mamm Seas 4(5):497-508

Vazzoler AEAM (l971) Diversificação fisiológica e morfológica de *Micropogonias furnieri* (Desmarest 1822) ao sul de Cabo Frio, Brasil. Bolm Inst Oceanogr (S Paulo) 20(2):1-70

Vazzoler AEAM (1991) Síntese de conhecimentos sobre a biologia da corvina *Micropogonias furnieri* (Desmarest, 1823), da costa do Brasil. Atlantica (Rio Grande) 13(1):55-74

Vazzoler G (1975) Distribuição da fauna de peixes demersais e ecologia dos Sciaenidae da plataforma continental brasileira entre as latitudes 29^0 21'S (Torres) e 34^0 44'S (Chuí). Bolm Inst Oceanogr (S Paulo) 24:85-169

Vazzoler G, Iwai M (1971) Relatório sobre prospecção e pesca exploratória na plataforma continental do Rio Grande do Sul (abril de 1968 a março de 1969). Contr Inst Oceanogr (S Paulo) Ser Oceanogr Biol 25:1-75

Vazzoler G, Kawakami E, Zaneti EM (1973) Distribuição e abundância de peixes demersais. Publ Esp Inst Oceanogr (S Paulo) 3(1):483-488

Vidal EAG (1994a) Relative growth of paralarvae and juveniles of *Illex argentinus* (Castellanos 1960) in southern Brazil. Antarct Sci 6(2):275-282

Vidal EAG (1994b) Paralarvas e juvenis de *Illex argentinus* (Cephalopoda: Ommastrephidae): descrição, crescimento, alimentação, distribuição e abundância em relação aos processos oceanográficos do sul do Brasil. MSc Thesis, Univ Rio Grande, Brazil

Vieira EF (1983) Rio Grande. Geografia física, humana e econômica. Ed Sagra, Porto Alegre

Vieira EF, Rangel SRS (1988) Planície costeira do Rio Grande do Sul: geografia física, vegetação e dinâmica sócio-demográfica. Ed Sagra, Porto Alegre

Vieira JP (1985) Distribuição, abundância e alimentação dos jovens de Mugilidae no estuário da Lagoa dos Patos e movimentos reprodutivos da tainha (*Mugil platanus*, Gunther 1880) no litoral sul do Brasil. MSc Thesis, Univ Rio Grande, Brazil

Vieira JP (1991a) Ecology of estuarine fish assemblages in Patos Lagoon, Brazil (32° S), and York River, USA (37° N), with review of zoogeography of fishes in western Atlantic warm-temperate and tropical estuaries. PhD Thesis, College of William and Mary, School of Marine Science, USA

Vieira JP (1991b) Juvenile mullets (Pisces: Mugilidae) in the estuary of Lagoa dos Patos, RS. Brazil. Copeia 2:409-418

Vieira JP, Musick JA (1993) Latitudinal patterns in diversity of fishes in warm-temperate and tropical estuarine waters of western Atlantic. Atlantica (Rio Grande) 15:115-113

Vieira JP, Musick JA (1994) Fish faunal composition in warm-temperate and tropical estuaries of western Atlantic. Atlantica (Rio Grande) 16:31-53

Vieira JP, Scalabrin C (1991) Migração reprodutiva da "tainha" (*Mugil platanus* Gunther, 1980) no sul do Brasil. Atlantica (Rio Grande) 13(1):131-141

Vieira PJC (1990) Biologia populacional de *Cynoscion striatus* (Pisces: Scianidae) no litoral sul do Brasil. MSc Thesis, Univ Rio Grande, Brazil

Vieira PJC, Haimovici M (1993) Idade e crescimento da pescada olhuda *Cynoscion striatus* (Sciaenidae) no sul do Brasil. Atlantica (Rio Grande) 15:73-91

Vilas Boas DF (1990) Distribuição e comportamento dos sais nutrientes, elementos maiores e metais pesados na Lagoa dos Patos, RS. MSc Thesis, Univ Rio Grande, Brazil

Vilela MJ (1990) Idade, crescimento, alimentação e avaliação do estoque de bonito listado (*Katsuwonus pelamis*) (Scombridae: Thunnini), explotado na região sudeste-sul do Brasil. MSc Thesis, Univ Rio Grande, Brazil

Vilela MJ, Castello JP (1991) Estudio de la edad y crecimiento del barrilete (*Katsuwonus pelamis*) en la región sur y sudeste de Brasil. Frente Mar (Montevideo) 9:29-35

Vilela MJ, Castello JP (1993) Dinámica poblacional del barrilete (*Katsuwonus pelamis*) explotado en la región sudeste-sur del Brasil en el período 1980-1986. Frente Mar (Montevideo) 14:111-124

Villwock JA (1978) Aspectos da sedimentação na região nordeste da Lagoa dos Patos: Lagoa do Casamento e Saco do Cocuruto, RS, Brasil. Pesquisas (Porto Alegre) 11:193-223

Villwock JA (1984) Geology of the coastal province of Rio Grande do Sul, southern Brazil: a synthesis. Pesquisas (Porto Alegre) 16:5-49

Villwock JA (1987) Processos costeiros e a formação das praias arenosas e campos de dunas ao longo da costa sul e sudeste brasileira. Publ Acad Ciencias Est (S Paulo) 54(1):380-398

Villwock JA (1994) A costa brasileira: geologia e evolução. Publ Acad Ciencias Est (S Paulo) 87:1-15

Villwock JA, Martins LRS (1972) Depósitos lamíticos de pós-praia, Cassino, RS. Pesquisas, Porto Alegre 1:69-85

Villwock JA, Tommazelli LJ (1989) Sea-level changes and holocen evolution in the Rio Grande do Sul coastal plain, Brazil. Sp Pub 1 International symp on global changes in south America during the Quaternary, Cong Brasil Assoc Brasil Estudos Quaternário (ABEQUA), 8-12 May, Rio de Janeiro, S Paulo, pp 192-196

Villwock JA, Tommazelli LJ, Loss EL, Horn NH, Bachi, FA, Dehnhardt BA (1986) Geology of the Rio Grande do Sul coastal province. In: Rabassa J (ed) Quaternary of South America Antarctica Peninsula, vol 4. Balkema, Rotterdam, pp 79-97

Virnstein RW (1979) Predation on estuarine infauna: patterns of component species. Estuaries 2:69-86

Virnstein RW, Nelson WG, Lewis FG III, Howard RK (1984) Latitudinal patterns in seagrass epifauna: do patterns exist, and can they be explained? Estuaries 7(4A):10-330

von Ihering H (1885) Die Lagoa dos Patos. Dtsch Geogr Bl 8:182-204

von Ihering H (1887) Die Vögel der Lagoa dos Patos. Z Gesammte Ornithol (Budapest) 4:142-165

von Ihering H (1892) Os mamíferos do Rio Grande do Sul. Anais Est Rio Grande do Sul (Porto Alegre) 9:1-11

von Ihering H (1896) Os peixes da costa do mar no Estado do Rio Grande do Sul. Anuário do Estado do Rio Grande do Sul para o anno de 1897, Porto Alegre, pp 98-124

Vooren CM (1983) Seleção pela malha na pesca de arrasto da castanha *Umbrina canosai*, pescada *Cynoscion striatus* e pescadinha *Macrodon ancylodon* no Rio Grande do Sul. Doc Tec Oceanografia, Univ Rio Grande, Rio Grande 4:1-32

Vooren CM (1992) Strategies reproductives comparées de huit espèces de selaciens vivipares du sud du Brésil. Bull Soc Zool Fr 117(3):303-312

Vooren CM, Chiaradia A (1990) Seasonal abundance and behaviour of coastal birds on Cassino beach, Brazil. Ornitol Neotrop 1:9-24

Vooren CM, Fernandes AC (1989) Guia de albatrozes e petréis do sul do Brasil. Ed Sagra, Porto Alegre

Vooren CM, Silva KG (1991) On the taxonomy of the angel sharks of southern Brazil, with the description of *Squatina occulta* sp. n. Rev Brasil Biol (Rio de Janeiro) 51(3):589-602

Vooren CM, Haimovici M, Vieira PC, Duarte VS, Ferreira BP (1988) Pesca experimental na margem externa da plataforma e no talude continental do Rio Grande no inverno de l986. Anais V Cong Brasil Eng Pesca, Fortaleza, pp 435-447

Vooren CM, Araujo MLG, Betito R (1990) Análise da estatística da pesca de elasmobrânquios demersais no porto de Rio Grande, de 1973 a 1986. Ciencia Cult (S Paulo) 42(12):1106-1114

Voss GL (1964) A note on some cephalopods from Brazil with a description of a new species of octopod, *Eledone massyae*. Bull Mar Sci Gulf Carib 14(3):511-516

Voss GL (1988) The biogeography of the deep sea Octopoda. Malacologia 29(1):295-307

Waechter JL (1985) Aspectos ecológicos da vegetação de restinga no Rio Grande do Sul, Brasil. Comun Mus Ciencias PUCRS Ser Bot (Porto Alegre) 33:49-68

Wasielesky WJ (1994) Tolerância do linguado *Paralichthys orbignyanus* (Valenciennes, 1839) (Pleunoctiforme, Paralichthyidae), à parâmetros físico-químicos. MSc Thesis, Univ Rio Grande, Brazil

Watanabe Y (1985) Histological changes in the liver and intestine of freshwater goby larvae during short-term starvation. Bull Jpn Soc Sci Fish 51(5):707-709

Weber L, Conceição M, Ruffino M, Levy JA (1993) Population genetics of *Artemesia longinaris* (Crustacea, Penaeidae) at the South-West Atlantic Coast. Comp Biochem Physiol 106B(4):1015-1020

Weiss G (1981) Ictioplancton del estuario de Lagoa dos Patos, Brasil. PhD Thesis, Univ Nac de La Plata, Argentina

Weiss G, Krug LC (1977) Características do desenvolvimento e metamorfose de *Lycengraulis olidus* (Engraulidae) e *Brevoortia pectinata* (Clupeidae) no estuário da Lagoa dos Patos, Brasil. Atlantica (Rio Grande) 2(1):83-117

Weiss G, Souza JAF (1977a) Estudo comparativo preliminar de pós-larvas e juvenis das três espécies de Engraulidae da costa sul do Brasil, Uruguai e Argentina. Atlantica (Rio Grande) 2(1):1-20

Weiss G, Souza JAF (1977b) Desova invernal de *Engraulis anchoita* na costa sul do Brasil em 1970 e 1976. Atlantica (Rio Grande) 2(2):5-24

Weiss G, Souza JAF, Santos A (1976) Contribuição ao conhecimento do ictioplâncton marinho da plataforma sul do Brasil. Atlantica (Rio Grande) 1(1/2):1-99

Weiss G, Hubold G, Bainay ACD (1987) Larval development of the zeiform fishes *Antigonia capros* (Lowe 1843) and *Zenopsis conchifer* (Lowe 1852) from the southwest Atlantic. Cybium 11(1):79-91

Weiss G, Hubold G, Bonecker ACT (1988) Eggs and larvae of *Maurolicus muelleri* (Cymelin, 1789) (Teleostei, Sternoptychidae) in the southwest Atlantic. Meeresforschung 32:53-60

Westhoff V (1985) Nature management in coastal areas of Western Europe. Vegetatio 62:523-532

White BN, Lavemberg RJ, McGowen (1983) Atheriniformes: development and relationships. In: Moser HG (ed) Ontogeny and systematics of fishes. Spec Publ Am Soc Icthyol Herpet, pp 355-362

Wiedemann AM (1984) The ecology of Pacific north west coastal sand dunes: a community profile. US Fish Wildl Serv FWS/OBS-84/04, Washington DC

Windom HL (1975) Heavy metal fluxes through salt marsh estuaries. In: Cronin LE (ed) Estuarine research. Academic Press, New York, pp 137-152

Wolfe DA, Kjerfve B (1986) Estuarine variability: an overview. In: Wolfe DA (ed) Estuarine variability, Academic Press, New York, pp 3-17

Wood E (1987) Subtidal ecology. Arnold, London

Wourms JP, Grove BD, Lombardi J (1988) The maternal-embryonic relationships in viviparous fishes. Fish Physiol XIB:1-134

Wright LD, Short AD (1984) Morphodynamic variability of surf zones and beaches: a synthesis. Mar Geol 56:93-118

Wüst G (1932) Das ozeanographische Beobachtungsmaterial. Wiss Erg Dt Atlant Exped Meteor 1925-1927 (Berl) vol 4(2)

Wüst G (1935) Die Stratosphaere. Wiss Ergebn Dt Atlant Exped Meteor 1925-1927 (Berl) vol 4(1/2)

Yamaguti N (1967) Desova da pescada foguete *Macrodon ancylodon*. Bolm Inst Oceanogr (S Paulo) 16(l):101-106

Yamaguti N (1979) Diferenciação geográfica de *Macrodon ancylodon* (Blach and Shneider, 1801) na costa brasileira, entre as latitudes 18° 36'S e 32° 10'S. Bolm Inst Oceanogr (S Paulo) 28(1):53-118

Yamaguti N, Santos EP (1966) Crescimento da pescada foguete *Macrodon ancylodon*: aspecto quantitativo. Bolm Inst Oceanogr (S Paulo) 15(1):75-78

Yesaki M (1973) Sumário dos levantamentos de pesca exploratória ao largo da costa sul do Brasil e da biomassa de peixe demersal e potencial pesqueiro. Doc Tecn SUDEPE PDP Bras 1:1-27

Yesaki M, Bager KJ (1975) Histórico da evolução da pesca industrial em Rio Grande. Programa de Pesquisa e Desenvolvimento Pesqueiro do Brasil PNUD/FAO, Min Agr SUDEPE. Ser Doc Tec (Rio de Janeiro) 11:1-15

Yesaki M, Barcellos BN (1974) Desenvolvimento da pesca do pargo róseo ao longo da costa sul do Brasil. Ser Doc Ocas SUDEPE PDP Bras 6:1-8

Yesaki M, Rahn E, Silva G (1976) Sumário das explorações de peixes de arrasto de fundo ao largo da costa sul do Brasil. Doc Tecn SUDEPE PDP (Rio de Janeiro) 19:1-37

Yunes JS, Suzuki MT, Silveira AG, Camargo M, Werner VR (1990) Cianobactérias fixadoras de nitrogênio do estuário da Lagoa do Patos, RS: *Nostoc muscorum*. Ciencia Cult (S Paulo) 42(5/6):375-383

Yunes JS, Niencheski LFH, Salomon PS, Parise M, Beattie KA, Raggett SL, Codd GA (1994) Development and toxicity of cyanobacteria in the Patos Lagoon estuary, southern Brazil. Intergov Oceanogr Comm Workshop Rep 101[Annex III]:14-19

Zaneti Prado E (1979) Bionomia, ciclo de vida de *Umbrina canosai*, Berg (1895). Bolm Inst Oceanogr (S Paulo) 28(1):119-165

Zavala-Camin LA (1974) Ocorrência de atuns no sudeste e sul do Brasil. Bolm Inst Pesca (S Paulo) 3(3):37-52

Zavala-Camin LA (1978) Algunos aspectos sobre la estructura populácional del rabil (*Thunnus albacares*) en el sudeste y sur de Brasil (1969-1977) con presentación de la hipótesis de la migración semestral. Bolm Inst Pesca (S Paulo) 5(11):1-25

Zavala-Camin LA (1981) Hábitos alimentares e distribuição dos atuns e afins (Osteichthyes: Teleostei) e suas relações ecológicas com outras espécies pelágicas das regiões sudeste e sul do Brasil. PhD Thesis, Univ S Paulo, Brazil

Zavala-Camin LA, Antero da Silva (1991) Histórico da pesquisa e pesca de atuns com espinhel no Brasil. Atlantica (Rio Grande) 13(1):107-11

Zavala-Camin LA, Tomas ARG (1990) A pesca de atuns com espinhel no Atlântico Sudoeste por barcos japoneses e brasileiros (1959-1979). Bolm Inst Pesca (S Paulo) 17:61-75

Zeltzer F (1976) Geologia e paleogeografia da restinga da Laguna dos Patos, RS. MSc Thesis, Univ Fed Rio Grande Sul, Brazil

Zembruscki SG, Barreto HT, Palma JJC, Milliman JD (1972) Estudo preliminar das províncias geomorfológicas da margem continental sul-brasileira. 26 Congr Brasil Geol, Belém. Anais Soc Brasil Geol (S Paulo) 2:187-209

Zenkovitch VP (1959) On the genesis of cuspate spits along lagoon shores. J Geol 67(3):269-277

Zenkovitch VP (1967) Processes of coastal development. Oliver and Boyd, London

Subject Index

SPRINGER NATURE

GPSR Compliance

The European Union's (EU) General Product Safety Regulation (GPSR) is a set of rules that requires consumer products to be safe and our obligations to ensure this.

If you have any concerns about our products, you can contact us on ProductSafety@springernature.com

In case Publisher is established outside the EU, the EU authorized representative is:

Springer Nature Customer Service Center GmbH
Europaplatz 3
69115 Heidelberg, Germany